AF750324

Studies in the Exact Sciences in Medieval Islam

Studies in the Exact Sciences in Medieval Islam

By
Ali A. Al-Daffa
and
John J. Stroyls

University of Petroleum and Minerals
Dhahran
Saudi Arabia

UNIVERSITY OF PETROLEUM AND MINERALS
Dhahran Saudi Arabia

and

JOHN WILEY & SONS
Chichester · New York · Brisbane · Toronto · Singapore

-4 APR 1986

509.1767
DAF

PK ~~2890~~
[al-DAFFĀʿ]

0|149,054

Copyright © 1984 by John Wiley & Sons Ltd.

All rights reserved.

No part of this book may be reproduced by any means, nor transmitted, nor translated into a machine language without the written permission of the publisher.

Library of Congress Cataloging in Publication Data:

Daffāʿ, ʿAlī ʿAbd Allāh.
Studies in the exact sciences in Medieval Islam.
Bibliography: p.
1. Science—Islamic Empire—History—Addresses, essays, lectures. 2. Mathematics—Islamic Empire—History—Addresses, essays, lectures. I. Stroyls, John J. II. Title.
Q127.A5D34 1984 509.17′671 83-21847

ISBN 0 471 90320 5

British Library Cataloguing in Publication Data:

Al-Daffa, Ali A.
Studies in the exact sciences in medieval Islam.
1. Science and Islam—History
I. Title II. Stroyls, John J.
509′.17′671 Q128

ISBN 0 471 90320 5

Photosetting by Thomson Press (India) Limited, New Delhi and Printed in Great Britain by Page Brothers (Norwich) Ltd.

Contents

Preface

The essays on the Exact Sciences in Medieval Islam in this collection represent a selection from those written by us over the past decade, and presented at conferences in America, Europe, the Middle East and Asia. They deal with problems in transmission of science and technology between East and West, historiography, as well as the more usual fare: works of individual scholars, and surveys on the development of a single scientific method. We offer no philosophical overviews, or theories on the development of Medieval Near Eastern science. The selection of topics is rather electric; astronomy is not represented, and physics appears only in so far as it relates to mathematics. We hope to deal with these apparent imbalances in a later collection.

We have tried to make each essay reasonably self-contained; in any case, we have provided the reader with a fair bibliography of related work drawn from Arabic, as well as most European languages. The secondary references in Russian are particularly noteworthy, since many of them are not available in translations.

It is with great pleasure that we acknowledge the support of the University of Petroleum and Minerals in Dhahran. Its Research Committee not only made it possible for us to write these essays and to present them at international conferences, but also recommended the publication of this collection. Our sincere thanks go to the University Press for agreeing to arrange for the printing.

On a more personal note, we thank Professors Ali Kettani, A. I. Sabra, Ahmed Saidan and Aydin Sayili for years of friendly encouragement; Mr David K. Easton of the University Library for patiently and efficiently processing our endless requests for source material, and Mr Nisar Ahmed Qureshi who carefully prepared the final typescript of these essays.

To our respective families who gave us support with their understanding and encouragement, a heartfelt thank you.

Ali A. Al-Daffa
John J. Stroyls
Dhahran, November 1982

Introduction

These essays reflect two of our main interests in Islamic and Arabic science during the Middle Ages: its development and transmission. The lead essay is devoted to some aspects of the latter, in the wider context of transmission of science and technology. More than a half century ago C. H. Haskins, in *The Renaissance of the Twelfth Century* and in his comprehensive *Studies in the History of Medieval Science*, observed that Arabic science reached the West through Spain and Sicily rather than by way of the Crusader Kingdoms. Moreover, he noted that during the time of the Crusades, this transmission of science was one way: from East to West. We re-examine Haskins' conclusions in the light of scholarship of the past 50 years, not only in the area to which he restricted himself (manuscripts in the quadrivium and medicine) but also in areas of applied science and technology, for which there are, apparently, no manuscripts to serve as agents of transmission. We find some need to modify Haskins' conclusions, particularly, if we extend classical science to include the latter two areas. In the more restricted sense of science transmitted by manuscript, however, his conclusions are still well-founded.

The next two essays, though about mathematics, are really concerned with a problem in the historiography of science: the preservation of mumpsimus, an adherence to exposed, but customary errors, in science history. One essay illustrates this process with the historical treatment of Pythagorean theorems. The other deals with the recurrent myth of logarithms in Medieval Near Eastern mathematics.

Our fourth essay deals with the thirteenth century mathematician Naṣīr al-Dīn al-Ṭūsī's attempt to prove the parallel or fifth postulate of Euclid as a theorem. Typical of our approach to such material, we trace the genesis of al-Ṭūsī's work in that of his predecessors al-Jawharī, al-Nayrīzī, Thābit ibn Qurra, ibn al-Haytham, al-Khayyāmī, and look for the possible influence of al-Ṭūsī's work on the European development of non-Euclidean geometry through John Wallis. The last section of the essay offers a novel way of classifying Medieval Near Eastern work on the parallel postulate into logically equivalent groups based on geometric equivalence theory developed in 1900 by Max Dehn. This provides an alternative to the more orthodox method of historical inference, which depends on internal evidence and "direct citation" for its conclusions.

After this comes an essay with the unlikely title, "Ibn Sīnā as a Mathematician".

Abū Ali al-Husayn ibn ᶜAbdallah ibn Sīnā (Avicenna) [980–1037] is known chiefly as a philosopher and physician. However, in his autobiography, preserved in ibn Abī Uṣaybiᶜa's *ᶜUyun al-abāᵓ fī ṭabaqāt al-aṭibbāᵓ*, ibn Sīnā makes numerous references to his early studies in mathematics, starting with lessons in the "Indian arithmetic", and continuing through Euclid and the *Almagest.* He also maintained a scientific correspondence and rivalry with al-Bīrūnī who is known, among other things, for contributions to mathematical geography and numerical approximation. Consequently, it is of interest to examine ibn Sīnā's own mathematical work, particularly since no comprehensive study has been devoted to it. Limited accounts do appear in a short thesis of Lokotsch (1912), which discusses ibn Sīnā's commentaries on the planimetric books of Euclid, and in notes of Woepcke and Tannery on the "method of casting out nines". To fill this gap, we examine the extant treatises of ibn Sīnā, in particular, *Kitāb ash-Shifāᵓ*, *Dānish Nāmā'i*, and *an-Najāt* to offer a comprehensive picture of his mathematical accomplishments, and thereby, to assess the validity of Sarton's claim that "ibn Sīnā's interest in mathematics was philosophical rather than technical".

Our penultimate essay also fills a gap in the existing literature. It deals with numerical analysis in ninth through fifteenth century Middle Eastern writing. Lest the reader accuse us of dealing in anachronisms, we certainly agree that as a distinct mathematical discipline, numerical analysis came to be recognized a mere thirty years ago, but as a part of mathematics proper, it has developed since Babylonian times. H. Goldstine's recent book, *A History of Numerical Analysis*, traces that development, as a part of the rekindling of interest in science in Europe, beginning in the sixteenth century. However, since his work makes only passing reference to numerical methods in the Middle East, e.g. al-Bīrūnī and al-Kāshī, we felt it fitting to supplement Goldstine's work with a survey of these methods, and to compare the "state of the numerical art" in the Middle East and in Europe in the fifteenth century. A synthesis of older work of Suter and Luckey with newer results of Kennedy, Youschkevitch, Saidan and Rashed makes such a comparison possible. We conclude the essay with comments on the failure of this Near Eastern work to have a direct effect on the development of numerical analysis in Europe.

The last essay of this collection expands on a paper we prepared for the 1981 International Congress on the History of Science in Bucharest. It treats the geometric theory of equations in Medieval Middle Eastern writing, as opposed to the competing numerical school of Ḥabash al-Ḥāsib, as-Samaw'al and Naṣir al-Dīn, and looks for possible Medieval links between the two approaches.

1 Transmission of Science and Technology Between East and West During the Period of the Crusades

In the 1920's, first in his collected essays *Studies in the History of Medieval Science*,[1] and shortly afterward in *The Renaissance of the Twelfth Century*,[2] Haskins advanced a thesis which has stood up remarkably to the scholarship of the past sixty years. He pointed out, on the basis of manuscript evidence, that the transmission of science between East and West was an activity contemporaneous with the Crusades, but owing very little to them. His *Studies* established that Arabic treatises in medicine, alchemy, and the quadrivium were translated into Latin, primarily in Spain, with Sicily a distant second, in the twelfth century. Moreover, the actual production of translators working in Syria during the Crusades, Stephen of Antioch and Philip of Tripoli, was small compared with that of centres like Toledo and Palermo. In fact, Haskins implied that the transmission of science during this period was one way: from Near Eastern and Arabic source to the Latin West.[3] He also attempted to give extra dimension to his analysis by discussing the activities of the school of Chartres,[4] the introduction of Arabic science into England by Robert of Chester and others,[5] the pre-Humanist translations from Greek by North Italian scholars, and closed his main account with Fredrick II's study of falconry, *De arte venandi cum avibus*. This latter essay, though quite relevant in the context of Haskins' work, retrospectively, gives *The Studies* an unfinished look, and even prompts the question: if falconry is relevant, what of agriculture, in fact, the whole of technology, from glass making and metallurgy to norias and mills, and machines of war? Of course, to open the door to such questions rapidly takes one beyond the scope of Haskins' intended study in at least two ways. First, it attaches technology and applied science to science proper, a conjunction quite uncommon before the middle, or even latter part of the nineteenth century,[6] and certainly at variance with the Medieval classification of science to which Haskins adheres. Secondly it forces one to leave the security of inference based on manuscript tradition, for example, the Medieval transmission of Euclid's *Elements*, or the influence of Ibn Sina's *Qānūn fiʾ l-ṭibb*, for the study of an area notorious for its lack of documentation. Nonetheless, questions such as these must be broached if one is to gain a more accurate picture of the transmission of science and technology between East and West during the Crusader period. In what follows, the authors offer a summary of some of the ways in which Haskins' thesis has been broadened and modified over the years, and then turn to the questions posed above. Given the limitation of space, however, we shall give only a small part of the evidence currently available, of transmissions of technology between East and West. Finally we offer some observations on the causes for the rather disappointing role of Outremer in these exchanges.

I. Some modifications of Haskins' thesis

In some respects Haskins' *Studies* is a reaction to popular histories of European science,[7] and even more elaborate works such as Taylor's *The Mediaeval Mind*,[8] which either pass over Arabic science in silence or give an undue role to the intellectual stimulus of the Crusades on the European mind. Moreover, Haskins' main theme is rather more the one developed in detail in *The Renaissance of the Twelfth Century*, namely the existence of such an early European intellectual stirring, than the simple consideration of the transmission of Eastern science. It is perhaps this latter theme, with its emphasis on a particular period of time, which caused Haskins to argue so strongly for twelfth century beginnings of the appearance of Arabic influences in Latin works on science. He had, for example, examined the famous Ripoll ms 225, dating from the late tenth century, and allowed for clear Arabic influence, but argued that with this single exception "Arabic influence [throughout the eleventh century] is limited to the technical terms of the astrolabe and the names of stars."[9] It remained for Millás Vallicrosa to study the manuscript with more care, and to establish not only that it contains a treatise on the quadrant translated from Arabic to Latin in the mid-tenth century, but that the Ripoll scriptorium is the first known centre for transferring the Arabic–Greek scientific tradition into Latin.[10] These claims are enhanced by the fact that the monastery of Ripoll also contained, for example, a Latin translation of Māshāʾallāh's treatise[11] on the astrolabe, and Boethius' *Arithmetic* with Arabic marginalia. Furthermore, Millás places the treatise on multiplication and division of Joseph the Wise (Joseph of Spain) requested in 984 by Gerbert of Aurillac from Miro Bonusfilius of Gerona well within the tradition of al-Khwārizmī, and his influence.[12] The very year that Millás' seminal work appeared, Welborn[13] made a case for Lotharingia as a centre for Arabic scientific influence in the eleventh century, extending the earlier work of Thompson on the introduction of Arabic science in the Lorraine in the tenth.[14] In consequence, it now appears that Arabic science made inroads into the Latin tradition much earlier than first postulated, and that in some cases, transmission was nearly contemporary with discovery.

At the other extreme, owing to the work of Millás Vallicrosa,[15] Menédez Pidal,[16] Bossong,[17] Proctor,[18] and Wegner,[19] one has a much clearer picture, than that available to Haskins, of the mechanism of translation in its final days in Spain: the Arabic science translations into the vernacular made during the time of Alfonso the Wise. The same is true of our knowledge of transmissions through Hebrew in Medieval France. Of perhaps greater importance, however, is our enhanced knowledge of the mechanism of the translational "impulse", the styles and work habits of particular translators, and the general level of linguistic competence of their times and the effect it had on the transmission process. Of equal interest are an element which Norman Daniel has dubbed the "cultural filter[20]," and another which, borrowing from Glick,[21] on might call "modalities of change". Let us now consider these items in order.

The first, which concerns the translators themselves, may be illustrated rather strikingly by comparing the approach to translation of Ḥunain ibn Isḥāq (808–873), the greatest and one of the earliest translators of Greek science into Syriac and Arabic with that of Gerard of Cremona (ca. 1114–1187), the most prolific of the translators from Arabic into Latin. Of course, such a comparison leads to numerous simplifications, but it seems worthwhile, nonetheless.[22]

According to the picturesque account of ibn Abī Uṣaibiᶜa (*ᶜUyūn al-anbāʾ fī ṭabaqāt al-aṭibbāʾ*), Ḥunain, who was born in Ḥīra, went to Baghdad to study medicine under the famous physician Yuḥannā ibn Māsawayh. After a short time ibn Māsawayh found his pupil's questions irksome, and perhaps professionally threatening, for he flew into a rage and dismissed him with the advice that he should return to Ḥīra to be a merchant.[23] Undeterred, Ḥunain, who already had a thorough command of Syriac and Arabic, went off to "bilād al-Rūm", probably, as Anawati conjectures,[24] Byzantium, and mastered Greek and Greek medicine as well. Upon his return to Baghdad, he began translations of Galen, and for a time became reconciled with ibn Māsawayh. Here he also pursued his work as a physician, but suffered loss of time and personal harassment owing to the jealousy of the Bukhtīshūᶜ family. Nonetheless, Ḥunain, his son Isḥāq, a nephew Ḥubaysh, and a pupil, ᶜIsā ibn Yaḥyā continued the translation of the Galenic corpus, in addition to works of Aristotle, Plato, Hippocrates, and Dioscorides. The medical and philosophical translations from Galen alone account for more than 100 titles.[25]

In his missive to ᶜAlī ibn Yaḥyā, he speaks at length of his view of the translator's art,[26] while at the same time elaborating on Galen's works.[27] Here are a few relevant excerpts:

Speaking of Galen's *περί αἱρέσεων τοῖς εἰσαγομένοις* he says:

> "A man named ibn Sahda from the neighborhood of Karkh̲ translated it for me into Syriac, but he was weak as a translator; then, I myself translated it as a young man of twenty, or perhaps a bit older, from a very defective Greek manuscript for a general practitioner from around Gondēshāpūr named Shīrīshoᶜ ibn Quṭrub. Then, later, when I was close to forty, my student Ḥubaysh begged me to correct it after we had collected a number of Greek manuscripts. When I'd collated these with one another, so that I now had a single correct manuscript, I then compared this with the Syriac text and corrected it. Thus was my procedure with all that I translated."[28]

Equally revealing is his account of the work on *θεραπευτική μέθοδος*:

> "Sergius had translated this book into Syriac, and, as a matter of fact, the translation of the first six parts took place while he was still a weak and fumbling translator. He then translated the next eight parts after he was more accomplished, and thus translated

them better than he had the first six parts. Salmawayh had forced me into correcting this second half for him, and had imagined that this would be better than making a (new) translation. He and I then collated a part of the seventh portion, that is, he held the Syriac and I the Greek, and while he read it to me, each time there seemed to me to be a divergence from the Greek, I told him, and he made the correction. Finally, however, this just got to be too much for him, and it became clear to him that it would be more convenient and more successful to make a (new) translation...[29] Some years later, I translated this book from the beginning for Jibrāʾīl ibn Bukhtīshūʿ at which time, I had a number of Greek manuscripts of its last eight parts. So I collated this material, produced from it a corrected copy and translated it with the greatest possible exactness and elegance at my command. But as far as the first six parts are concerned, I had but a single manuscript, and it was error ridden at that. For this reason, it was impossible for me to treat this part as carefully as necessary. Then (later) I came across another manuscript, collated it and made corrections of those things which could be corrected (through its help); I would gladly collate the book a third time, should I find a third manuscript. The manuscripts of this book are rather scarce, and to be sure, because it was not among those (texts) read in the School of Alexandria."[30]

Ḥunain continues, in later entries, to explain that it was his usual custom also to translate from Greek to Syriac, and then from Syriac to Arabic, occasionally leaving the last step to his colleagues,[31] and even, in some cases, correcting their efforts.[32] One final excerpt from the "missive" further highlights his active search for manuscripts, in this case Galen's *περι ᾿αποδείξεως*:

"Up to this time not a single complete Greek manuscript of this book has come to light, although Jibrāʾīl went to a great deal of trouble to find it. I myself searched for it, travelled all of Mesopotomia and Syria, Palestine and Egypt, until I reached Alexandria, but I found nothing except for about a half of it in Damascus, and even this was out of sequence and incomplete..."[33]

These claims are certainly not exaggerated; Bergsträsser, in his careful study of Ḥunain and his school even remarks that his translations are studiously correct, but that one is "nonetheless... left with the impression that this is not the result of anxious effort, but of a free and sure mastery of the language. This is seen in the easier adaptation to the Greek original and the striking exactness of expression obtained without verbosity.... This... constitutes the eloquence of Ḥunain."[34] Needless to say, Ḥunain procedures conform quite well in principle with current ideas regarding the editing of classical texts.[35]

Now let us turn to Gerard of Cremona, whom Haskins treats at less length than his predecessor Adelard of Bath. Unlike the wealth of biographical detail about Ḥunain, that concerning Gerard is sparse: a eulogy appended (possibly)[36] by his colleagues at the end of his last translation, Galen's *Tegni*, and a vivid description by Daniel of Morley of his encounter with Gerard in Toledo: "I who am speaking am a king (said Gerard).... With clear irony, I asked him (Gerard) where he reigned, and he responded 'in animo', I serve no man...."[37] But Daniel appears to have been something of a wandering heckler, given to temper, so one must take his comments with care.[38] Indeed, they are quite at variance with the biographical fragment on Gerard following the *Tegni*:

"Although he scorned fame, although he fled from praise and the vain pomp of this world; although he refused to spread his name in quest for empty, insubstantial things, the fruit of his works diffused through the world makes plain his worth.... He always remained in a state constancy. Hostile to fleshy desires, he clove to spiritual ones alone. He worked for the advantage of all, present and future, mindful of Ptolemy's words: when approaching your end, do good increasingly."[39]

Surely the truth must have been somewhere between this cloying, saintly picture, and Daniel's fiery one. The biographical fragment, however, gives one valuable piece of information regarding Gerard's impulse toward translation:

"He was trained from childhood at centers of philosophical study and had come to a knowledge of all of this that was known to the Latins; but for the love of the *Almagest*, which he could not find among the Latins, he went to Toledo; there, seeing the abundance of books in Arabic on every subject, and regretting the poverty of the Latins in these things, he learned the Arabic language, in order to be able to translate. In this way, combining both language and science... he passed on the Arabic literature... to the Latin world...."[40]

So we may conclude that what drew Gerard to translate Arabic was also of a very personal sort, though his reasons may, perhaps, have been more altruistic than Ḥunain's. Unfortunately, there appears to be no account either in his own hand, or from contemporary sources of his translation procedure, barring Daniel of Morely's assertion that Gerard used the services of a Mozarab named Galippus.[41] However, as Lemay observes, Galippus' name is mentioned only in connection with the *Almagest*, and no collaborators, including him, are listed as such in the bibliography of Gerard's translations appended to the eulogy.[42] Consequently one must fall back entirely on inference based on a study of the translations themselves. In this regard, the remarks of Lemay and Kunitzsch[43] are extremely valuable.

Lemay's conclusions, drawn primarily from a study of Gerard's reworkings of John of Seville's translation of Abū Ma ͨshar's *Greater Introduction*, and another of his similar "reworking" of Gundissalinus' translation of Al-Fārābī's *On the Sciences*,[44] are roughly as follows. Early in his career, whenever possible, Gerard tried to rework an existing translation, comparing Latin with Arabic, keeping what seemed passable, standardizing vocabulary in a way consistent with contemporary Latin scientific usage, and above all, restoring faithfully any deletions of previous editors.[45] Of these earlier editors, John of Seville seems to have become his favourite. In fact, as Lemay remarks, the form adopted by both is very similar: "closeness to the Arabic original, preservation as far as possible of the construction of the Arabic sentences, and scrupulous rendering of nearly every word contained in the Arabic."[46] In fact, Gerard's method amounts to word-for-word translation.

Kunitzsch concurs with this characterization of Gerard's end procedure, which indeed is common to a high proportion of all medieval Arabic to Latin translations: "ist die geradezu sklavische Wörtlichkeit, mit der sie dem arabischen Original in Worthwahl und Syntax folgt",[47] and goes on to make valuable comparisons between Gerard's renderings, his Arabic source, and the Greek originals, in the case of the *Almagest*. His selections show rather conclusively that Gerard's adherence is to the ideal of the medieval translator, that of "fidus interpres", that for this reason his translations must be judged on linguistic–philological grounds, and that their actual information value (astronomy-wise) depends, in many caes, on the previous knowledge of the reader.[48]

Notably missing from all of these analyses, however, is any reference to collation by medieval Arabic to Latin translators of *Arabic* manuscripts, to prepare a clear text for translation, or even the comparison of manuscripts to ameliorate the effects of scribal errors that would make passages devoid of sense. On the other hand, a word-for-word translation approach would seem to make such refinements less necessary—in a naive sense. Still this matter is puzzling. The principle text for Gerard's translation of the *Almagest*, to take one example, was, with only the exception of parts of certain books, the Arabic version made by Ḥajjāj ibn Yūsuf ibn Maṭar, in 827/828.[49] As Kunitzsch notes, Gerard seems to have come on this text first, then found a version of Isḥāq (son of Ḥunain) to complete his translation, and later, used the version of Isḥāq to make corrections.[50] Clearly, the underlying idea was simply to establish a "complete" translation of the *Almagest*, regardless of version, rather than a "complete" translation of a particular version.[51]

One might be inclined to explain this apparent failure to collate texts, initially, on simple historical grounds: perhaps the Latin translators, working for the most part among contemporary Arabic speakers, looked upon the texts as nearly contemporary themselves. But, of course, the reverse was actually the case. Hugh of Santalla was by no means isolated in confusing the temporal order of Greeks and Babylonians, of placing Abū Maʿshar among the "Indians" in Egypt, and in counting Arabs among the ancients.[52] A final point should be remembered, however. The history of textual criticism in Europe does not even begin until the time of Angelo Poliziano (1454–94).[53]

This brings us to our final three remarks in this all too brief survey of modifications of Haskins' thesis; these deal with linguistic competence, "cultural filters", and "modalities of change".

There is, of course, ample evidence of competency in Greek among the Syriac and Arabic speaking translators in the time of Ḥunain. The claim for Latin, on the other hand, is slight. One searches Sezgin's *Geschichte* in vain for Latin to Arabic translations in medicine, veterinary science, zoology, pharmacy, botany, mathematics, and astronomy.[54] In fact one of the few known translations from Latin to Arabic was of Paul Orosius' *Historiarum adversum paganos libri VII*, made in Spain, probably during the time of al-Ḥakam (961–976). This "universal history", detailing events from creation to 417, and dedicated to St. Augustine was highly regarded throughout the Middle Ages.[55] On the other hand, glossaries and dictionaries are extant from the time of the Crusades, or perhaps earlier, from Latin to Arabic. These include Leiden, Codex 231 Scal., a Latin–Arabic glossary of (possibly) the tenth century[56] and the *Vocabulista in Arabico* composed in

Catalonia around 1275.[57] However, even the various studies on the Muslim madrasah are silent on language teaching, apart from the teaching of Arabic.[58]

Of more interest is the European knowledge of Greek during the late Middle Ages, for it helps to shed some light on the passage of Greek science via Arabic into Latin.[59] Apart from Sicily and south Italy, which Haskins discussed at length for their role as centres for direct Greek to Latin science translations,[60] knowledge of Greek in Europe during the period was abysmal. As a case in point, Bischoff gives a lovely example of the progressive corruption of the Greek maxim: Know Thyself—ΓΝΩΘΙ ΣΕΑΥΤΟΝ into ΓΝΟΤΟΣΟΛΙΤΟΣ, with the contemporary scholarly explanation: "ΛΙΤΟΣ equals firstly, 'the stone', and secondly, 'oneself'."[61] His next remark is equally telling: "This Greek of the Western schools and their textbooks and lexica was full-grown in the twelfth century and then lived on to the very end of the Middle Ages, unharmed by the criticism of... Roger Bacon, who himself wrote a Greek and a Hebrew grammar...."[62]

Our penultimate remark is a mere mention of an interesting theory of cultural barriers and cultural filters put forth by Normal Daniel in the early seventies to explain why certain ideas and blocks of knowledge are transmitted between cultures and others are not. Basically the argument has to do with a society's need to preserve its identity. Daniel illustrates this interesting view with examples of Greek to Arabic and Arabic to Latin science transmissions.[63]

Finally we mention some points made by Glick[64] in the context of the spread of science and technology in Medieval Spain, points which we have loosely called "modalities of change". After studying the cultural process in Spain, utilizing the works of Millás Vallicrosa, Menéndez Pidal, Vernet, and others, Glick shows how tangled a tale the story of transmission of simple technological changes can be. For example Arabic and Latin cultures may have duplicate terms for the same piece of technology. How does one establish priority? Which technologies are shared, i.e. have a common origin, as for example Roman surveying practices? Which are products of bilateral diffusion, such as military technology? What mechanisms governed the flow of technology from East to West?

In the next section, we keep these questions in mind while discussing modifications to Haskins' thesis when technology is added to his more traditional view of science and its flow, during the Crusades.

II. Technology—East and West

Sources for the transmission of technology in the Crusader period are quite diverse; some evidence stems from manuals such as the five books on clocks in the *Libros del saber*, al-Ḥasan al-Rammāḥ's *Kitab al-Furūsīya wal-munāṣab al-harbīya*, Alfred of Sareshel's translation of the alchemical parts of *K. al-Shifā*ᵓ, the *K. al-falāḥa* of Ibn al ᶜAwwām; from artistic depictions: a mid-thirteenth century, illuminated manuscript showing a water clock at the court of St. Louis at Paris,[65] the illustrations of mechanical devices in the treatises of the Banū Mūsā, Al-Jazarī, and Aḥmad b. Jalaf al-Murādī;[66] from archaeological sites and finds:

dams in Spain and the Near East, fortifications in the same, terracotta grenades, swords, glass; from philology: the Italian "cerbottana" from the Arabic "zabaṭāna",[67] the Medieval Castilian "alambique" and the Arabic "al-inbīq";[68] from the wealth of Geniza documents studied by Goiten, which indicate the dispersion of various raw materials and manufactured goods from Fusṭāṭ throughout the Mediterranean (950–1250);[69] from contemporary chronicles and histories of the Crusades.[70] The latter, as expected, provide a considerable wealth of information on the technology of warfare, for example: Ibn al-Athīr's description of a very effective form of Greek fire used against the Frankish siegetowers of Acre (1190).[71] (Equally expected, and justified is Usama ibn Munqidh's (1095–1188) account of the generally low level of Frankish medicine.[72] Yet, as Holmes points out, Usama is quite willing to praise as well, as when he mentions a Frankish remedy of glasswort and vinegar for a skin disease.[73])

It would, of course, be impossible to do justice to all this material, so we shall simply sketch a case or two from each of the first four categories.[74] Needless to say, the whole subject of technology in the Middle Ages is still in flux, to say nothing of the even more delicate subject of transfer and transmission of technology. Lynn White, Jr's delineation of the development of the trebuchet is a classic example; appearing in China at the beginning of the eleventh century, as a rope activated sling-beam for hurling stones, it is described in twelfth century Mozarab manuscripts.[75] So far so good, but then, as White notes:

> "The development of this machine was rapid and international. An Arabic treatise written in Syria between 1187 and 1192 not only speaks of Arab, Turkish, and Frankish forms of it, but also describes and illustrates a much more complex version powered by a swinging counterweight which is credited to Persia. In Europe it was known by 1199 under the name trebuchet. It is strange that this remarkable substitution of gravity for manpower should be ascribed by a Syrian to Iran, since from c. 1220 onward oriental sources frequently call such engines magribī, i.e., 'Western' and probably 'Frankish'. Moreover, while trebuchets spread with great rapidity in Europe and quickly displaced the older torsion machines, the new and more powerful artillery appears to have become dominant in the Mameluke army only in the second half of the thirteenth century."[76]

Medieval technical manuals also seem to provide evidence for transmission of certain types of technology from West to East, although the picture is far from clear. For example, al-Haṣan al-Rammāḥ's treatise mentioned above contains a description of the preparation of saltpeter, along with directions for the manufacture of various pyrotechnic devices, using it.[77] This probably reflects a Chinese influence, since saltpeter was known as "Chinese snow" in mid-thirteenth century Egypt, while al-Rammāḥ, writing in Syria, refers to his rockets as "Chinese arrows".[78] His various recipes for flares call for mixtures of (by weight) saltpeter, sulfur and charcoal in the ratios of 10, $1\frac{1}{2}$, $2\frac{1}{2}$; 10, 1, 1; 10, 2, 2; 11, $3\frac{1}{4}$, $3\frac{1}{4}$; 10, $1\frac{1}{2}$, 3.[79] The first of these is equivalent to 13.6, 2, 3.4, which comes near the usual modern proportions of 15, 2, 3[80] for gunpowder; however, all are quite far from the usual fifteenth century composition of 4, 5 or 6 to 2 to 1.[81] The high proportion of saltpeter would have probably rendered the material useless as an explosive in al-Haṣan's time owing to the hygroscopic properties of saltpeter itself. A final recipe of al-Haṣan, translated by Berthelot as "Formule de la graine franque de Patrices", has led to the inference that all the preceding recipes were also of Frankish origin.[82] Whatever the truth of this matter, a contemporary of al-Haṣan, Marcus Graecus (*c.* 1300) does include a recipe with the proportions 6, 1, 2 in his book *Liber ignium ad comburendos hostes*, but this is not really much closer to fifteenth century gunpowder than al-Haṣan's 10, $1\frac{1}{2}$, 3.[83] Their sources may well have had something in common, but Marcus, too, gives no hint that his mixture may be used to propel a projectile.[84] The real uncertainty is how this connects with what next transpired: the earliest actual use of gunpowder and cannons in Europe seems to have come about 40 to 50 years after the fall of Acre, and to have taken another 30 to 40 years to reach Egypt.[85]

One would hope that the situation would be rather more clear cut in our illustration taken from illuminated text, in this case those in the treatises on mechanical devices by the Banū Mūsā, Al-Jazarī, and (possibly) al-Murādī. However, the evidence for the influence of the Banū Mūsā on Medieval European technology is very indirect, in fact, inferential. As Hill points out, their *Book of Ingenious Devices* was highly regarded and well circulated in the Medieval Islamic world, and may possibly have been known to the bilingual population of Al-Andalus.[86] Nevertheless, there is no evidence that it was translated, either into Latin[87] or Castilian. Moreover, their work cannot have exerted a direct influence on the composition of the books on water clocks in *Libros del Saber*, since the *Book of Ingenious Devices* does not treat them.[88] On the other hand, water clocks and candle clocks (which are also treated in Alfonso's text)[89] are considered at some length by al-Jazarī. However, the jewel of the Alfonsine treatise, the mercury clock, is independent of al-Jazarī's work.[90] Again there is no evidence of a separate Latin or Castilian translation of al-Jazarī, but indirect influence cannot be ruled out until both works are compared in detail. Hill, himself, seems to hold more hope for making a connection through an Arabic manuscript of Andalusian origin, in a copy dated 1266, and recently analyzed by him.[91] Originally, it was believed that this anonymous tract, which describes "thirty machines, including water clocks"[92] was the work of the eleventh century scholar Ibn Muʿadh, however, there now seems to be evidence for attribution to one Ahmad b. Jalaf al-Murādī.[93] The date of the manuscript copy, however, still does not rule out possible influence on the water clock treatise in *Libros del Saber*, since these were composed by Isaac ibn Sid before 1282,[94] and possibly during or after 1277 when he translated the book on quadrants from an Arabic original. Yet, since the mercury clock seems to have been his own invention, this may well be gratuitous.[95] More puzzling is the fact that many of the devices and techniques described by the Banū Mūsā and al-Jazarī do reappear, possibly independently in Renaissance Europe.

This brings us to the third of our categories: archaeological finds and sites. Let us briefly examine military fortifications as an example.[96] As White notes, some knowledge of Western military technology was disseminated through parts of the Near East even before the first crusade by Frankish mercenaries.[97] In fact, in 1087, the distinctive kite shaped Norman shield appears as a decoration on

Cairo's Bāb an-Naṣir, probably owing to the services of Armenian architects who may have encountered the Norman mercenaries of Hervé, or those employed by the Byzantines.[98] As Boase[99] and others have pointed out, however, a principal problem in assessing the direction of flow of fortification technology in the Crusader era, particularly in Outremer, is that of establishing the date of various sites. This problem is coumpounded by the lack of contemporary documentation on the sites themselves, as well as the fact that many make use of ruins from pre-existing structures, e.g. the castle at Jubail employs masonry from Phoenician Bybols,[100] Chastel-Rouge (Qalᶜat Yaḥmur),[101] near Ṣāfīthā is Byzantine, Ilan-kale is Armenian,[102] and so on. Moreover, the fortunes of war led to temporary occupations, with the result that most sites were modified and extended in Crusader times, to say nothing of later Turkish rebuilding, and more recent dismantling of structures as sources of local building material. Even the distinctive stone tooling with toothed hammer that Clermont-Ganneau took as evidence of Crusader work is not an infallible test, since Arab builders used similar instruments with like results.[103] Moreover, the general tendency of Crusaders to use Syrian builders, and the prevalence of Armenian architects on both sides further complicates matters. Carbon-14 dating methods, theoretically useful in determining the age of mortar samples that are uncontaminated by shell fragments, seem, as White has observed, to provide only very muddy dates for the period 1050–1225.[104]

Villena, on the other hand, observes that much can be learned about transfer of fortification technology during this period by studying its development in Islamic Spain between the ninth and fifteenth centuries.[105] Moreover, structural dating problems here are minimal owing to stone inscriptions at sites, and abundant contemporary records, and "over building" is minor at numerous fortifications, e.g., "'Alcazabas' (citadels) of Almeria, Malaga, Gaudix, . . . , the castles of Merida, Gormaz, Tarifa, Banos de la Encina, El Vacar, etc."[106] After studying these sites, he concludes that Muslims introduced numerous fortification techniques into Spain, some as early as the eighth century, and most developed on a pre-existing eastern model. These include the barbican (Persian, "barbah-hana"), portcullis (existing, in a stone form in Pharonic Egypt and used by Arab builders in seventh century Iraq), brattices (introduced by the Byzantines in Syria in the sixth century, and by Arabs into Spain in the eleventh), machicolations (possibly existing at Nineveh, and introduced to Spain by Syrian builders).[107] These all seem to have become part of general European fortification technology after the Crusades. Other eastern innovations in Spain such as detached towers, vertical defences and cutting walls were not transferred.[108]

Spain, again, provides our best evidence for transfer of technology in the last category—the evidence of philology. Some of it, like the long parallel word lists in Spanish and Arabic for plants introduced into Al-Andalus by Arab agronomists is well known.[109] Moreover, the actual cultivation procedures for many of these plants are detailed in ibn al-ᶜAwwām's *K. al-falāḥa.*[110] Other lists, such as those for Medieval Castilian chemical terminology, are equally indicative of the strong dependence of early Spanish chemistry on Arabic models.[111] Equally striking are those for such diverse areas as textile production,[112] pottery,[113] net fishing,[114] and leather work.[115]

Hence, it would seem, that the principal transfer of technology between East and West during the Crusader period (and even earlier), was channelled through Spain, just as the transfer of science via translations from manuscripts was. This is not to say, however, that no direct transfers in these areas were made from Outremer itself. James of Vitry, for example, mentions the glass and paper manufacturing sites of Tripoli, the glass production of Acre and Tyre, and the process for preparing sugar from cane.[116] But, the establishing of direct links to the West is difficult.

III. Failures in Outremer

There is really little to add to Haskins' account of the two principal translators from Syria during the Crusades: Stephen of Antioch, a Pisan who went to Syria before 1127 to translate ʿAlī ibn al-ʿAbbās' *Kitāb al-mālikī*, and Philippus Clericus Tripolitanus who translated *Sirr al-asrār*, a compendium of wisdom attributed during the later Middle Ages to Aristotle.[117] The former, translated by Stephen as *Regalis dispositio* (*Liber regius*) was intended to replace the earlier translation by Constantine the African, a translation regarded by Stephen as incomplete.[118] The later, known as the *Secretum secretorum* was done sometime between the middle third and the first half of the thirteenth century. Both had considerable impact, one in medical science, the other on the popular imagination.[119] In light of our earlier comparisons of Ḥunain and Gerard, it may be of interest to remark that Stephen's translation is "slavishly literal and verbose",[120] while Philip's close and usually accurate translation is, in his own words, done, "sometimes literally and sometimes according to the sense, for the Arabs have one idiom and the Latins another."[121] Regardless of the impact of these two works, however, they represent little when held up against the nearly two hundred year life span of Outremer.

Summing up the intellectual life of the Crusader Kingdoms, Runciman admits that in comparison with that of Spain and Sicily it was disappointing.[122] Indeed, if *La Roman de Renart* is to be believed, one can not be too surprised. In an episode between Renart and Grimbert, five months is mentioned as the average time that a knight, and perhaps even other visitors, might have given to the entire journey to Outremer, including transit and stay.[123] This is certainly not enough time to learn Arabic, even from the travellers' manuals noted by Bischoff,[124] let alone to become involved in cultural interchanges based on language. Such interchanges were left to the long term residents, and of these, it seems that only James of Vitry, a thirteenth century bishop of Acre had any great interest in local science.[125] The major thrust of intellectual life was directed toward jurisprudence, the writing of chronicles and histories,[126] and romances—occupations that engaged but a tiny part of the population. Moreover, these people were, for the most part, educated in France. The majority of the population, on the other hand, was made up of soldiers and merchants, with little interest, or time for scientific curiosity.

In Section I, we gave some idea of the traditional Arabic schools of the time, and discussed Norman Daniel's theories of "cultural filtering", so there is no need to dwell further on these two elements and the role they played in damping possible scientific exchanges. One should also be cognizant of the simple chronology: Constantine the African had already completed his translations of Galen nearly a decade before Urban II called the first crusade, and Gerard of Cremona was dead close to 90 years before the fall of Acre in 1291. In other words, much of early Arabic science had already been translated, and, in fact, a steady, parallel set of translations from Greek, notably James of Venice's translations of Aristotle had begun by the first third of the 12th century.[127]

This is not to say that Arabic science was dead before the fall of Acre, leaving nothing worth translating. This is certainly far from true; one has only to recall, for example, that the physicist Abuʾl-Barakāt al-Baghdādī was a contemporary of Gerard of Cremona, that the Damascene physician Ibn al-Nafīs died in Cairo in 1288, and that the chemist Abū-l-Qāsim al-ʿIraqī and the engineer and astronomer Muʾayyad al-Dīn al-ʿUrḍī al-Dimishqī were his contemporaries. Moreover, the geographer Abuʾ l-Fidā of Damascus was only eighteen years old at the fall of Acre, and the noted Syrian astronomer ibn al-Shātir and his colleague al-Khalīlī produced some of their best work forty years afterward. It is of course true, however, that the center of Eastern science, by the middle of the Crusader period, had shifted from Baghdad, Damascus and Cairo into Iran. Many Eastern scientists of this period flourished in places like Tabriz, Merv, and Ṭūs. Sadly, this shift is also nearly contemporary with the decline and death of the impulse for translations from Arabic into Latin.

Also, one must not forget that the Frankish base of intellectual life in the Latin Kingdoms would have provided little in the way of science from Europe to be translated into Arabic, even if such translators had been available. One has only to consider the preoccupation of the schools of Paris and Montpellier with Arabic science in the second half of the thirteenth century to see the soundness of this claim.[128] Equally important factors are territorial, and historical. In the first place, Outremer was never more than a coastal entity. At no time did it hold a grip on centres like Baghdad or Damascus. Finally, from a historical point of view, the Crusader incursion was but a short and rather weak force in the main currents of the time: the rise of the Ayyubids, Ṣalāh al-Dīn, and then successively, the Mamlūks, the Saljūqs in Anatolia, the Qara-Khitai and Khwārizm-Shāhs, the Ghūrids, and finally the fall of Baghdad to the Mongols in 1258.[129]

Perhaps, one should take stock of this point, and express wonderment that Outremer produced any intellectual interchange at all.

Notes

1. Haskins (57).
2. Haskins (58).
3. This is only an inference. Haskins, himself, speaks of "indebtedness of the Western world to the Arabs", etc. [See (58) in Lewis' reprint, p. 84].

4. Three studies by Schipperges (125)–(127) add considerably to our current understanding of the role of Chartres.
5. See also Crombie (20)–(21); Hunt (68); Metlitzki (94).
6. See Grant (51), 111 who cites a series of essays of Lynn White, Jr, including, "Medieval uses of air", *Scientific American* **223** (1970), 92–100.
7. There is no need to be specific; the reader is invited to make his own list. Even very recent histories of the exact sciences suffer this defect.
8. Taylor (143), II, 335.
9. Haskins (57), 8–9. He had examined Ripoll ms 225 as early as 1913, and the essay "Translators from the Arabic in Spain" of (57), in which he makes these remarks, was read before the American Philosophical Society in 1923, but not published.
10. See Millás Vallicrosa: *Assaig d'historia de les idees fisiques i matemàtiques a la Catalunya medieval*, Barcelona, 1931; Glick (44), 250. Compare with Haskins (57), 8–9, especially notes 18 and 21.
11. Sezgin VI, 127–9.
12. Millás Vallicrosa (98), and more recently Vernet (154), especially 56–7.
13. Welborn (160).
14. Thompson (145).
15. In particular, see Millás' *Estudios sobre Azarquiel*, Madrid–Granada, 1943–50 and his *Las traducciones orientales en los manuscritos de la Biblioteca Catedral de Toledo*, Madrid, 1942, as well as (97)–(98).
16. Pidal (108)–(109).
17. Bossong (11).
18. Proctor (112).
19. Wenger, A., "Die astronomischen Werke Alfons X", *Bibliotheca mathematica* **3** (1905), 129–85, omitted by Haskins. We should also mention G. J. Toomer's, "A study of the Toledan Tables", *Osiris* **15** (1968), 5–174.
20. Especially Daniel (25), 151–77.
21. Glick (44), in particular part two: "Movement of ideas and techniques", 217–99.
22. It is, as a matter of fact, inspired by Walzer's informative comparison of Ḥunain with Cassiodorus Senator (540 A.D.). See Walzer (157), 162ff.
23. Anawati (5), 230 gives a translation of the event from A. Müller's edition of ʿUyūn al-anbāʾ (p. 185):

> "What makes the people of Ḥīra want to study medicine? Go away and find one of your friends; he will lend you fifty dirhems. Buy some little baskets for a dirhem, some arsenic for three dirhems, and with the rest buy coins of Kūfa and of Qādisiyya. Coat the money of Qādisiyya with arsenic and put it in baskets and stand by the side of the road crying: 'Here is true money, good for giving alms and for spending....' Sell the coins; that will earn you much more than the science of medicine."

24. Anawati (5), 230.
25. Walzer (157), 163. For more details on the list of translations, themselves, see G. Bergsträsser, ed., *Ḥunain ibn Isḥāq*: *Über die syrischen und arabischen Galen-Übersetzungen*, Leipzig, 1925 (Kraus Reprint, 1966), Iskandar (70), and Ullman (150), especially 25–68. Also, of value is Bergsträsser's *Neue Materialen zu Ḥunain ibn Isḥāq's Galen-Bibliographie*, Leipzig, 1932 (Kraus Reprint, 1966).
26. Our translations follow the text of Bergsträsser edition cited in note 25.
27. There is also a commentary on this text by Meyerhoff, "New light on Ḥunain ibn Isḥāq and his period", *Isis* **8** (1926), 685–724.
28. Bergsträsser, text 4, 18–5, 8; translation (German), 4.
29. Bergsträsser, text 17, 15–18, 2; translation (German), 14. See also Iskandar (70), 235–6.

30. Bergsträsser, text 18, 7–18, 15; translation (German) 14. See Iskandar (70), 235, note 13 for an interesting detail.
31. See, for example, entry 24 of Bergsträsser, 20 (text), 14 (German), etc.
32. This follows in the continuation of our excerpt from Bergsträsser, text 18, 15.
33. Bergsträsser, text 47, 12–17, translation (German), 38–9.
34. Cited by Meyerhoff, *The Book of the Ten Treatises on the Eye Ascribed to Ḥunain ibn Isḥāq*, Cairo, 1928, p. xxv, and repeated by Anawati (5), 231, from Bergsträsser's *Ḥunain ibn Isḥāk und seine Schule*, Leiden, 1913.
35. Nonetheless, one wonders what he would have made of the later subtleties of Lachmann and Timpanaro. Indeed, one might speculate on whether an eventual update of Sandys might not devote an entry to Ḥunain.
36. Lemay (84), 173–4 presents some interesting arguments against the socii theory. He also offers (p. 175) a role for the socii as manuscript buyers offering supporting citations from Levi-Provençal.
37. Ego qui loquor rex sum.... Cum vero ironice interrogarem ubi regnaret, respondit, in animo, quia nemini mortalium servirem.... Cited by Lemay (84), 174, who gives references to more detailed excerpts from Daniel's *Philosophia*; e.g., Rose (115), 349, etc.
38. Metlitzki (94), 36–7, and M. Müller, "Die Stellung des Daniel von Morley in der Wissenschaft des Mittelalters", *Philosophisches Jahrbuch* **41** (1928).
39. M. McVaugh's translation, in Grant (50), 35.
40. *Loc. cit.*
41. Rose (115), 347–9; Haskins (57), 15.
42. Lemay (84), 174–5.
43. Lemay (84) and Kunitzsch (79), especially pp. 104–12. A third study, I. Opelt, "Zur Übersetzungstechnik des Gerhard von Cremona", *Glotta* **38** (1959/60), 135–70, receives Kunitzsch's comment, "Leider sehr unzulänglich".
44. Lemay (84), 175–6, 181–2.
45. Such a procedure is also consistent with a learning process, but Gerard's steady application seems to have turned it into a method.
46. Lemay (84), 175–6.
47. Kunitzsch (79), 104.
48. *Ibid.*, 105. He offers, for example, Gerard's rendering:

    Et erit illud inuentum breuiter secundum considerationem que est inter loca earum scripta ut comparetur ad diuersitatem super quam significant stelle;

    of Hajjaj's Arabic:

    wa-yakūnu ḏālika mawǧūdan bi-l-yasīr bi-ʿtibār mā bayna mawāḍi ʿihā l-maktūba an yuqāsa ilā ḫtilāf ma tadullu ʿalayhi l-kawākib;

    the Greek text:

    πρόχειρον μέντοι γενοιτ᾽ ἂν αὐτόθεν δι᾽ αὐτῆς τῆς κατὰ τὰς ἀναγραφομένας αὐτων ἐποχὰς συγκρίσεως ἐπιβάλλειν τοῖς διαφόρως σημαινομένοις τῶν ἀστερων.

    Gerard's text is a literal (though hardly understandable) rendering of the Arabic, which itself is only an approximation of the clear Greek: it would be very easy to identify differently designated stars by direct comparison of their recorded positions.
49. *Ibid.*, 98–102.
50. *Ibid.*, 102 and 104.
51. Quality and style don't really enter the picture, since Kunitzsch's selections seem to indicate that the generally superior version is the Isḥāq translation, e.g. Kunitzsch (79), 104–5.

52. Daniel (24), 268–71, with additional citations from Raymund of Marseilles and Hermann of Carinthia.
53. See the interesting account of its history: E. J. Kenney, *The Classical Text*, (University of California Press, 1974), 5ff.
54. Sezgin (128), *passim*. Nonetheless, J. Vernet's work on ibn Juljul's *Ṭabaqāt* (DSB VII, 187–8) indicates that St. Isidore's *Etimologiae*, as well as works of some Christian physicians were translated into Arabic from Latin in Córdoba in the eighth and ninth century. See also Vernet's "Al-Bakrī", DSB I, 413.
55. Sarton (123), **1**, 395; Hourani (67), 99; Rosenthal (116), **1**, 95n, and G. Levi della vida's "La traduzione arabe delle storie di Orosio", Miscellanea G. Galbiati (3), 185–203, Hoepli, Milan, 1951.
56. Evidence for this early dating is given by W. M. Whitehill, "The date of the earliest Latin–Arabic glossary", *Isis* **26** (1937), 370–2, who argues against Seybold's post eleventh century dating.
57. Sarton (123), **1**, 783. Given the penchant of both Eastern and Western Medieval scholars for producing handbooks and encyclopedias, e.g. Ibn Sīnā's *K. al-Shifā*ʾ, al-Jāḥiẓ's *K. al-Ḥayawān* to mention only two in Arabic, one might speculate on why some of Pliny's *Natural History* did not pass into Arabic, at least in Andalusia. [See Stahl (136).] Interesting, in this respect, are ibn Khaldun's remarks on the detrimental effects of "encyclopedias" on the general level of education, Rosenthal (116), **3**, 290 ff. His particular references are to summaries like Ibn al-Ḥājib's works on jurisprudence, Ibn Mālik on philology, etc.
58. Dodge (28)–(29); Goldziher (48); Nakosteen (101), Totah (149).
59. For the earlier period, Traube's observations are still valid: the appearance of Greek tags in pre-tenth century north European manuscripts, for example, does little more than show a possible Irish authorship or influence. See Laistner (81), 238–50, who goes on to point out (citing Siegmund) that monks in the Byzantine enclaves in Rome and south Italy during the seventh century spent considerable time translating "Greek saints' lives, homilies, ..., into Latin." (Laistner (81), 250). See also Bischoff (9), 27–55. Kristeller (78), 112–13, 137–50 comments at length on the Humanist period, with its direct dependence on Byzantium.
60. Haskins (57), 141–54, where mention is made of an anonymous translation of Ptolemy's *Almagest* (1160), James of Venice's translations of Aristotle's *Topics, Prior Analytics, De Sophisticis Elenchis, Posterior Analytics*, and Burgundio of Pisa's translations of ten works of Galen.
61. Bischoff (10), 216.
62. *Loc. cit.* Very amusing, too, is his account of Arnold von Harff's Arabic, Turkish, Hebrew word list for pilgrims, compiled in the late fifteenth century. (p. 219–20).
63. Daniel (25); also Daniel (24), 282–98, and Schipperges, H., "Ideologie und Historiographie des Arabismus", *Sudhoffs Archive*, Beiheft 1 (1961), especially 6–26.
64. Glick (44), 217–47.
65. White (161), endplate no. 11; see 120, note 3 for commentary. As White points out, however, "iconography lagged behind actuality, and innovations were seldom reflected in objects of art until their novelty had worn off and they were taken for granted." (p. 25). See also his paper "Natural Science and naturalistic art in the Middle Ages", *American Hist. Rev.* **52** (1947), 421–35.
66. Hill (61)–(63), and his papers, "A treatise on machines", *J. History of Arabic Science* **1** (1977), 33–44; "Medieval Arabic mechanical technology", 222–37 in (60). He remarks (Bibliography, item 22, 250 in (62)) that it is more likely that the author of the manuscript discussed in "A treatise on machines" was Aḥmad b. Jalaf rather than Ibn Muʿadh to whom he originally assigned it. On this point, see King (72), note 8, 288–9.
67. White (161), 95.

68. Glick (44), 271.
69. There is also philological material, e.g., the Italian term "scala" (plank used in lading of ships) taken into Arabic as "isqāla", by Ibn ʿAwkal of Alexandria as early as the middle third of the eleventh century. Goiten (45)–(46). The "scala" reference is from his paper, "Mediterranean trade in the eleventh century: some facts and problems", 51–62 in Cook (18). See also Lewis (88), and for a fascinating collection of tenth century Arabic materials, Mez (95), especially chapters 25–9.
70. Many of these are listed, for example, in Runciman (117), C. Cahen's article, "Crusades", in *The Encyclopaedia of Islam* (new edn.) **2**, 63–6, and his text, *La Syrie du Nord à l'époque des Croisades*, 33–94.
71. Gabrielli (40), 198–200.
72. *Kitāb al-Iʿtibār*; see Gabrielli *supra*, 76–7.
73. Holmes (64), 26 note 175.
74. Perhaps we may be able to return to this subject and survey the last two categories in a future study.
75. White (161), 102. See also Needham (102) **4** (3), *passim.*, and (102), §30.
76. White (161), *loc. cit.*, and his paper, "The Crusades and the technological thrust of the West", 97–112 in Parry and Yapp (104), especially 102–3. Hill (61), 253, finds evidence only of the *manjanīq*, i.e. the sling-beam form of the trebuchet in al-Jazarī's *Book of Knowledge of Ingenious Mechanical Devices* (Kitāb fī maʿrifat al-ḥiyal al-handasiyya) of 1204 or 1206. See also his paper, "Trebuchets", *Viator* **4** (1973). Needham (102), §30 collects Chinese sources for the appearance of the counterweight.
77. Sarton II (2), 1040.
78. White (161), 97.
79. Berthelot (8), **2**, 198.
80–81. A. R. Hall, "A note on military pyrotechnics", 374–382 in Singer (130) **2**. Especially 381 and 382.
82. Berthelot, who cites a treatise of Reinaud and Favé *Le Feu gregeois*, 23 ff. (unavailable to us), seems to indicate that all these recipes are from al-Ḥasan, while Sarton II (2), 1040 cites "Formule de..." as though it came from an entirely different treatise. See also White (161), 97; Lippmann (90), 394.
83–84. Hall, *supra*, 380.
85. White (161), 99–100.
86. Hill (63), 74.
87. Gerard of Cremona did translate their teatise on the measurement of plane and spherical figures, however. See J. Al-Dabbagh, "Banū Mūsā", DSB I, 444.
88. Hill (63), and (62), 23.
89. Hill (61), 17–93, 272, who refers to Wiedemann and Hauser's "Über die Uhren in Bereich der Islamischen Kultur," *Nova Acta Abh. der Kaiserl. Leop. Carol Deutschen Akad. d. Naturforscher* **100** (1915) 1–272 (especially 17ff.).
90. Kurz (80), 18.
91. Hill, "A treatise..." *supra*, note 66.
92. Hill (63), 74.
93. See note 66.
94. Kurz (80), 18.
95. Sarton II (2), 837.
96. We could also examine the rich history surrounding the astrolabe as an archaeological "find". See Destombes (27), for example.
97. White in (104), 99.
98. *Loc. cit.* White also notes that these emblems appear in a Greek manuscript of 1066.
99. T. S. R. Boase, "Military architecture in the Crusader states in Palestine and Syria", 140–64 in volume 4 of K. M. Setton's *A History of the Crusades.*

100. Boase, *supra*, 144.
101. *Ibid.*, 145.
102. *Ibid.*, 163.
103. *Ibid.*, 143.
104. White, "The Crusades and the technological thrust of the West", 104–5. (See note 76.)
105. L. Villena, "Fortification technology in Islamic Spain", preprint of paper from *Proceedings of International Symposium on the History of Science* (Hamdard–UNESCO), *Rawalpindi*, 1979, and his essay, "Eastern influences in Spanish fortifications", *Bulletin de l'Institut Hellenique des Chasteaux Historiques* (1973), 151–8.
106. Villena, "Fortification", 2.
107. *Ibid.*, 2–4.
108. A. H. Thompson's essay, "Military architecture", 773–84 in volume 6 of (54) provides a summary picture of (principally) Northern European fortification technology in the late Middle Ages.
109. Glick (44), 77, for example, offers the following, by no means, exhaustive table:

Edible crops introduced by the Arabs

Spanish derivation	Meaning	Arabic form
aceituna	olive	al-zaitūna
albaricoque	apricot	al-barqūq
alcachofa	artichoke	al-kharshuf
algarrobo	carob	al-kharrūba
arroz	rice	al-ruz
azafrán	saffron	al-za ʿfarān
azúcar	sugar	al-sukkar
azufaifa	jujube	al-zufayzaf
berenjena	egg plant	bādhinjāna
chirivia	parsnip	jiriwiyya
limón	lemon	laimūn
naranja	orange	nāranjā
toronja	grapefruit	turunja
zanahoria	carrot	isfannāriya

110. See, for example, J. J. Clément-Mullet's translation (2 volumes in 3), *Le Livre de L'Agriculture*, 2nd edition (reprint), Editions Bouslama, Tunis, 1977.
111. Again, Glick (44), 271 provides a handy table:

Medieval Castilian chemical terminology

Medieval Castilian	Arabic	Meaning
	A. The seven spiritis	
1. alcrebite	al-kibrīt	sulfur
2. almocrate, almohatre	al-noshadīr	sal ammoniac
3. atutía	al-tūtiyāʾ	zinc oxide
4. azarnefe	al-zirnīkh	arsenic
5. azogue; sebic	al-zāʾuq; zibaq	quicksilver
6. magnesia	—	magnesia
7. marcasita	marqashītā	marcasite

B. Chemical Apparatus

1. alambique	al-inbīq	upper part of distillation apparatus
2. alculla	al-qulla	large glass decanter, cruet
3. alcuza	al-kuza	oil jug
4. alquitara	al-qiṭāra	distillation vessel
5. aludel	al-ʿutal	open tube to receive products of sublimation

Similar lists can be found in E. J. Holmyard, *Makers of Chemistry* (Oxford, 1921), 101–2, illustrating the relationship between Arabic and Latin alchemical terminology.

112. *Ibid.*, 242–4.
113. *Ibid.*, 238–41.
114. *Ibid.*, 244.
115. *Ibid.*, 245.
116. Holmes (64), 14–16.
117. Haskins (57), 130–40.
118. McVaugh (92) offers very sensible comments on this and the relative merits of the two translations. See also Schipperges (126), 35–7 (Stephen) and 17–43, 45–54, etc. (Constantine).
119. Haskins (57), 137 who notes that nearly two hundred manuscripts survive.
120. Talbot (142), 38.
121. Haskins (57), 140, 137.
122. Runciman (117) **3**, 489.
123. The episode is quoted by Holmes (64), 5. Of course, one must not forget that Adelard of Bath may have collected Arabic scientific manuscripts during his short stay in Palestine. See Haskins (57), 31–2.
124. Bischoff (10). Specific citations are given in his notes. See also Altaner (3).
125. Runciman (117) **3**, 489–90, where Palestine Pilgrims' Text Society, London 1896–7, **1**, 91–2 is given as a citation. As Runciman goes on to note, however, James' interest (local theories of earthquakes) was devoid of any contact with Muslims or local Christians owing to his religious bias. Usama also leaves us an interesting picture of "Orientalized Franks", although it has nothing to do with science; see Gabrieli (40), 78–9 for a translation of the passage.
126. Of historians, William of Tyre is the giant. He seems to have been exceptional in having his early education in Palestine, and was fluent in Latin, Arabic, Greek, French, and perhaps, had some Hebrew. Krey (77), 151 argues that his later training was completed in Italy. See also Smalley (133), especially 121–57, and Huygens (69), and the interesting chapter, "Books and their illustrations", 164–81 in Smail (132).
127. See the interesting time line in Hourani (67), 109–13.
128. Sarton (122); Sarton (123) II (2), 985–1104; Haskins (59), and the series by Schipperges (124)–(127).
129. Spuler (135), presents a rapid account of these events. See, in particular pages 88–100. Given the significance of Jerusalem to Islam, however, one must take his remark (p. 89) characterizing the Crusades for the East as "a side-show, running solely in Syria, Palestine and on the coast of Egypt", as ill chosen.

2 Pythagorean Theorems and Mumpsimus

Quite recently two historians of science have offered some didactic observations on their subject. Unguru[1] has argued the need to study the early history of mathematics within a cultural context, and Bruins[2] has pointedly shown the necessity for distinguishing between those early mathematical manuscripts which are "research papers" and those which are "instructive essays". The point of the former's argument is that the history of mathematics deserves more, than to be viewed as a linear progression whose terms are successively less anachronistic. The latter quite rightly asserts, that to assess mathematical progress—say at the time of al-Kashī—we must know whether his *Miftāḥ al-Ḥisāb* is intended as a text for beginners, or is the author's final statement on numerical methods.[3]

In the same vein, we would like to offer some personal observations on a fear that must strike the hearts of many who work in the history of science—the fear of "mumpsimus". Both of us have always had the fear, but we owe R. J. Gillings[4] a debt for having given us its name. In case the word "mumpsimus" is new to you, it means simply: "an adherence to exposed, but customary errors".[5] Sometimes the harsher perjorative "bigoted" preceds adherence, but we rather like the weaker definition, and perhaps should offer it for the shorter word "m-u-m-simus"!

In any case let us give a few illustrations of our meaning from the history, in particular some involving Pythagoras' right triangle theorem (*Elements* I, 47). Between 80 and 100 years ago, it was well known that Pythagoras discovered this theorem after conversations with certain Egyptian rope stretchers, whom the cognoscenti knew as "harpedonapti"! Moreover, others knew that Pythagoras' theorem had emerged in China in the *Chou Pei Suan Ching* dating to at least 100 B.C., and in India in the *Śulvasutras* a few centuries later. In the last 50 years these, and many other traditions relating to Pythagoras' theorem have been challenged and revised. We now distinguish cultures which studied "Pythagorean triples" distinct from Pythagorean geometry. The works of Neugebauer and Sachs[6] have called our attention to the Babylonians; Gillings[7] and R. A. Parker[8] to the Egyptians and possible influences of Babylonians on Egyptians, and the works of Needham,[9] Kaye,[10] and Pingree[11] have revised our notions on the antiquity of Chinese and Indian texts. However, in spite of this work, the old stories and traditions crop up again in various recent histories of science and mathematics—ever persistent victories for "mumpsimus".[12]

Let's take one more example, one which is much closer to our own area of interest: Thābit ibn Qurra's generalization of Pythagoras' theorem. This example has the additional charm of isolating features in the history of mathematics which call "mumpsimical tendencies" into play—namely gaps in the historical record, hypothetical proof-reconstructions, problems in determining an author's sources, and so on.

Thābit, as is well known, was born in Harran (Mesopotamia) around 836, and worked there as a money changer until going to Baghdad with Mūsā ibn Shākir, who was impressed by Thābit's linguistic talents.[13] While in Baghdad, he earned a considerable reputation for Arabic translations of Greek mathematical texts, including parts of Apollonius' *Conics*, the *Elements*, the *Almagest*, and Nicomachus' *Arithmetic*. The translations are clear and precise, and in some cases, became part of the Medieval Latin tradition, owing to retranslation by Gerard of Cremona. In addition, Thābit made a number of original mathematical contributions: a rule for construction of amicable numbers, theories for composition of ratios, a computation equivalent to $\int_0^1 \sqrt{x}\, dx$, an attempt to establish Euclid's Fifth postulate, a remarkable treatise on cylindrical sections, which uses notions akin to equi-affine transformations, and establishes propositions equivalent (in modern language) to reduction of elliptic integral of a more general type to those of first kind.[14] Of course, all this information has become relatively general knowledge only in the last century. As late as 1857, Thābit's entry in the Cambridge University Library catalogue made him a 13th century Jew of unknown origin.[15]

The particular contribution of Thābit's that interests us here is his *Risāla fi'l-Ḥujja al-Mansūba ilā Suqrāṭ fi'l-murabbaᶜ wa-quṭrihi* (Treatise on the proof attributed to Socrates on the square and its diagonal), which survives in two manuscripts, Ayasofia 4832/5 and Cairo, Dār, riyāḍ. 40 m. The first dates from approximately the fifth century A.H. and the second is dated 1159 A.H.[16] Aydin Sayili was the first to call attention to the Ayasofia manuscript in a long article in Turkish[17] in 1958, followed by an English summary in 1960.[18]

In this preliminary report, we shall refer only to Sayili's Arabic transcription of the Ayasofia manuscript. This manuscript is apparently a copy of a letter of Thābit to a friend who has expressed the wish for a proof of the general Pythagorean theorem that would be as immediately convincing as that described in Plato's *Meno* for isosceles right triangles.[19] Thābit's answer is to offer two such proofs, and three generalizations of the Pythagorean theorem. The style of the manuscript proceeds in the very best mathematical fashion—from very detailed, to very sketchy! It is this last element which will occupy most of our attention. To be fair, we must note, as Thābit does, that his discussion is ever increasing in its generality.

Thābit's first two proofs follow Figures 1 and 2 respectively, and depend, in the Socratic tradition, on dissection and composition. (In Figures 1 and 2 the $\triangle ABC$ is right, with the right angle at C.) Thābit's proof is clear and careful in each case. As Sayili[20] points out the first proof is also documented in al-Nayrīzī's tenth

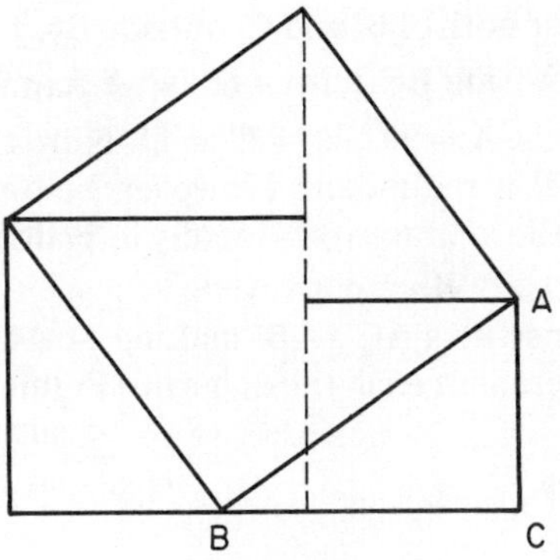

Figure 1

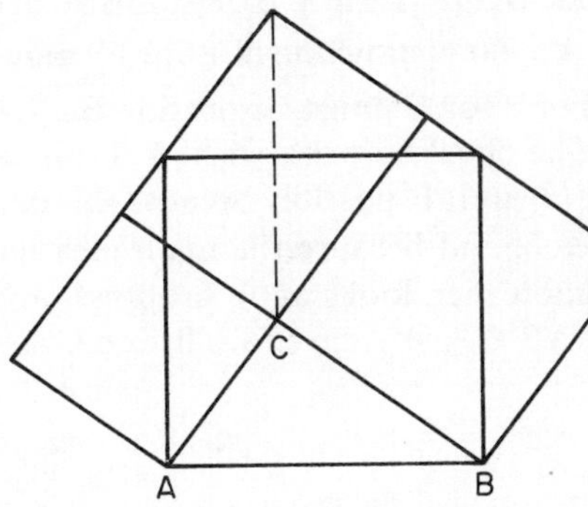

Figure 2

century commentary on Euclid, and is there attributed to Thābit. Heath,[21] in a minor slip, implies that Thābit's first proof may have been suggested by Pappus' generalization of Pythagoras' theorem.[22] But an easy check shows this is not the case. Thābit's second proof has no mention in al-Nayrīzī, though it has been rediscovered, in various forms several times in the last 80 years.

Thābit then mentions Euclid's own generalization of the theorem: *Elements* VI, 31: "In right-angled triangles the figure on the side subtending the right angle is equal to the similar and similarly described figures on the sides containing the right angle".[23] As Thābit notes, this removes the restriction of only having squares inscribed on the sides; then he goes on to consider the case most interesting to us, namely, removal of the right angle condition. He lets $\triangle ABC$ be arbitrary, and breaks his discussion into cases. First, if the vertex angle at A is acute (Figure 3), draw AC', AB' to base AB making $\measuredangle AC'C = \measuredangle AB'B = \measuredangle A$. Of course C', B' or both may lie on $\overline{BC}$ extended, and Thābit shows due concern for their relative locations. This care is quite usual in Thābit's geometric constructions, as anyone will testify who has studied, for example Thābit's *Qawl fī Taṣḥīḥ masā'il al-jabr bi 'l-barāhī al-handasiyya*,[24] which sets al-Khwārīzmī's discussion of solutions of quadratics on a firm Euclidean basis. The cases are enumerated: (1) if $\measuredangle A < 60°$, [then both $\measuredangle B$ and $\measuredangle C$ could exceed $60°$, with both B', C' outside $\overline{BC}$, or, for example, $\measuredangle B < 60°$, $\measuredangle C > 60°$, giving C' inside, B'

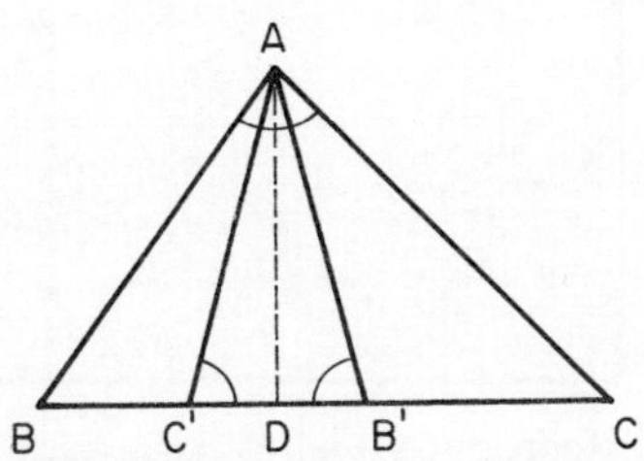

Figure 3

outside $\overline{BC}$];[25] (2) if ∢B or ∢C is 90°, [one or both of B′ and C′ outside $\overline{BC}$]; (3) if ∢A = 60°, [only one of B′ or C′ may be outside $\overline{BC}$]; (4) if 60° < ∢A < 90°, [one of B′ or C′ must be outside $\overline{BC}$]; (5) if ∢A = 90°, B′ = C′ = D. Note that since he chose to work with ∢A, his case (2) is redundant. However, he omits cases (3) and (4), possibly because this part of the manuscript is merely intended as a sketch, and because the main idea has already been conveyed.

Thābit then looks at ∢A obtuse, and constructs AC′, AB′ making ∢AC′B = ∢AB′C = ∢A; in effect, B′ and C′ are interchanged in the diagram (Figure 4),

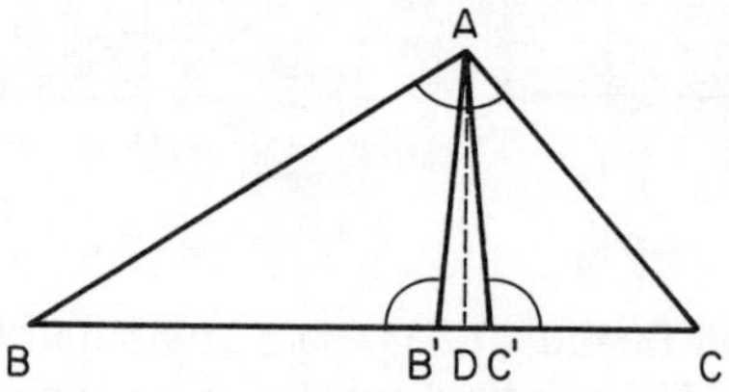

Figure 4

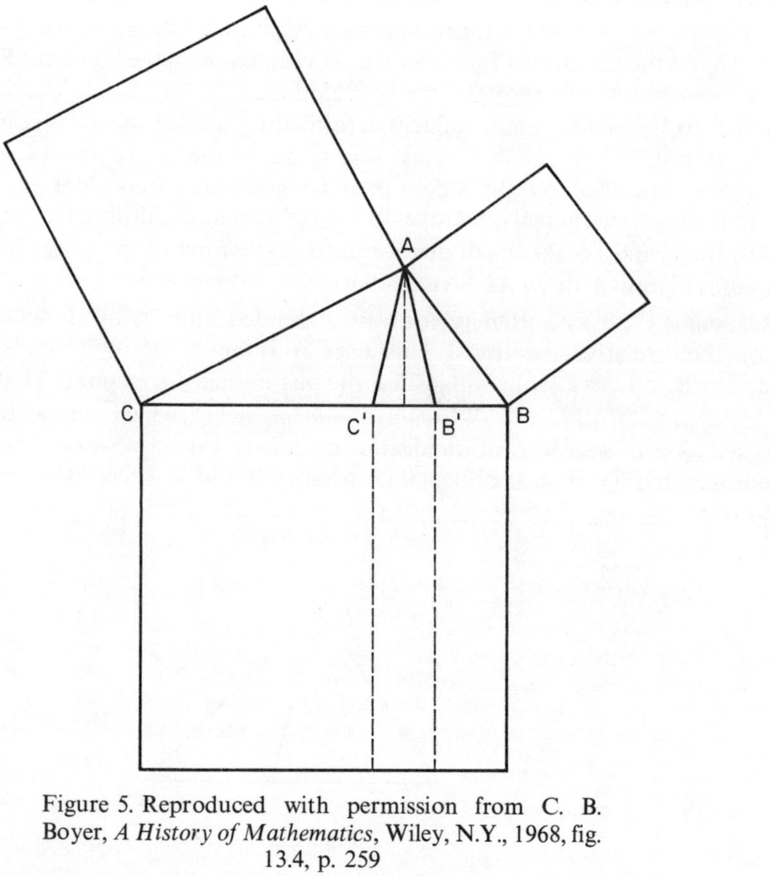

Figure 5. Reproduced with permission from C. B. Boyer, *A History of Mathematics*, Wiley, N.Y., 1968, fig. 13.4, p. 259

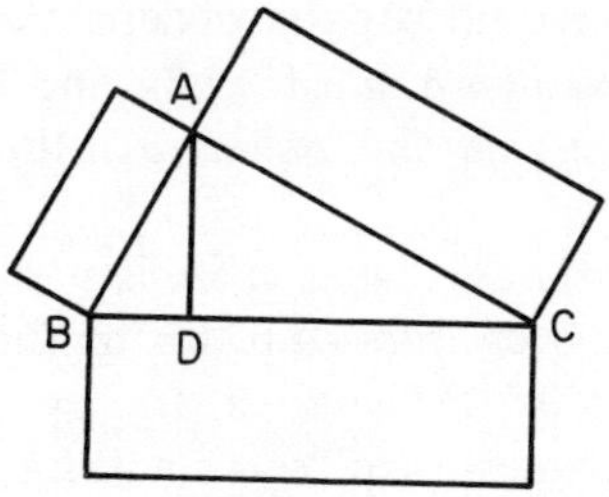

Figure 6

and there's only one case, namely BB′ + CC′ < BC, since both B′ and C′ lie inside $\overline{BC}$. Thābit then remarks that the generalization he has in mind is that (BB′ + CC′)BC = $\overline{AB^2}$ + $\overline{AC^2}$, and that the proof follows from Euclid. Since none is actually offered, however, a number of modern commentators have supplied their own, usually invoking the law of cosines, and then gone on to assert that Thābit probably used it himself, in the form of *Elements* II, 12 or II, 13.[26] This would seem more likely if Thābit's drawing included the dotted line AD perpendicular to BC, but it doesn't. Boyer, who states the theorem but only parenthetically mentions Thābit's concern for the locations of B′, C′, remarks that a proof is easily supplied by similar triangles.[27] This is certainly true; for example, in figure 4, △BAC ~ △BB′A ~ △AC′C. Hence BB′/AB = AB/BC and CC′/AC = AC/BC. Thus (BB′ + CC′)BC = $(AD)^2 + (AC)^2$, as Thābit claims. In fact, an examination of Thābit's various cases shows that such an argument is generally applicable. Moreover, as Thābit was adept at handling composite ratios—witness his manuscript on them: *Kitāb fī Taʾlīf al-nisab*—such a proof-reconstruction is possible. In addition, Boyer gives a nice drawing showing where the "pieces" (BB′)(BC), (CC′)(BC), etc. fit (Figure 5). But there is another course open. Recall that Thābit had just been discussing Euclid's generalization, *Elements* VI, 31. The diagram for this is (Figure 6). Here the triangle is right, and the bounding figures are similar rectangles, similarly placed, and as Thābit undoubtedly saw, the proof "works" because there are three similar triangles, something he merely has to duplicate to get his theorem. In fact, if one works through VI, 31 with Thābit's diagram (Figure 4), it seems most likely that Thābit would have used the three triangle similarity in the form

$$\frac{BB'}{BC}=\frac{(BB')^2}{(AB)^2} \quad \text{and} \quad \frac{CC'}{BC}=\frac{(CC')^2}{(AC)^2},$$

i.e. following Euclid's prescription [VI, 19, porism] that, if for example, BB′/AB = AB/BC, then BB′ is to BC as the similarly placed similar figure on BB′ is to the similarly placed similar figure on AB.[28]

This latter reconstruction has the benefit of providing a rather more immediate possible motivation for Thābit's theorem, as well as being closer to a definite Euclidean model. Of course, the second consideration is not particularly decisive in Thābit's case, since he was quite capable of departures from Euclid, e.g. *Maqāla fī burhān al-musādara ᵓl-mashhūra min Uqlīdis*, and *Maqāla fī anna ᵓl-khaṭṭayn idhā ukhrija ʻalā zawiyatayn aqal min qā'imatayn iltaqayā*, two treatises on the parallel postulate.

Of course, one may rightly ask: what value is such a reconstruction anyway? Well, some have certainly been important for mathematics itself, e.g. Fermat's reconstruction of Apollonius' *Plane Loci*, to say nothing of everyone else's attempt to reconstruct Fermat's own "too small for the margin proof". But, in general reconstructions are of value in forcing the historian of mathematics to examine mathematics of a given time *within the context of that time*, and in allowing him to separate flashes of genius from routine mechanics.

Thābit closes his "instructive essay" with the remark that one might also wish to deal with general triangles, and bounding figures that are not squares. He immediately supplies the theorem: the sum of similarly placed similar figures on two sides of a triangle is equal to a figure whose ratio to a similar figure similarly placed on the third side is as $BB' + CC'$ to BC. No proof is offered, but it's immediate that one follows in a similar fashion to the proof reconstructed above.

Of course Pappus (Fl. A.D. 300–350) had included an elegant generalization of *Elements* I, 47 in his *Collection*, which allows arbitrary triangles and the similar figures to be parallelograms.[29] The form of this generalization, and the circle of ideas in its proof are quite different from Thābit's. In fact, as far as is known, the *Collection* never became a part of the Arabic mathematical tradition, and very few of Pappus' other works were rendered into Arabic. The only one of these with purely mathematical content is, possibly, the *Tafsīr al-maqāla al-ᶜāšira min Kitab Uqlīdis fī maqālatain* (*Commentary on Euclid*, X) translated by Abū ᶜUthman al-Dimīshqī (Fl. *c.* 908–932).[30] So, as Sayili quite rightly observes, Thābit makes no mention of Pappus' generalization.

Still, there is a detail worth noting. If one takes Pappus' generalization, restricts the triangle to be right angled, and the similar figures to be squares, the resulting figure is the same as Thābit's second "Socratic" construction. Lest one place too much significance in this, let's also note that Thābit's generalization itself has been independently rediscovered at least three times in less than 150 years: by Wallis (1665), and by Ozanam (1790), possibly through Clairaut,[31] and as noted earlier, the second "Socratic" proof itself has been rediscovered several times in the last 80 years.

And finally, two disquieting observations, and a suggestion. Let's not forget that the translatorium in which Thābit worked also produced translations from Indian sources as well. Also some mathematical notions, even specific examples from texts do get transmitted from one language to another even though the texts themselves may never be translated. This point was driven home to us in some studies we are currently doing on the mathematical sources of Ibn Sīnā's *Kitāb al-Shīfa*, and an earlier study on Arabic work on the parallel pustulate.[32]

The suggestion is simply a reiteration of one of Gilling's: perhaps even the name Pythagorean theorem is a "mumpsimus".[33]

Notes

1. Unguru (47).
2. Bruins (4).
3. Actually Bruins remarks in this paper are mainly concerned with Archimedes' "Measurement of the circle". The references to al-Kāshī, equally apt, appear in a paper he read at the 2nd International Congress on the History of Arabic Science, Aleppo, Syria, 1979.
4. Gillings (18), 1.
5. See entry "mumpsimus" in *Webster's Third International Dictionary.*
6. See Neugebauer and Sachs (31), and also (5)–(10), (17), (20), (23)–(24), (29)–(30), (35), (46), (48)–(49).
7. Gillings (18), (19).
8. Parker (31)–(32).
9. Needham (28), 21ff. See also Mikami (27).
10. Kaye (26). See also (1), (14)–(15).
11. Specifically, see Pingree's *Bibliography of Sanskrit Science.*
12. Compare with van der Waerden (49), I, 6.
13. Rosenfeld and Grigorian (36). See also Suter (45), 34–38; Sarton (37), 599–600; Daffa' (12), 101–4, and (13), 93–104.
14. See note 13.
15. Steinschneider (44), 331.
16. Sezgin (41), 269.
17. Sayili (38).
18. Sayili (39).
19. Plato (34), 82B–85B. As Gow rightly points out [*A Short History of Greek Mathematics*, 175], Plato is less interested in the Pythagorean theorem, than in the chain of reasoning. See also Heath (21), 297–303.
20. Sayili (39), 45–6, note 2.
21. Heath (22), I, 365.
22. Hultsch (25), I, 177. (Book IV, prop. 1)
23. Heath (22), II, 268–9.
24. See, for example P. Luckey, "Ṯābit b. Qurra über den geometrischen Richtigkeitsnachweis der Auflösung der quadratischen Gleichungen", *Berichte über die Verhandl. der sachs. Akad. d. Wiss., Leipzig, math.—phys. K*l 93 (1941), 93–114.
25. Brackets in (1)–(5) are editorial additions.
26. Sayili (39), and R. Schloming, "Thābit ibn Qurra and the Pythagoren Theorem," *Math. Teacher* **63** (1970), 519–28.
27. See C. Boyer, *A History of Mathematics*, Wiley, New York, 1968, especially, 259. But see also Scriba (40), 62. Actually, all references to the *Elements* used by Thābit, including figures, can be traced to the Isḥāq-Thābit translation (Bodleian, Nicoll 279), rather than to Heath. But Heath is used as a convenience.
28. Of course, porisms are not part of the Arabic versions of the *Elements*. See Heath (22), I, 81. But the intent of VI, 19 which suggests the porism would, evidently, not have been lost on Thābit. See also Appendix I(B) of our essay on Ibn Sīnā.
29. Hultsch (25), I, 177.
30. Bulmer-Thomas (11), 299–300. Also see Sezgin (41), 174–6: Brockelmann (3), *passim*; Steinschneider (42), (43); Dodge (16).
31. Scriba (40) and Boyer (2).
32. These are essays 5 and 4 respectively in this collection.
33. Gillings (18), 1.

3 Some Myths about Logarithms in Near Eastern Mathematics

The careful work of Thureau-Dangin (37), Neugebauer (26), Sachs (28), Bruins and Rutten (71) has established that the Babylonians knew something of logarithms, and fixed the degree of their knowledge. The evidence is preserved in at least three different types of cuneiform tablets. The first, of which MLC 2078 is an example,[1] is merely a list of powers of an integer. In modern notation, it reads:

$$
\begin{array}{ll}
16^{1/4} = 2 & 2 = 2^1 \\
16^{1/2} = 4 & 4 = 2^2 \\
16^{3/4} = 8 & 8 = 2^3 \\
16^{1} = 16 & 16 = 2^4 \\
16^{5/4} = 32 & 32 = 2^5 \\
16^{3/2} = 64 & 64 = 2^6
\end{array}
$$

Of course "logarithms" are not mentioned, but the columns in MLC 2078 are equivalent to statements about logarithms: $\log_{16} 2 = \frac{1}{4}, \ldots, \log_2 2 = 1, \ldots$, etc. However, as Neugebauer and Sachs note,[2] the use of identical terminology (íb–si$_8$) to describe the relations between the terms in the two columns of this table-text seems to indicate an awareness of the connection between "exponentiation" and "taking logarithms". In fact, two problem-texts bear this out. Each is concerned with computation of interest. One (YBC 4669)[3] does little more than find the value of a for which $1 = a(\frac{6}{5})^3$. The other (AO 6770) is more to the point. It reads, "Now borrow 1 kur at interest. After how many years shall they be equal? You, operating, you shall compute for 4 years. How much is it 'more' than 2 kur? What shall I have to put to that, which resulted in 3 years, in order to obtain what I have subtracted from 4 years for 2 kur? This gives me 2.33, 20: from 4 years I subtract 2 [months]. 33.20."[4] In other words, when is $(\frac{6}{5})^n = 2$?[5] The scribe's language suggests that the solution is gotten by linear interpolation. In fact, if we set up the usual table:

$$
\begin{aligned}
(\tfrac{6}{5})^4 &= 2.0736 \\
(\tfrac{6}{5})^n &= 2.0000 \\
(\tfrac{6}{5})^3 &= 1.7280,
\end{aligned}
$$

linear interpolation gives

$$n = 4 - \frac{2.0736 - 2.0000}{2.0736 - 1.7280} = 4 - 0.2129 \quad \text{years.}$$

However, 0.2129 Babylonian years = 76.644 days = 2.555 months of 30 days, or 2.33, 20 months in sexagesimal notation.[6]

These three texts probably date from the Old-Babylonian period, perhaps somewhere around 1700 B.C..[7] The last (AO 4669) comes tantalizingly close to the discovery of logarithms as we know them. What is lacking in the Babylonian corpus, however, is the choice of a fixed base and the computation of tables of suitably small increment.[8]

Nonetheless, such texts seemingly had no impact on the mathematical history of the later Near East. Time may have been the major factor; for, as we have noted, the Old-Babylonian period extended from roughly 1800 to 1600 B.C..[9] On the other hand, a considerable body of Old-Babylonian mathematics was still known in the Seleucid-period (300 B.C.–1 A.D.).[10] Still, one must remember that the latter period, from which most Babylonian astronomical texts date, played almost no role in shaping the content of astronomical studies in Islam that began seven centuries later.[11] So the lack of Babylonian influence on Near Eastern mathematical history is, perhaps, less surprising.

However, one does find juxtaposed the sequences 1, 2, 3, 4,... and 2, 4, 8,... in late Arabic mathematical compendia, just as one finds such things in pre-Naperian mathematical literature in Europe. A case in point is a passage in 'Ali ibn Walī ibn Ḥamza al Mağribī's *Tuḥfat al-aʿdād liḏawi ʾr-rušd was-sadād* (999/1590).[12] Of course, the intent is much less clear than that of a comparable passage in Stifel's *Arithmetica Integra* (1544).[13] In any case, although Napier may have been influenced by considerations similar to those entertained by Stifel, his logarithms are formulated in kinematic terms.[14]

A second "inspiration" for logarithms, the method of prosthaphaeresis is often credited by historians of mathematics to the eleventh century Arab astronomer Abu 'l-Ḥasan ibn Aḥmad ibn Yūnus al-Ṣadafī (1009). This method uses one of four formulae involving products of sines and cosines, e.g. $\cos\theta\cos\psi = \frac{1}{2}\{\cos(\theta-\psi)+\cos(\theta+\psi)\}$, to do multiplication via addition and extensive sets of trigonometric tables. For example, suppose one wishes to multiply 35.84 by 54.46. Since $\cos(69°) = 0.3584$ and $\cos(57°) = 0.5446$, $35.84(54.46) = 100^2\{\cos(12°)+\cos(126°)\}/2 = 100^2(0.9781 - 0.5878)/2 = 10{,}000(0.19515) = 1951.5$.

Prosthaphaeresis certainly enjoyed considerable use by astronomers in Europe well into the seventeenth century. However, the crediting of ibn Yūnus with the discovery of one of its basic formulae, and the blaming of later Arab mathematicians for failure to exploit it, are parts of a modern myth. King, who did a careful study of ibn Yūnus' *Ḥākimī zīj*,[15] and the earlier commentaries of Schoy[16] and Caussin de Perceval,[17] showed conclusively that none of the four product formulae appear in the zīj. In fact, as King observes,[18] it was the astronomer Delambre who found, in 1819, that ibn Yūnus' two methods of finding the "time from solar or stellar altitude",[19] are related by the prosthaphaeresis formula for $\cos\theta\cos\psi$.[20] In fairness to Delambre, however, one should point out that his own astronomical expertise often made him impatient with the work of his predecessors. Occasionally his own "short-cuts" even get mingled with the historical record. Yet, in this instance he is much less at fault than von Braunmühl, a later interpreter. In the relevant passage, Delambre leads off with a red herring, by writing: "In a continuation of the same subject, he [ibn Yūnus] tells us, take

$$\frac{\cos(H-D)+\cos(H+D)}{2},$$

and you will have sin verse arc semi-diurne

$$=\frac{\cos(H-D)}{\frac{1}{2}\{\cos(H-D)+\cos(H+D)\}}=\frac{\cos(H-D)}{\cos H\cos D}$$
$$=1+\operatorname{tang}H\ \operatorname{tang}D."^{21}$$

But then he says, "Ibn Jounis does not give these developments, but they prove that his rule is identical to ours. He then employs the ortive amplitude . . ., etc."[22] By 1896, von Braunmühl,[23] was asserting that the formula is, indeed, in ibn Yūnus, at least, in verbal form. From that point, the prosthaphaeresis myth took a life of its own, finding its way, with varying degrees of embellishment, into texts of Boyer,[24] Nasr,[25] and even Juschkewitsch.[26]

One can, of course, speculate that the formula for prosthaphaeresis *might* have been discovered. For example, Abū'l-Wafā al-Būzjānī (940–997/8), a near contemporary of ibn Yūnus, does include proofs of addition formulae for $\sin(a+b)$ in Chapter 5 of his thirteen chapter commentary on *al-Majisṭī*, and of course, such formulae form the basis for the product formulae. For example $\sin(a+b)=\sin a\cos b+\cos a\sin b$ and $\sin(a-b)=\sin a\cos b-\cos a\sin b$ imply that

$$\tfrac{1}{2}\{\sin(a+b)+\sin(a-b)\}=\sin a\ \cos b.$$

At least two manuscripts of Abū'l-Wafā's commentary are known to exist: Bibliotheque Nationale, ar. 2497, and Istanbul, Carullah, 1479.[27] Only the Paris manuscript has been studied; it served as the basis for Carra de Vaux's analysis of 1892.[28] Abū'l-Wafā conceives of the sine of a central angle θ of a circle as being one half the chord of double the arc cut by θ on the circle, a convention prominent in early Indian mathematics,[29] and appearing in Arabic commentaries on the *Almagest* at least as early as the time of al-Fārābī (d. 950).[30]

Abū'l-Wafā offers two proofs of the "addition formula for the sine"; the first presents the result in the extremely cumbersome form:

$$\sin(a\pm b)=\sqrt{\sin^2 a-\frac{\sin^2 a\sin^2 b}{R^2}}\pm\sqrt{\sin^2 b-\frac{\sin^2 a\sin^2 b}{R^2}}$$

(deducible from the verbal presentation).[31] A second proof yields the classical formulae directly:

$$\sin(a\pm b)=\sin a\cos b\pm\sin b\cos a$$

(again verbally).[32] In fact, according to de Vaux's transcription, the Paris manuscript, at this point reads:[33]

حساب جيب مجموع القوسين وجيب تفاضلهما اذا كانت كلّ واحد منهما

معلوما اذا أردنا ذلك ضربنا جيب كلّ واحد منهما في جيب تمام الاخر

دقايق فما حصل جمعناها أن أردنا جيب مجموع القوسين واخذنـــا

تفاضلهما أن أردنا جيب تفاضلهما •

that is,

> "Calculation of the sine of the sum or difference of two arcs, when each of them is known. We multiply the sine of each of the two arcs by the sine of the complement of the other, expressing them sexagésimally in minutes, and we add the two products if we wish to know the sine of the sum; we subtract if we want to know the sine of their difference."

Curiously enough, as de Vaux points out, Abūʾl-Wafā makes no attempt to simplify his first version of the formula, which reduces to the second with $R = 1$ and a formal taking of roots. What is even more telling, however, is the fact that he actually uses the first version in his computations.[34] One gets the feeling that Abūʾl-Wafā believes that the second form is merely "pretty", while the first has true computational value. This puts him a long way from "discovery" of prosthaphaeresis, even though suitable trigonometric tables were available. Yet one must not lose sight of the simple fact that formulae easily derived in a suitably chosen notation are difficult in a poor one, and often impossible in "rhetorical algebra". Also, the notion of multiplication by doing addition is certainly not a "natural" one. Even al-Kāshī's *Miftāh al-Ḥisāb*, a veritable fifteenth century treatise on numerical methods, complete with a section on the laws of exponents, does not have anything to say on this point.[35]

Still, the myth of prosthaphaeresis does not die easily in the history of Arabic mathematics. Not only will uncorrected editions of older texts continue to appear, but new papers may add further details. In fact, the Second International Symposium on the History of Arabic Science (Aleppo, 1979) witnessed the rumour of evidence for the occurrence of one of the four product formulae in the astronomical manuscripts of Ḥabash al-Ḥāsib (d. 864–874).[36] To date this study has not appeared in print.

Finally, let us observe that regardless of the true disposition of what we call the "prosthaphaeresis myth", the history of Near Eastern mathematics is rich enough in real accomplishments to survive quite well.

Notes

1. Neugebauer and Sachs (28), 35.
2. *Ibid.*, 36.
3. Neugebauer (26), I, 516.

4. The translation is from E. M. Bruins' study, "Interpretation of cuneiform mathematics", 289.
5. In most problem-texts, the interest rate is 20%.
6. Neugebauer (26), II, 40–1, and III, 63. Goetsch (14), 150 essentially follows Neugebauer, omitting details. Bruins (above), 290, presents a more convincing reconstruction of the actual computation.
7. Neugebauer and Sachs (28), 1. See also his note 5, and the additional discussions of the texts in MKT.
8. *Ibid.*, 36.
9–10. Neugebauer (27), 14–15.
11. Even Al-Battanī's astronomical works contain little more of direct Babylonian stamp than a short section on sexagesimals. See, for example, C. A. Nallino's edition of *Opus Astronomicum*, G. Olms reprint, 1977, I, 6–9.
12. See Salih Zéky Efendi (12), 37–8, and Ḥajji Khalīfa's *Lexicon* (ed. Flügel), **2**, 221–2.
13. See Boyer (2), 342; Kline (24), 256. Boyer construes similar meaning from Archimedes' essay "The sand-reckoner", Heath (16), 221–32. A perusal of Steinschneider (34) indicates that this work was probably not translated into Arabic.
14. Smith (33), 148–55.
15. King (19), (21)–(23).
16. Schoy (29)–(31).
17. Caussin de Percival (10).
18–19. King (20), 218.
20. Delambre (11), 108, 128.
21–22. Delambre (11), 108.
23. von Braunmühl (3), 105: "so teilt Delambre mit, dass IBN YÛNOS die Formel gekannt habe:

$$\cos\psi \cos\delta = \tfrac{1}{2}\{\cos(\psi - \delta) + \cos(\psi + \delta)\},$$

das will sagen, dass er den Inhalt dieser Formel, welche eine der prosthapharetischen ist, in Worten angab." In (5), 62–4, von Braunmühl even conjectures a possible mode of proof.
24. Boyer (2), 264.
25. Nasr (25), 84.
26. Juschkewitsch (17), 300: "Bei dem in Kairo tätigen Astronomen und Mathematiker Abū-l-Hasan ʾAlī ibn abī Saʿīd ʿAbd ar-Raḥman ibn Aḥmad ibn Yūnus (gest. 1009), einem hervorragenden Beobachter und Schöpfer astronomischer Tabellen, finden wir eine mit Hilfe der Orthogonalprojektion hergeleitete Beziehung, die der Formel

$$\cos\alpha \cos\beta = \tfrac{1}{2}[\cos(\alpha + \beta) + \cos(\alpha - \beta)]$$

äquivalent ist. Diese Formel wurde von Tycho Brahe und anderen benutzt, um die Multiplikation durch eine Addition zu ersetzen. Später diente die dazu, die Summe zweier Kosinus oder Sinus in eine logarithmierbare Form zu bringen."
27. Youschkevitch (38), 43. See also, Sezgin VI.
28. Carra de Vaux (9). See also, Bond (1), 308.
29. Kennedy (18), 346.
30. Hairetdinova (15), 30. See section V of our paper on Ibn Sīnā, and its notes for details.
31. Carra de Vaux (9), 417–19.
32–33. *Ibid.*, 419. Since both proofs are readily available in Carra de Vaux, we omit them in this discussion.
34. *Ibid.*, 418.
35. Numerous Arabic editions exist; for the Russian reader, there is also a fine edition by Rosenfeld.
36. Tekeli (36), 612.

4 Naṣīr al-Dīn al-Ṭūsī's Attempt to Prove the Parallel Postulate of Euclid

Τοῦτο καὶ παντελῶς διαγράφειν χρὴ τῶν αἰτημάτων θεώρημα γάρ ἐστι,...

Nasīr al-Dīn al-Ṭūsī (1201–74) is only one of many Near Eastern mathematicians to follow an injunction similar to Proclus' to establish the parallel postulate as a theorem of Euclidean geometry.[1] Yet his work deserves singular attention for at least four reasons: his attempts indicate an awareness of the connection between the postulate and the sum of the angles of a quadrilateral; he seeks to refute the acute and obtuse angle hypothesis by the method of contradiction; he is the last Near Eastern mathematician, during the four centuries of most intense interest in geometry, to criticize and modify the work of his predecessors and contemporaries on Euclid's fifth postulate; among all Near Eastern mathematicians, including Al-Samarkandī and Qāḍī Zādā al-Rūmī, his work is the only one to reach the Latin West, thereby effecting, directly, the geometrical studies of Wallis, Saccheri, and others.[2]

Born in Ṭūs, Persia during the tumultuous times of the Mongol advance across Central Asia, al-Ṭūsī is known for contributions to mathematics, mineralogy, astronomy, logic, philosophy, ethics, and theology.[3] In Nīshāpūr, where he later gained a reputation for scholarship, he studied philosophy under Farīd al-Dīn al-Dāmād, who considered himself a philosophical descendant of Ibn Sīna, medicine with Qutḅ al-Dīn al-Masrī, a prominent student of al-Rāzī, and mathematics, notably geometry, with Kamāl al-Dīn ibn Yūnus.[4] His mathematical works are mainly in the form of commentaries and recensions (taḥrīr), and include the works of Apollonius, Archimedes, and Menelaus as well as Euclid.[5] He treats the parallel postulate (Postulate V, *Elements* I) in three places: a short tract written before 1251 entitled "Discussion designed to remove all doubts concerning parallel lines" (*Ar-risāla aš-safiya ʿan aš-sakk fi-l-ḥutūt al-mutawāziya*), and in two editions of the *Elements*.[6] These editions are not actually translations of the *Elements*, but rather reworking of Euclid, based on Arabic translations. Several printed versions are extant. One, called the larger by Suter, and actually wrongly attributed to al-Ṭūsī, was published in Rome in 1594 under the title *Euclidis elementorum geometricorum libri tredecim, Ex traditione doctissimi Nasiridini Tusini nunc primum arabice impressi*; some copies actually do contain 13 books, others 12.[7] The other, called the smaller by Suter, and correctly attributed to al-Ṭūsī, was printed in its entirety (15 books) in 1801 in Constantinople, and later (1824), partially (books 1–6) in Calcutta.[8] Al-Ṭūsī died in Kadhimain, near Baghdad in 1274, two years after the completion of the astronomical work *Zij i-īlkhānī*, and following 13 years of work at the Marāgha observatory, which he designed and constructed under the patronage of his benevolent "jailer" Hūlāgū.[9]

In what follows, we discuss the Greek tradition relating to the parallel postulate, the work of al-Ṭūsī in the context of that of his Near Eastern predecessors, contemporaries, and successors. We also include an outline of a scheme based on work of Max Dehn,[10] which makes it possible to arrange these Near Eastern parallel theories in a logical hierarchy. The paper concludes with some remarks on the influence of al-Ṭūsī's work in the West, and comments on a recent paper of Gray,[11] relating to the historiography of non-Euclidean geometry. More specifically our account of the Greek tradition examines the pre-Euclidean work mentioned in the *Corpus Aristotelcium*, the contributions of Ptolemy, Posidonius, Geminus, Proclus, Simplicius, and Aganis (Sabra's philological Agapius, Rosenfeld's Agan, an Armenian student of Gazar Farpezi). The Near Eastern predecessors of al-Ṭusi include: Al-Jawharī (*c.* 830), Al-Nayrīzī (fl. *c.* 897, d. 922), Thābit ibn Qurra (836–901), Al-Qass (*c.* 950), Ibn Sīnā (980–1037), Ibn al-Haytham (965–1039), al-Khayyāmī [Omar Khayyam] (1048 (?)–1131 (?)). Among the contemporaries of al-Ṭūsī who contributed to the theory of parallels are Qaiṣar ibn abi-l-Qāsim (d. 1251), Muḥyi 'l-dīn al-Maghribī (fl. *c.* 1260–65), Athīr al-dīn al-Abharī (d. 1265; the al-Abahrī of Suter), Al-Samarqandī (fl. 1276). The Near Eastern successors include the mathematician Qāḍī Zādā al-Rūmī (*c.* 1364–*c.* 1436), a contemporary of Ulugh Beg. Another, closer in time to Al-Ṭūsī, but born in France, is Levi ben Gerson (1288–1344), variously known as Gersonides and Leo Hebraeus. Those later European mathematicians whose work shows a plausible link with al-Ṭūsī's work on parallels are Clavius (1537–1612), Cataldi (1552–1626), Wallis (1616–1703), Saccheri (1667–1733), Lambert (1728–1777).

I. The Greek Tradition

Euclid writes (*Elements* I, Definition 23): "Parallel straight lines are straight lines which, being in the same plane and being produced indefinitely in both directions, do not meet one another in either direction".[12] His famous Postulate V, or parallel postulate, then reads: "If a straight line falling on two straight lines makes the interior angles on the same side less than two right angles, the two straight lines, if produced indefinitely, meet on that side on which are the angles less than two right angles".[13] It is used for the first time to prove Proposition I, 29: "A straight line falling on parallel straight lines makes the alternate angles equal to one another, the exterior angle equal to the interior and opposite angle, and the interior angles on the same side equal to two right angles".[14] The *Elements* appeared around 300 B.C., and there is some evidence to suggest that Postulate V was Euclid's own device to remove a *petitio principii* in the theory of parallels inherited from earlier geometers.[15] It is also reasonable to suppose that the theory of parallels appears in the *Elements* after a long chain of mathematical evolution. In fact, Toth, in lengthy paper, attempts to reconstruct this evolution, by citing more than 30 quotations from the *Corpus Aristotelicum*.[16] But much of this is tenuous and speculative, and parts, such as Toth's discovery of a theorem in *De Caelo* I 12, 28 lb 5–7, are doubtful.[17] In any case, the late use of the postulate in the *Elements*, its cumbersome wording compared to that of the other postulates, and the fact that it is, technically, the converse of Proposition I, 17 caused it to be attacked by later commentators.[18] For these reasons, we restrict our attention to the successors of Euclid in our discussion of the Greek Tradition. Most of our knowledge comes from the commentaries of Proclus and Simplicius. The former evidently never became a part of the Arab lore, but we include it in our discussion for completeness. The latter, however, comes down to us through an Arabic translation by Al-Nayrīzī. Proclus, an Athenian astronomer, philosopher and mathematician (410 [412?]–485),[19] in his *Commentary on the First Book of Euclid's Elements*, records the objections of Posidonius (b. Apameia, Syria *c.* 135 B.C.–d. *c.* 51 B.C.) and Geminus (fl. Rhodes [?] *c.* 70 B.C.) to the Euclid's theory. The former wished to substitute his own definition for parallel lines, namely: "parallel lines are those which, being in one plane, neither converge nor diverge, but have all the perpendiculars equal which are drawn from the points of one line to the other".[20] As we will see, this "equidistance" theory of parallels reappears in Simplicius and Aganis, and the concomitant notion of divergence and convergence in several attempts to prove the parallel postulate. It is not recorded whether Posidonius used this definition to produce his own proof.[21] Geminus, while also subscribing to Posidonius' definition,[22] voices much stronger objections. He observes that Postulate V is a converse to Proposition 17,[23] and insists that it should also admit a proof.[24] Then he admits that the postulate is plausible,[25] but calls attention to the hyperbola and conchoid and their asymptotes to show "that some lines exist which approach indefinitely but yet remain non-secant".[26] No proof of Postulate V by Geminus is known, although Heiberg, Heath, and others once identified him with Aganis and credited him with Aganis' proof.[27–28]

The first recorded attempt is that of Alexandrian astronomer Ptolemy (fl. *c.* 125–160);[29] it appears in Proclus, with extensive commentary.[30] Ptolemy's approach is to prove *Elements* I, 29 without recourse to Postulate V. Unlike Posidonius and Geminus, he keeps Euclid's definition of parallels, but as Proclus points outs, the "proof" is marred by Ptolemy's unconscious use of what eighteenth century mathematicians called Playfair's axiom.[31] Since the Fifth Postulate follows directly from I, 29, as Ptolemy indicates, we merely sketch his attempt at this proposition. Let AB and CD be parallel, and EF be a transversal (Figure 1). FB and DE are no more parallel than AF and EC. Hence if the sum of the interior angles on one side is less than (greater than) π, so will the sum on the other side. But this is impossible since the sum of the four angles must be 2π.[32]

The second proof, chronologically, is due to Aganis (Aghanis), who was once

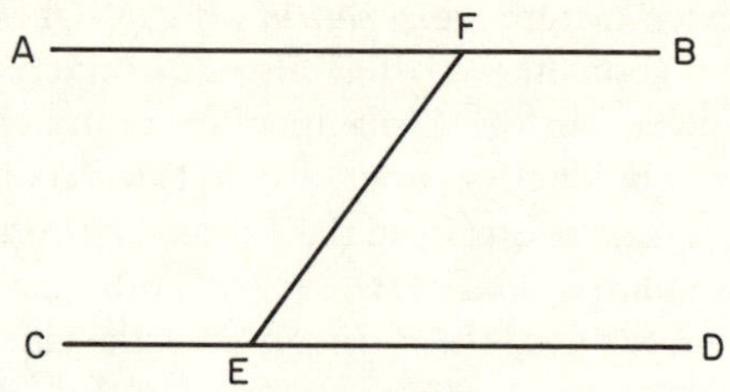

Figure 1

identified with Geminus by Heiberg; but Tannery offered very strong arguments against this.[33] Two identifications are now current: Sabra claims that Aganis is the Greek philosopher Agapius (fl. 511 A.D.), a student of Proclus;[34] Rosenfeld and Petrosian, on the other hand, say that he is an Armenian, Agan (fl. end of 5th c.), a student of Gazar Farpezi.[35] His proof appears in that part of Simplicius commentary on Euclid reproduced by al-Nayrīzī. It begins with Aganis' definition of parallels, which closely resembles those of Geminus and Posidonius. Then he establishes four propositions: (1) the 'distance' between parallels lies along a segment perpendicular to them; (2) if a straight line is perpendicular to, and intersects two straight lines, they are parallel, and the 'distance' between them is the length of the intercepted segment; (3) if a straight line falling on two straight lines make the exterior angle equal to the interior and opposite angle on the same side,..., the straight lines will be parallel. [This is Euclid I, 28. Aganis also proves I, 27 here]; (4) the converse of (3), i.e. *Elements* I, 29. He then proves *Elements* I, 30, 31 (on parallels), and *Elements* I, 33, 34 (on parallelograms). This sets the stage for his proof of the Fifth Postulate which actually tries to specify the intersection point of the two lines: let AB and CD be two straight lines which are intersected by a third line at E and F respectively (see Figure 2), and suppose the sum of the interior angles $\angle BEF + \angle EFD < \pi$. Let G be any point on FD. Draw GI parallel to AB. Bisect EF at J and JF at K, and so on until $\overline{KF}$, for example, is less than $\overline{IF}$. (This is a direct appeal to the axiom of Archimedes.) Draw FL, KMN parallel to AB. Extend FM to O and let FO be to FM as FE is to FK. Then Aganis claims that AB and CD intersect in O.[36] Unfortunately, his rather ingenious proof breaks down at the very end, owing to an unconscious use of Playfair's axiom:

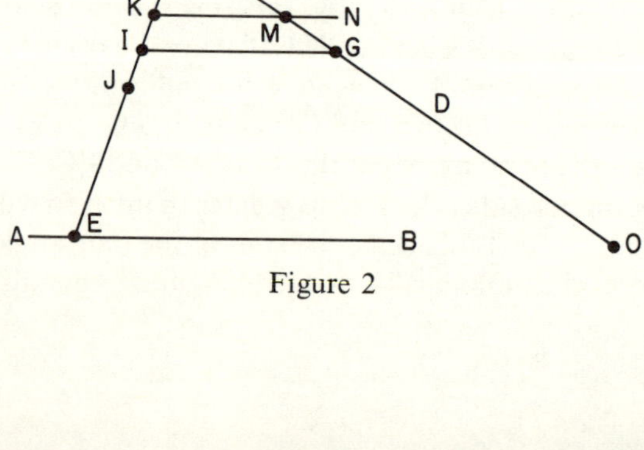

Figure 2

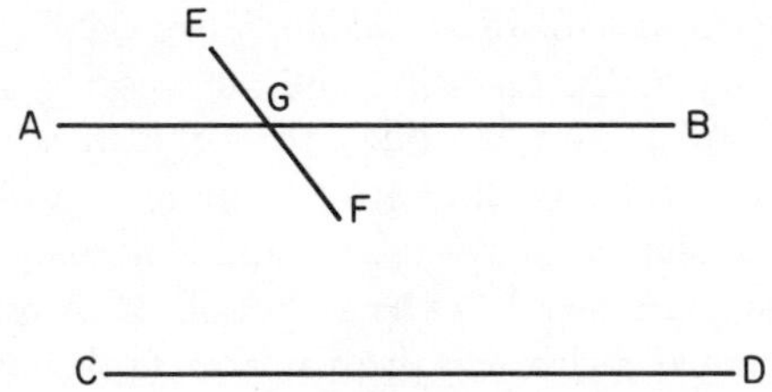

Figure 3

through a given point only one parallel can be drawn to a given line. Actually, as we noted, this axiom goes back at least to Proclus.[37] Of course, as Saccheri pointed out, the proof also runs into difficulty at the very beginning, since Aganis has to prove that the locus of points equidistant from a given line is a straight line, before he can use his definition of parallels.[38] Let us note that we record these flaws for the purpose of information, and not as criticism. For the reader's convenience we include a table of these in the section on Max Dehn's classification scheme mentioned above. (See Section VI, especially Figure 28.)

The third and last known proof of Postulate V in the Greek tradition, is Proclus' own. He bases his argument on the axiom: "if from one point two straight lines forming an angle be produced indefinitely, the distance between the said straight lines produced indefinitely will exceed any finite magnitude.[39] He off-handedly cites Aristotel [*De Caelo* I 5, 271 b 28] for a proof of "something similar".[40] He then uses the axiom to prove a lemma: if a straight line cuts one of two parallels, it cuts the other also. We sketch this proof, since his next result, a deduction of Postulate V via the lemma, is correct. He says: let AB and CD be parallel (Figure 3) with EF cutting AB at G. Now GB and GF are two straight lines from F, and if produced indefinitely will have a greater distance than the distance between the parallels. Hence GF must cut CD also.[41]

Clavius (1537–1612) argued that Proclus' axiom requires proof, and Saccheri tried to supply one.[42] However, the flaw in Proclus' argument is the assumption that two parallels are so related that, however far produced, the perpendicular from a point on one onto the other remains finite.[43]

Simplicius, the sixth century Neoplatonist philosopher,[44] whose commentary al-Nayrīzī reproduced, mentions two other scholars who offered proofs of Euclid V: "Anthisathus" or "Abthiniathus" and Diodorus. Nothing is known of these proofs, but Heath does make a conjecture on the identity of Diodorus.[45] This concludes our sketch of the Greek tradition.

II. The Near Eastern Predecessors of al-Ṭūsī

Al-ʿAbbas ibn Saʿīd al-Jawhrī (fl. Baghdad, *c.* 830) was an astronomer, an expert in *tasyīr* (the art of predicting life span based on Ptolemy's *Tetrabiblos* III, 10), and a geometer. He served the ʿAbbasīd Caliph al-Maʾmūn, and took part in observations made in Baghdad and Damascus during the early third of the ninth century.[46] Ibn al-Nadīm ascribes two works on mathematics to him: *Kitāb tafsīr kitāb Uqlīdis* and *Kitāb al-ashkāl allatī zadahā fi ᵓl-maqāla ᵓl-ūla min Uqlīdis*,[47] the first being a commentary on the *Elements*, and the second a set of propositions (50 in number according to al-Ṭūsī)[48] added to the *Elements*. Al-Jawharī's is the earliest known proof in Arabic of the Fifth Postulate; it probably appeared in his *Commentary on Euclid*. But the *Commentary* and *Additions* are now lost, and we have only al-Ṭūsī's account of al-Jawharī's proof. This appears in al-Ṭūsī's *al-Risāla ᵓl-shāfiya* mentioned earlier. His proof makes implicit use of the proposition: if a transversal makes equal alternate angles with two lines, then so does every other transversal of the two lines.[49] In fact, he uses this in the first of the five propositions he needs to establish the parallel postulate; in particular, the first proposition is: two such lines do not intersect and are equidistant. The remaining four propositions are: (2) a line segment joining the midpoints of two sides of a triangle is half the length of the third side; (3) for any angle, any number of bases may be drawn; (4) if ABC (Figure 4) is an angle; BJ a divider of $\langle$ABC and DE any base, and $\overline{FD} = \overline{BD}$, while $\overline{GE} = \overline{BE}$, then $\overline{BK} = \overline{KL}$; (5) from any interior point of an angle a line may be drawn which intersects both sides. The next proposition he proves is Euclid's Fifth postulate. Al-Jawharī does this by using (1)–(4) and the Archimedian axiom.[50] Since this axiom and (3) appear in Aganis' proof, preserved in Simplicius commentary, and since (5), according to ᶜAlam al-Dīn Qayṣar (13th century) was originally proposed by Simplicius, Sabra places al-Jawhari's proof among those directly influenced by Simplicius' work.[51] This seems to imply that the Simplicius commentary used by al-Nayrīzī was known to the Arabs much earlier.[52] However, this may well be offset by the fact that al-Jawharī's notion of distance between parallels is markedly inferior to that of Aganis, (Posidonius and Geminus). Specifically, he says (part 2 of proposition (1), following Sabra's own account): "if [two lines are] parallel to one another, then the distance from every point on one to the corresponding point (nazira) on the other is always the same, that is, the distance from the first point in the first line to the first point in the second line is the same as that from the second point in the first line to the second point in the second line, and so on."[53] The fifth proposition of al-Jawharī appears in a strengthened form as an assumption in one of A. M. Legendre's (1752–1833) attempts at proving the Fifth Postulate. In the third edition of his *Elements de Geometrie* (1800), Legendre assumes: through an interior point of an angle of

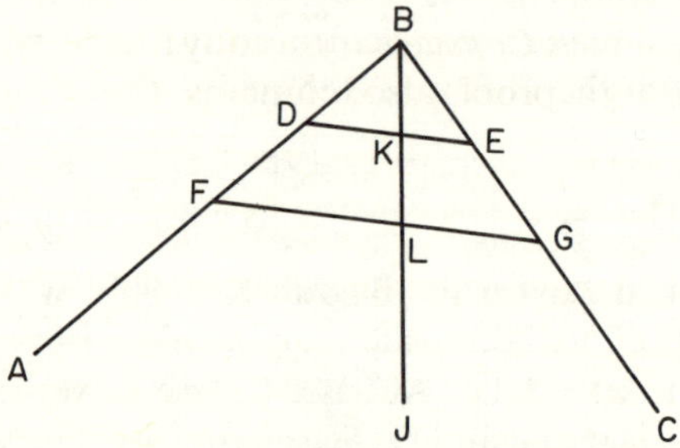

Figure 4

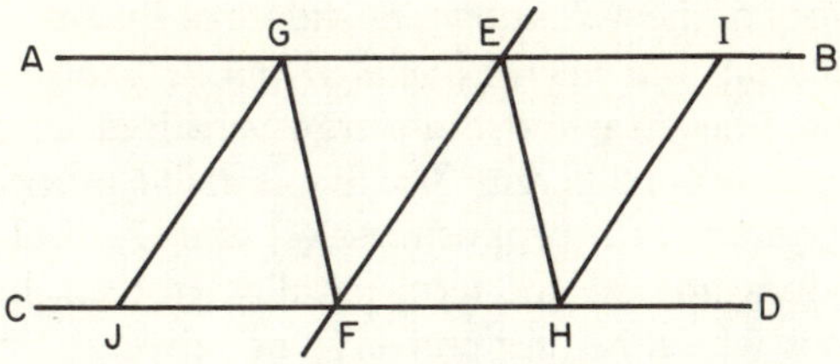

Figure 5

measure less than $\pi/3$, one can draw a straight line which intersects the sides. He abandoned this apprach in the ninth edition. There is, of course, no evidence to suggest Legendre knew anything of al-Jawharī.[54]

Since the primary flaw in al-Jawharī's proof is in proposition (1), let's examine it in detail. The first part is just Euclid I, 27 and hence is independent of the parallel postulate. In the second part he has to prove that parallel lines are "equidistant", in his sense. He takes AB, CD parallel (Figure 5) EF *any* transversal, $\overline{GE} = \overline{FH}$. Since $\langle GEF = \langle EFH$, $\triangle GEF$ and $\triangle EFH$ are congruent. Let $\overline{FJ} = \overline{IE}$. Then $\triangle GJF$ and $\triangle EHI$ are congruent, and $\overline{GJ} = \overline{HI}$. But GJ and HI join the endpoints of the equal segments $\overline{GI}$ and $\overline{JH}$. Hence they join "corresponding" points.[55] As al-Ṭūsī himself points out in the *al-Risala ʾl-shāfiya,* the pairs G and J, I and H aren't "corresponding" in al-Jawharī's sense, and even if $\overline{GE} = \overline{EI} = \overline{FJ} = \overline{FH}$, the proof doesn't show that $\overline{GJ} = \overline{FE} = \overline{HI}$.[56] In effect, al-Jawharī's failure to prove that parallels are "equidistant" in his sense, is tantamount to his adopting this as an axiom.

The work of Abuʾl-ʿAbbas al-Faḍl ibn Ḥātim al-Nayrīzī (fl. Baghdad, *c.* 897)[57] is not chronologically next in the sequence, but we consider him briefly before analysing the more extensive work of Thābit. Al-Nayrīzī, a geometer and astronomer is best known for his commentary on Euclid's *Elements* based on a translation of the *Elements* by al-Hajjaj ibn Yusuf ibn Maṭar, and the now lost commentaries of Heron and Simplicius.[58] As we have noted, this *Commentary* is of special value for its recording of the work of Aganis. Although it does not contain a proof of the parallel postulate by al-Nayrīzī himself, one is extant in manuscript, *Risala fī Bayān al-muṣādara al-mašhūra*, in at least four copies.[59] But the manuscript so closely follows Aganis' proof, that it may merely be a draft of that proof for inclusion in the *Commentary*, a copy from it, or a reworked version. For example, al-Nayrīzī's proof also contains the same three preliminary propositions as does Aganis', and the proof of the fourth (the parallel postulate), "closely follows Aganis".[60]

Thābit ibn Qurra (836–901) spent his youth as a money changer in Harran in Mesopotamia, and went to Baghdad at the request of one of the Banū Mūsā, who was impressed by Thābit's linguistic ability.[61] There he became known as an eminent translator, as well as a mathematician, astronomer, physician and philosopher. His mathematical work is phenomenal and includes two attempts to prove the Fifth Postulate: *Maqāla fī burhān al-maṣādara ʾl-mashhūra min Uqlīdis*

(Treatise on the proof of the well known postulate of Euclid), and *Maqāla fī anna ᵓl-khaṭṭayn idhā ukhrijā ᶜalā zawiyatayn aqal min qāᵓimatayn iltaqayā* (Treatise on the fact that two lines drawn (from a transversal) at angles less than π will intersect).[62] Both proofs are different. The first is in five propositions, its first and third agreeing in content with propositions (1) and (2) of al-Jawharī's proof.[63] Thābit, however, does not use the term parallel. Instead, his first proposition speaks of "two lines which neither converge nor diverge from one another in either direction when they are cut by a third line which forms equal alternate angles with them".[64] This has something in common with Posidonius' definition of parallels as given by Proclus, but is much vaguer.[65] Thābit, in proposition one, shows that if two such lines intersect, and converge on one side of the intersection, they must converge on the other. Similarly, if they diverge on one side they diverge on the other. But to do this he assumes that two lines which diverge on one side of a third, which they cut, must converge on the other side. In the third proposition, he establishes the existence of parallelograms, from which he deduces al-Jawharī's proposition (2) in his own fourth proposition. He then completes the proof of the parallel postulate, using (1)–(4) and the Archimedian axiom.[66]

In his second proof Thābit, possibly on the basis of his careful study of Archimedes' works and his knowledge of mechanics, argues that the concept of motion is necessary in geometry[67]—a point of view echoed by Helmholtz in the nineteenth century.[68] He goes on to criticise Euclid for his timid and sparing use of kinematical considerations in the *Elements*, and then offers a proof of the Fifth Postulate in seven propositions. He begins by postulating that parallel translation ("one simple motion") preserves straight lines.[69] By proposition two, Thābit has used this notion, and kinematic principles to prove that a plane quadrilateral with two angles and two sides adjoining the base equal, has the other two angles equal. When the two base angles are $\pi/2$, this so-called isoceles birectangular quadrilateral appears again, in the work of al-Khayyāmī, al-Ṭūsī, and Saccheri. In proposition (4), Thābit makes this very assumption on the base angles and concludes that the quadrilateral must be a rectangle. He uses this result and the Archimedian axiom in proposition seven to complete the proof.[70] These kinematical ideas also carry over into the works of al-Haytham, without explicit reference to Thābit.[71] Al-Haytham, on the other hand, justifies their use by citing Euclid's definition, in the *Stereometry*, of a sphere as the locus of a semicircle rotated about its diameter.[72]

The next recorded proof is one by Yūhannā ibn Yūsuf ibn al-Ḥariṯ ibn al-Biṭriq al-Qass (the priest) (fl. *c.* 950).[73] He was a translator of Greek works, and an authority on Euclid.[74] Ibn al-Nadīm cites his proof of the parallel postulate in the *Fihrist: Maqāla fi l-burhān ᶜalā annahū matā waqaᶜa ḫaṭṭ mustaqīm ᶜalā ḫaṭṭain mustaqīmain mauḍū ᶜain fī saṭh wāḥid ṣaiyara z-zāwiyatain ad-dāḫilatain allatī fī ǧiha wāḥida angas min zāwiyatain qāᵓimatain*, but it does not survive, even among those parts of al-Qass' work copied by al-Sijzī (*c.* 945, d. *c* 1020).[75] Passing mention is made of it in al-Ṭūsī's *al-Risāla*, specifically in the letter to ᶜAlam al-Dīn appended to the manuscript.[76]

After the time of al-Qass, the famous physician, and philosopher Abū ʿAli al-Ḥusayn ibn ʿAbdallāh ibn Sīnā (Avicenna) (980–1037)[77] also wrote briefly about the theory of parallels: in the *Kitāb al-shifāʾ* (the Cure [of ignorance]) which is primarily an encyclopedic treatment of Aristotelianism, and in the *Dānish Nāma-i ʿAlāʾi* (Book of Science), a popular precis of logic, metaphysics, physics, and mathematics. The latter contains a section on the properties of "lines separated from one another",[78] and distinguishes two classes, those which converge, and those which "are separated in such a way that the distances between the two 'extremities' are equal, that is, if a perpendicular is dropped from one line to the other, it is perpendicular to the second line also".[79] Ibn Sīnā continues, in this way, with little attention to rigour, and without reference to postulates, to give an account of the Euclidean theory of parallels. The passage ends with a statement of *Elements* I, proposition 30, although it is not identified as such.[80] There is, however, no evidence to suggest that Ibn Sina presented any proof of the parallel postulate.

This brings us to a consideration of the work of Abū ʿAli al-Ḥasan ibn al-Ḥasan ibn al-Haytham (965–*c.* 1040),[81] who, with al-Jawharī and al-Khayyāmī plays a considerable role in the formation of al-Ṭūsī's work on parallels.[82] Ibn al-Haytham (Alhazen) was born in Iraq, and spent a considerable part of his life in Fāṭimid Egypt.[83] Justly famous for his work on optics (*Kitāb al-manāẓir*), which presents his theory of light and vision based on experimental and physiological principles, as well as a solution of the reflection problem (Alhazen's problem) for plane, spherical, cylindrical, and conical surfaces, he also wrote several treatises on geometry.[84] These include *Commentary on Propositions not Proven in Euclid's Elements* (*Kitāb šarḥ muṣādarāt Uqlīdis fi-l-ʾ Uṣūl*) and *Book on the Solution of Difficulties in Euclid's Elements and the Explanations of his Ideas* (*Kitāb fī Ḥall šukūk Kitāb Uqlīdis fi-l-Uṣūl wa-šarḥ maʿānīhi*).[85] Apparently al-Haytham considered these two works together as his definitive statement on the *Elements*;[86] at least eight other geometrical treatises of his are known.[87] The *Kitāb ṣarḥ muṣadārāt* is especially concerned with the Fifth Postulate. As we noted above, his work reflects the influence of Thābit's kinematics, although Thābit is never directly mentioned. Al-Haytham begins by claiming that Euclid uses the axiom: two straight lines do not enclose a space.[88] This does not appear implicitly in the *Elements*, although Euclid does assume it in Proposition I.4,[89] and Proclus in his *Commentary* on I.4 not only mentions this: ("therefore two straight lines do not enclose a space and it was with knowledge of this fact that the writer of the *Elements* said in the first of his Postulates, 'to draw a straight line from any point to any point', implying that it is one straight line which would always join the points, not two"),[90] but also attempts a proof.[91] Al-Haytham then considers Euclid's definition of parallels, and argues that since one can not represent an "indefinitely extended line", the existence of parallels must be proved.[92] To this end, he uses the Archimedian axiom (implicitly) to construct a line segment whose length is an arbitrary multiple of a given segment, and then introduces a new definition of parallels, which he considers to be more evident than Euclid's.[93] It is kinematic and of the "equidistance" category, and appeals to geometric continuity: if a line segment slides along a fixed line with one end attached to the line, in such a way, that it always remains perpendicular to that line, its free end will trace a straight line parallel to the first.[94] Thus, from the very beginning, al-Haytham's work is marred by the same flaw that runs through Thābit's second proof, and all proofs in the "equidistance" category—the substitution of a definition of parallels equivalent to Euclid's parallel postulate. However, his deduction of the Fifth Postulate has historical importance in its own right, since it contains (as does Thābit's second proof) the analysis of a quadrilateral. This time, however, the quadrilateral is the so-called trirectangular one, associated with the work of J. H. Lambert on parallels.[95] We reproduce a paraphrase of the proof here.

Let ABDC be a quadrilateral with right angles at A, B, and D (see Figure 6). Al-Haytham seeks to show that the angle at C is also a right angle, by establishing, indirectly, that $\overline{CD} = \overline{AB}$. First he assumes that $\overline{CD} > \overline{AB}$. After extending CA to DE so that $\overline{CA} = \overline{AE}$,[96] he drops a perpendicular say DF from E onto the extension of DB; then he draws BC and BE. Now $\overline{FE} = CD$, since $\triangle EAB$ and $\triangle CAB$ are congruent ($\langle EAB$ is also right), making $\overline{BC} = \overline{BE}$, and $\langle CBD = \langle EBF$, and hence (since $\langle CDB$ and $\langle EFB$ are right) making $\triangle CDB$ congruent to $\triangle EFB$. However, by hypothesis $\overline{EF} > \overline{AB}$. Al-Haytham then allows EF to slide along DBF while always remaining perpendicular to it (his kinematic postulate). When F reaches B, E must coincide with H say, since $\overline{EF} > \overline{AB}$, and when F reaches D, E coincides with C. However, in the course of this motion, E describes a straight line. Consequently, the lines CAE and CHE enclose an area, thereby contradicting the axiom al-Haytham attributes to Euclid. Similarly, if $\overline{CD} < \overline{AB}$, so is $\overline{EF}$, and the same argument prevails. Al-Haytham then argues that ABDC must be a rectangle (for $\triangle DBA$ and $\triangle BDC$ are congruent, making $\triangle AKB$ congruent to $\triangle CKD$, $\triangle CKA$ and $\triangle DKB$ congruent isosceles, so $\langle KCA = \langle KAC$, and $\langle DCA$ is right). The existence of this rectangle is actually equivalent to the parallel postulate.[97] Al-Haytham argues the implication by

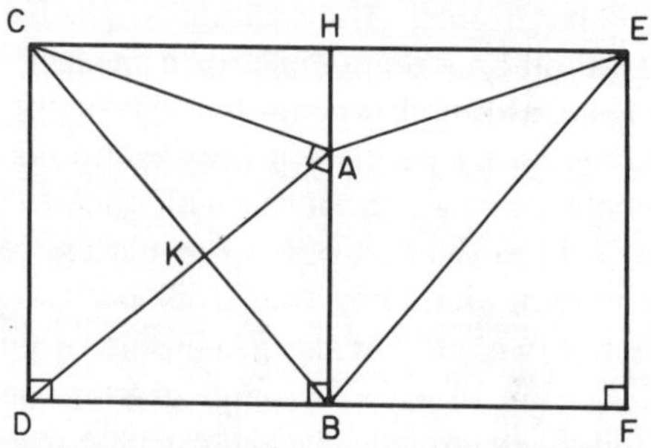

Figure 6. © American Council of Learned Societies. Reproduced with permission from C. C. Gillispie, ed., *Dictionary of Scientific Biography*, Charles Scribner's Sons, N.Y., 1970, vol. VI, Figure 4 p. 202

considering three cases: two lines are cut by a transversal and the sum of the interior angles is less than two right angles, and (a) one of these is right, (b) is actue, (c) obtuse. We omit al-Haytham's argument here, since we will present a similar one in discussing al-Ṭūsī's work. Two remarks of Youschkevitch are in order here.[98] First, in the case in which one line cuts the transversal at right angles, and the other cuts it obliquely, al-Haytham introduces another "obvious" proposition: a line which intersects a side of a triangle but does not pass through a vertex cuts another side also. This is an axiom of Pasch (1882)[99] which was also used by al-Ṭūsī (see below). Second, the quadrilateral CDFE is Saccheri's birectangular quadrilateral, so al-Haytham joins Thābit, al-Khāyyamī, al-Ṭūsī, and Saccheri in using it, albeit indirectly. Finally an observation of Sabra's is in order. The *Kitāb fī Ḥall šukūk*, written after the *Kitāb šarḥ muṣādarāt*, reformulates the Fifth Postulate as Playfair's axiom, and mentions the preceding proof.[100]

We now turn to the work of Ghiyāth al-Dīn Abu'l-Fatḥ 'Umar ibn Ibrāhīm al-Nīsābūrī al-Khayyāmī (Omar Khayyam, b. Nīshāpūr 1048 [?], d. Nīshāpūr 1131 [?]),[101] another who exerted a great influence on al-Ṭūsī. Known in the West today more as a poet, owing to the efforts of Fitzgerald, al-Khayyāmī was also an astronomer, philosopher, and mathematician of considerable stature. Book I of his *Commentary on Euclid* (*Sharḥ ma ashkala min muṣādarī Kitāb Uqlīdis*) contains his contribution to the theory of parallels.[102] He begins his account by chastizing al-Haytham for bringing "kinematical" principles into geometry, and argues that al-Haytham's use of Euclid as his authorization is specious.[103] In particular, he observes that Euclid could afford loose language in the *Stereometry* since it was presumably addressed to those who had already mastered the *Elements*;[104] then he clinches his argument by noting that in the *Elements* Euclid was careful to define a circle as a locus, rather than as a figure generated by the planar rotation of the end point of a segment round a fixed point.[105] Having said as much, he blithely gives Aristotle as his authority for a principle he will use to establish the parallel postulate: "two convergent straight lines intersect, and it is impossible that they diverge in their direction of convergence".[106] A perusal of the known *Corpus Aristotelicum* uncovers no such principle, but the reader is invited to compare this with the assumptions made by Thābit in proposition one of his first proof of the parallel postulate. After stating Archimedes' axiom, al-Khayyāmī uses the Aristotelian principle, and a sym-

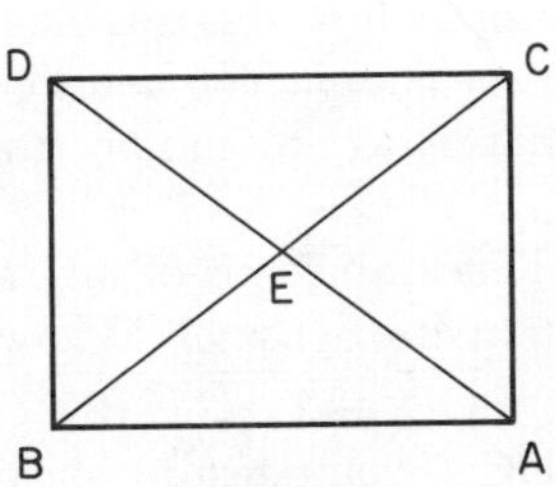

Figure 7

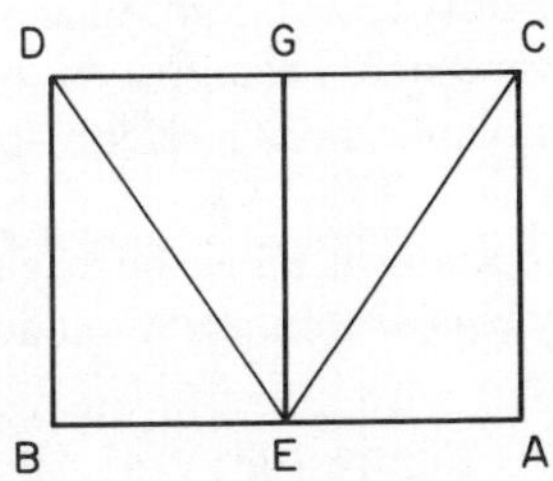

Figure 8

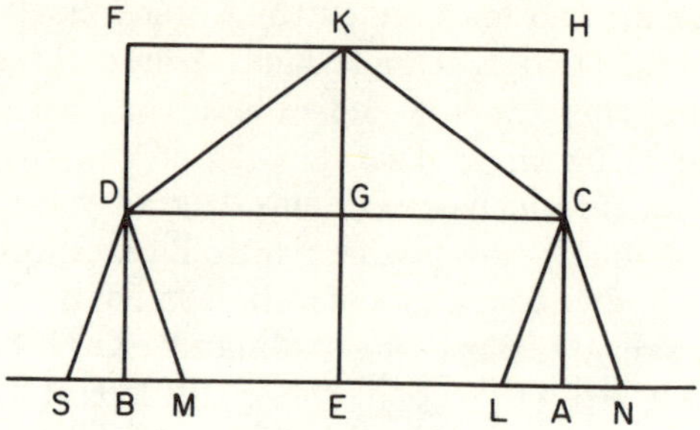

Figure 9

metry argument to show that perpendiculars to a given line can neither converge nor diverge, and so, are equidistant.[107] His proof of the Fifth Postulate follows in a sequence of eight propositions. First he considers a quadrilateral ACDB (Figure 7) with base AB, sides AC and BD equal, and right angles at A and B (Saccheri's quadrilateral).[108] In Proposition I he has shown that $\angle BDC = \angle ACD$. Figure 7, in fact suggests his proof. Then in proposition II, al-Khayyāmī proves (Figure 8) that if a perpendicular to BA is erected on the midpoint E, GE is perpendicular to DC and $\overline{DG} = \overline{GC}$. The proof again follows the suggestion of the figure. Proposition III establishes that both $\angle BDC$ and $\angle ACD$ are right angles. Al-Khayyāmī's accompanying figure is Figure 9. As in proposition II, E is the midpoint of BA, GE perpendicular to BA at E, and GE extended so that $\overline{GE} = \overline{GK}$. FKH is perpendicular to EK, and AC, and BD are extended, intersecting HKF in H and F respectively. Draw DK and GK. Then $\angle GDK = \angle GCK$, and consequently so are $\angle HCK$ and $\angle KDF$, and thus $\angle KHC$ and $\angle KFD$. Therefore, since $\overline{DK} = \overline{KC}$, $\overline{CH} = \overline{DF}$ and $\overline{HK} = \overline{FK}$. If $\angle ACD$ and $\angle BDC$ are right angles, there's nothing to do (case 1). Otherwise, both angles are acute (case 2), or obtuse (case 3). Al-Khayyāmī deduces a contradiction in the latter two cases by appealing to his Aristotelian axiom. Suppose the two angles are acute. Fold Figure 9 around the line CD. HF then coincides with SN, where $\overline{SN} > \overline{AB}$. (In the obtuse angle case, it coincides with ML, $\overline{ML} < \overline{AB}$.) Now undo the fold and lower CD to BA. Under the obtuse angle hypothesis, the perpendiculars BF and AH to AB diverge on both sides of AB, while under the acute angle hypothesis, they converge on both sides. But this violates the preliminary proposition that perpendiculars to a given line neither converge nor diverge. Hence only case 1 can hold. Notice again how this double convergence and double divergence is reminiscent of ideas of Thābit. Yet, in the course of this work al-Khayyāmī only refers to Thābit as a minor translator of Euclid.[109]

Since the proposition III is the key to al-Khayyamī's proof we merely record the remaining propositions and their accompanying figures.[110] These are: (IV) In the figure ACDB with all right angles, $\overline{AB} = \overline{CD}$ and $\overline{AC} = \overline{BD}$. (The accompanying figure is Figure 10.) (V) If AB and DC are equidistant lines (Figure 11), any line perpendicular to one is perpendicular to the other. (VI) Any two lines,

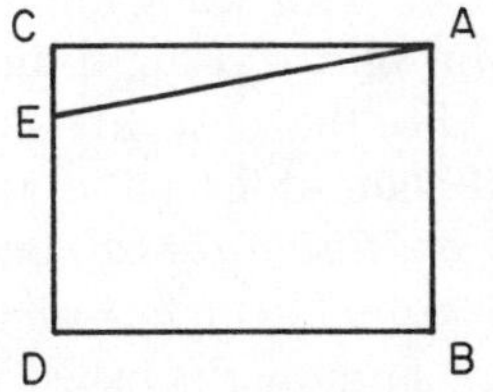

Figure 10

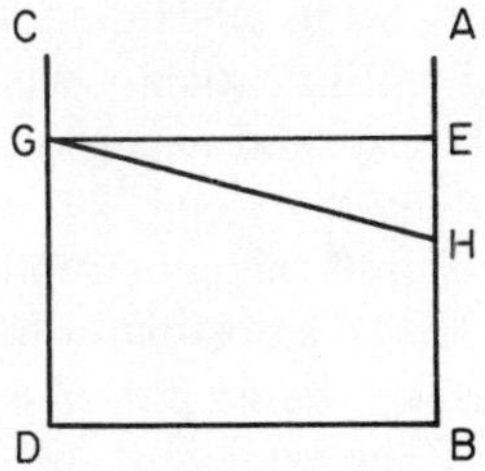

Figure 11

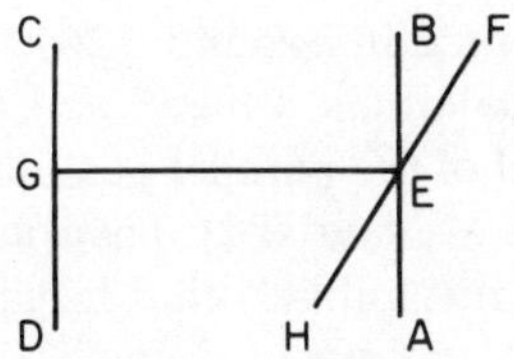

Figure 12

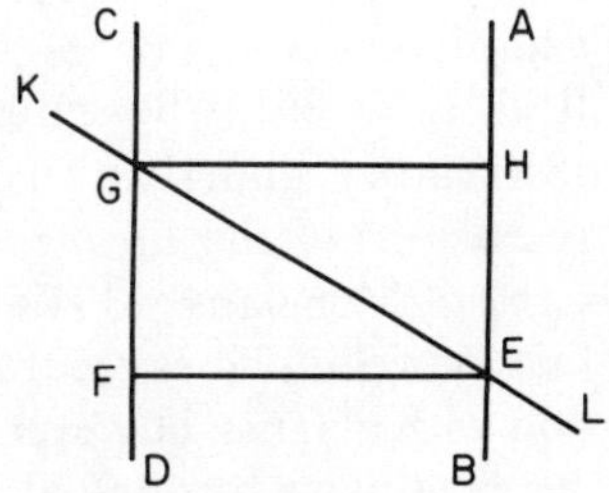

Figure 13

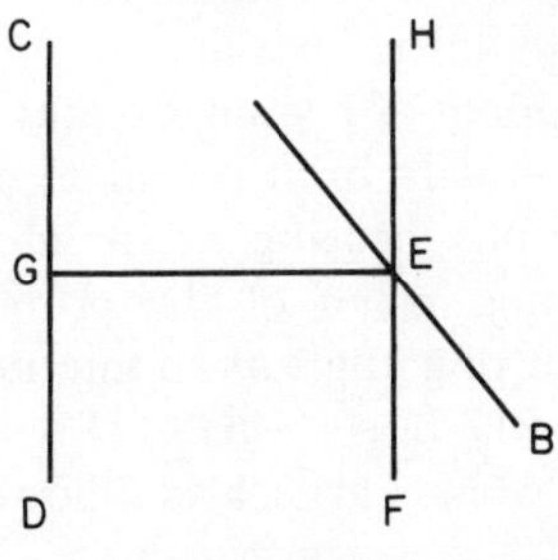

Figure 14

parallel in the sense of Euclid, are equidistant (Figure 12). (VII) Corresponds to *Elements* I, 29: "A straight line falling on parallel straight lines makes the alternate angles equal to one another, the exterior angle equal to the interior and opposite angle, and the interior angles on the same side equal to two right angles".[111] (VIII) This is al-Khayyāmī's deduction of the parallel postulate. The accompanying figure is Figure 14.

Before discussing the work of al-Ṭūsī and his contemporaries, it is well to record Youschkevitch's observation that al-Khayyāmī's treatment of the subject of parallels testifies to his familiarity with the *Commentary* of al-Nayrīzī.[112] In fact al-Khayyāmī cites the *Commentary* explicitly.[113]

III. The Near Eastern Contemporaries of al-Ṭūsî

Before analysing the work of al-Ṭūsī, let's consider that of his contemporaries. The first of these is the Egyptian Qaiṣar ibn Abī-l-Qāsim ibn ʿAbd-al-Ghanī ibn

Musāfir, ᶜAlam al-dīn, al-Ḥanafī (1178/79 [1168/69?]–1251), an engineer, astronomer, and mathematician, who studied with Kamāl al-dīn ibn Yūnus, as did al-Ṭūsī.[114] He worked in Syria for Muẓaffar II Taqī al-dīn Maḥmūd, who ruled Ḥamāh for fifteen years (1229–44).[115] For the most part this work had to do with fortifications, and the construction of water-mills (nā ᶜūra). He dedicated a commentary on Euclid's postulates to al-Ṭūsī: *Risāla fī Ma ᶜrifat ḫawāṣṣ al-ḫuṭūṭ al-mutawāziya wa-a ᶜrādiha ḏ-ḏātīya wa-i-Mutaqāti ᶜa*, but we know little more than the title.[116] [The same must be said of a number of other commentaries on the postulates, for example *Šarh musādarāt Uqlīdis*, mentioned by Ibn al-Qifṭī, and written by Faḫraddīn al-Rāzī (d. 1210)].[117]

There is extant, however, a proof of the parallel postulate by Muḥyi ᵓl-dīn al-Maghribī (fl. *c.* 1260–65), who came from the Islamic West to work at the Marāgha observatory under al-Ṭūsī's direction around 1264.[118] Among his works are recensions of the *Elements*, Apollonius' *Conics*, and the *Spherics* of Theodosius and of Menelaus.[119] His proof of the parallel postulate is related to the circle of ideas of Simplicius recorded in al-Nayrīzī's *Commentaries.*[120] Given his association with al-Ṭūsī, however, it is not unlikely that he came to know of the Simplicius postulates through al-Ṭūsī's critique of al-Jawharī's work, which also contains these postulates.

This brings us to Athīr al-dīn al-Muffaḍḍal ibn ᶜUmar al-Abharī, (d. *c.* 1265) a Persian philosopher, astronomer, logician, and mathematician who wrote in Arabic.[121] He was also a student of Kamāl al-Dīn.[122] Like al-Jawahrī, he bases his proof of the parallel postulate on a postulate of the "Legendre-type". We sketch this proof since it has bearing on a later one by Al-Samarqandī. Specifically, he begins with a proof of the proposition (1): a straight line perpendicular to the bisector of a angle at an interior point must intersect both sides of the angle.[123] Let BD bisect ⟨ABC (Figure 15). Construct FE perpendicular to BD. Then al-Abharī argues as follows:

> "If FE does not intersect BA and BC then any segment away from B would not intersect BA and BC. If A'MC', for example, which is perpendicular to BD, intersects BA and BC, then FE will also, since, in △A'MB, FE intersects BM and hence either A'M or A'B. But FE cannot intersect A'M because we cannot have a triangle with two right angles (Euclid I, Proposition 17)."

Of course, the "if A'MC'" statement is *petitio principii.* Al-Abharī goes on to

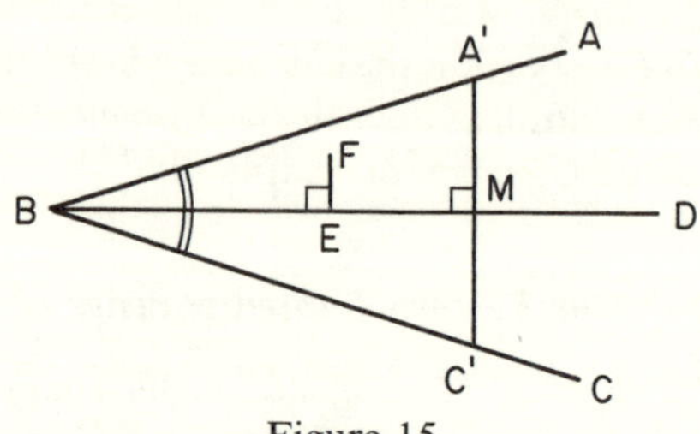

Figure 15

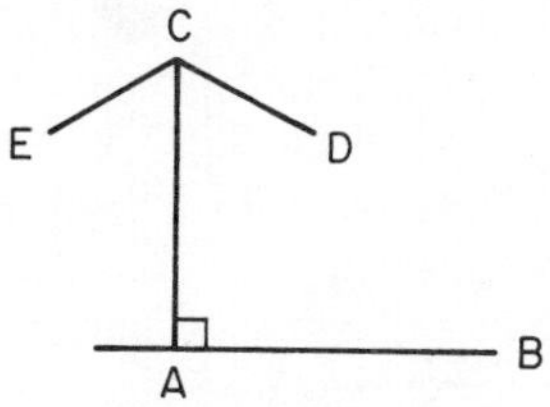

Figure 16

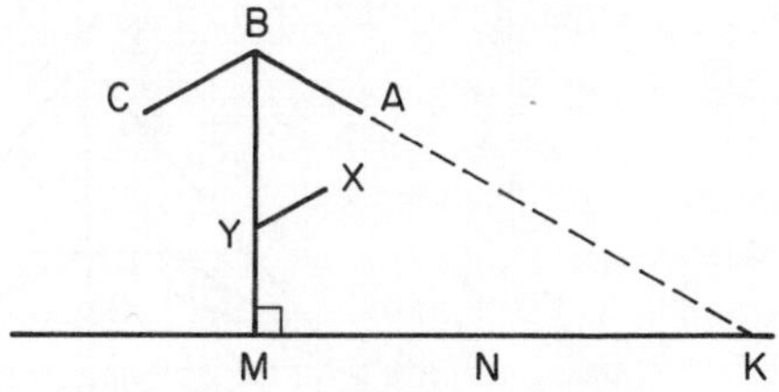

Figure 17

complete his short proof (four manuscript pages) by considering three cases. These are shown in Figures 16, 17, 18.

In the first case (Figure 16), he lets $\langle BAC = \pi/2$ and $\langle DCA$ be acute, and proves that DC and AB intersect, as follows. Draw CE so that $\langle ECA = \langle DCA$. Then CA is the bisector of $\langle ECD$, with BA perpendicular to it, hence the result follows by proposition (1). In the second case, he takes $\langle ABY$ and $\langle BYX$ to be acute, and shows the line segments BA and YX extended intersect. He draws BC so that $\langle ABY = \langle YBC$, and extends BY to M say, and draws MN perpendicular to BM. Again MN extended and AB extended meet by proposition (1), say at K. In $\triangle$BMK, YX intersects BK or MK. However, it can not intersect MK (Euclid I, proposition 17, again). The third case (Figure 17) is, of course that of an acute and obtuse angle.

Al-Abharī's argument goes as follows. Let $\langle AXY$ be acute, $\langle CYX$ obtuse, and $\langle AXY + \langle CYX < \pi$. Draw from N, the midpoint of XY, the perpendicular NF on CD. Thus $\langle YNF$ is acute (Euclid I, proposition 17), and consequently, so is $\langle MNX$. Now, since $\langle AXN$ is also, MN and AX intersect, say at K (case 2). Furthermore $\langle NKX$ can not be a right angle, for otherwise $\triangle$NXK is congruent to $\triangle$NYF (Euclid I, proposition 26). This implies that $\langle KXN = \langle NYF$, and in turn that $\langle KXN + \langle CYX = \pi$, a contradiction. Next, if $\langle NKX$ is acute AB and CD intersect (case 1). Finally, if $\langle NKX$ is obtuse the sum of $\langle NKX$ and $\langle NFY$ is more than two right angles, contradicting Euclid I, proposition 17. Hence, al-Abharī concluded, AB must intersect CD.

This brings us to the last of al-Ṭūsī's contemporaries known to have produced a proof of the Fifth Postulate, Shams al-Dīn Muḥammad ibn Ashraf al-Ḥusaynī al Samarqandī (fl. 1276).[124] He was a mathematician, logician and astronomer,

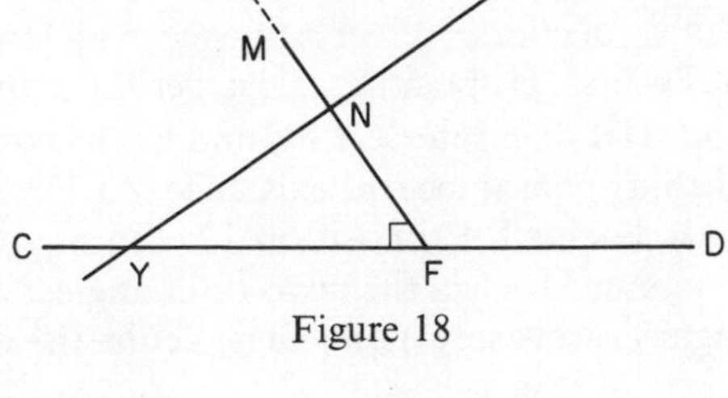

Figure 18

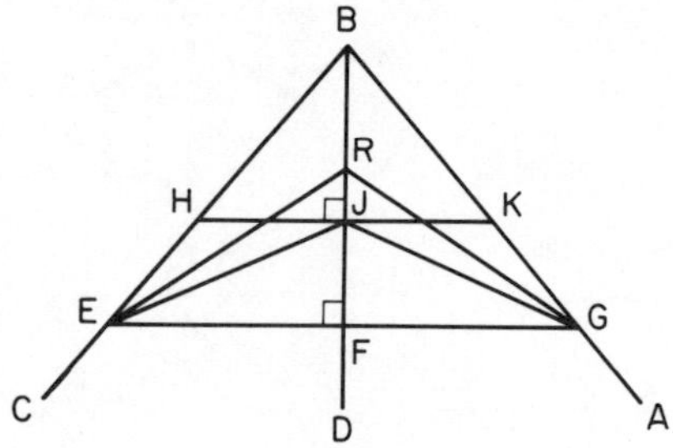

Figure 19

but did not belong to the Marāgha group working under al-Ṭūsī.[125] Al-Samarqandī's proof appears in his treatise *Aškal at-ta᾽sīs*, which contains short proofs of 35 propositions selected from *Elements* I.[126] It borrows heavily from al-Jawharī, al-Haytham, al-Khayyāmī, al-Abharī and al-Ṭūsī. Unlike al-Ṭūsī, however, he does not offer an extended critique of his predecessors, except to say he prefers to base his discussion of the parallel postulate on al-Abharī's because it is shorter than al-Ṭūsī's![127] The treatise enjoyed considerable success. Mīrak al-Buhārī (d. *c.* 1340) and Qāḍī Zādā al-Rūmī (d. *c.* 1436) wrote still extant commentaries on it, while Tāğ As-Sa ʿīdī (a student of al-Rūmī), al-Mašhadī (fl. 1420), al-Niẓāmī (d. 1513), and Mulla Čelebi (d. 1656) all provided glosses.[128]

The first proposition in al-Samarqandī's proof is: one can construct an infinite number of segments, one beneath the other (Figure 19), cutting the bisector BD of a given $\angle CBA$, so that on each of the segments, one can erect isosceles triangles, with one vertex on BD, and BD bisecting the vertex angle; e.g. segments EG, HK, ...[129] His proof runs as follows. Take a point K on side BA, and take BH = BA. The line through the points H, K forms two triangles: $\triangle BJH$ and $\triangle BJK$. Since $\overline{BH} = \overline{BK}$, BJ is common and $\angle HBJ = \angle KBJ$, the two triangles are congruent. The $\angle BJK$ is thus a right angle. (Here al-Samarqandī says that this also follows from a result of the ʿAbbāsid Caliph al-Ma᾽mūm, (r. 813–833))[130] Thus HK is perpendicular to BD. In the same manner, choose another point, say G on AB, take $\overline{BE} = \overline{BG}$. Al-Samarqandī then claims that the line EG cannot pass through J; for if it did, by the previous argument, $\angle BJG + \angle BJK = \pi$. But since BJ is perpendicular to HK, $\angle BJK + \angle BJH = \pi$. Hence $\angle BJG + \angle BJK = \angle BJK + \angle BJH$, which is absurd. Moreover, if EG cuts ER, we have a contradiction to Euclid's postulate six (al-Haytham's: two straight lines do not enclose a space). Hence, al-Samarqandī concludes, GE does not cut HK, and therefore crosses BD at same point F, below J.

This proposition shows, of course, direct influence of al-Jawharī's propositions (2)–(5), and al-Abharī's first proposition. The *petitio principii* is even more cleverly concealed, since HK does intersect BC and BA by construction (Euclid I, Postulate I),[131] but nothing guarantees the existence of J, H, ... Al-Samarqandī's next proposition is just Euclid I, Proposition 17: "In any triangle two angles taken together in any manner are less then two right angles."[132] The proof of the parallel postulate is again casewise: (1) one angle acute, the other right; (2) both angles acute; (3) one angle acute, the other obtuse. The arguments follow the order and arrangement by al-Abharī, so we do not reproduce them here.

IV. The Work of al-Ṭūsī

Since we have already recorded a biographical sketch, we may proceed directly to al-Ṭūsī's work, "Discussion designed to remove all doubts concerning parallel lines". This begins with a critical appraisal of the contributions of al-Jawharī, ibn al-Haytham and al-Khayyāmī.[133] We have also noted al-Ṭūsī's criticism of al-Jawharī;[134] his criticism of al-Haytham is based on *Kitāb fī Ḥall šukuk Kitāb Uqlīdis fi l-Usul wa-šarḥ ma ͨ ānīhi* (Book on the solution of Difficulties in Euclid's Elements), rather than the earlier *Kitāb šarḥ muṣādarāt*, which al-Ṭūsī says was not known to him directly.[135] As Jaouiche has pointed out, this criticism is basically the same that al-Khayyāmī levels at al-Haytham: kinematic principles are not admissible in geometry. The criticisms of al-Khayyāmī are much more detailed, since al-Ṭūsī builds his own theory primarily on this work; these are in particular: first, al-Khayyāmī's use of the principle of continuity (for angles), which is not a feature of the *Elements*; second, his incorrect use of the axiom of Archimedes; third, his confusion about the distance between two coplanar lines (in evidence by al-Khayyāmī's attempt to define the distance between intersecting lines). Finally al-Ṭūsī points out that all his predecessors have, implicitly, or explicitly used equivalents of the parallel postulate in their proofs.[136] He then sets himself the task of trying to establish the Fifth Postulate by using only the orthodox principles and definitions in the *Elements*. This, of course, is a new point of departure. However, all this careful work is foredoomed, since he implicitly uses a postulate that he had explicitly given as an alternate to Euclid's Postulate Five in one of his recensions of the *Elements*: "If two co-planar lines diverge from one another in a given direction, they can not, provided they do not intersect, converge in that same direction."[137]

As Youschkevitch notes, this is also one of Robert Simon's axioms (*Euclidis elementorum libri priores sex*..., Glasgow, 1756),[138] but it also has much in common with the circle of ideas implicit in proposition (1) of Thābit's first proof of the parallel postulate. With these remarks in mind, we examine al-Ṭūsī's proof, as set forth in the "Discussion".

This begins with a proposition I: "The shortest of the lines drawn from a point to any line which does not contain it, the so-called distance from the point to the line, is the perpendicular dropped from the point to the line."[139] Al-Ṭūsī's proof follows. Let DC be a given line, A a point off the line (Figure 20) and AB perpendicular to DC. Extend another line, say AE, from A to E on DC. Since $\angle$EBA is a right angle, $\angle$AEB is less than a right angle by *Elements* I, proposition 17.[140] Now since $\overline{AB}$ is opposite the smaller of these two angles, and AE is opposite the larger, $\overline{AB} < \overline{AE}$ by *Elements* I, proposition 19. ("In any triangle, the greater angle is subtended by the greater side.")[141] Hence AB is the shortest distance from A to DC as claimed. Al-Ṭūsī's second proposition is: "If two equal perpendiculars are dropped to a line (and intersect it), and the endpoints are

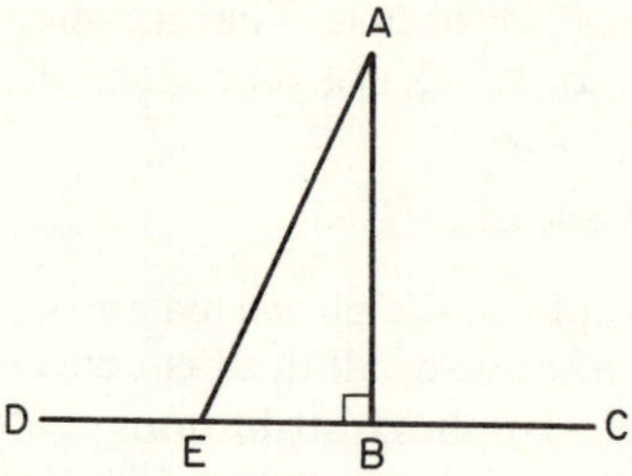

Figure 20

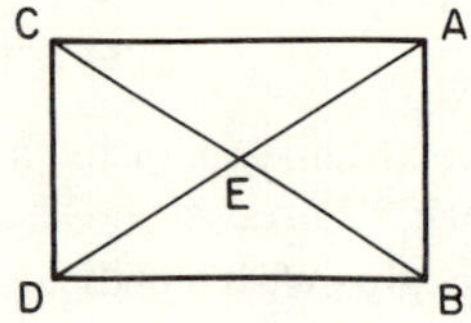

Figure 21

joined, the (upper internal) angles are equal angles" (Figure 21). For a proof, let $\overline{CD} = \overline{AB}$, with both perpendicular to DB. Draw lines AD and BC; they intersect in E. In $\triangle ABD$ and $\triangle BCD$, $\overline{AB} = \overline{DC}$ and $\overline{BD} = \overline{BD}$. Since both are right triangles, they are congruent by *Elements* I, proposition 4. Thus $\overline{AD} = \overline{BC}$ and $\angle BAD = \angle BCD$, $\angle ADB = \angle CBD$. Therefore $\overline{BE} = \overline{DE}$ ($\triangle BED$ is isosceles) by [Elements. I, proposition 6].[142] Hence $\overline{AE} = \overline{CE}$ and $\angle ECA = \angle EAC$ by *Elements* I, proposition 5.[143] Thus $\angle DCA = \angle BAC$. Of course this is al-Khayyāmī's proposition 1, with figure and proof.

Al-Ṭūsī's next proposition coincides with al-Khayyāmī's third; namely that $\angle DCA$ and $\angle BAC$ are right angles. However, the proof is different. Al-Ṭūsī proceeds in the following way (Figure 22). Drop equal perpendiculars to BD as in proposition 2. If $\angle BAC$ and $\angle DCA$ are not right angles, they are both obtuse or acute. Suppose the former, then by proposition 11 of *Elements* I,[144] erect a perpendicular AE to line AC. It is necessarily between AB and CD, and that $\angle AED$, the exterior angle to the right $\triangle ABE$, is greater than either interior and opposite angle by *Elements* I, proposition 16.[145] Hence it is also obtuse. Next erect a perpendicular to BD at E, say EG. Again, it necessarily comes between AE and CD, and $\angle EGC$, the exterior angle to $\triangle EAG$ is bigger than the interior opposite right angle GAE. Thus it is also obtuse. Now erect a perpendicular at G to line AC, say GH, and continue. Then the perpendiculars passing through those points situated on the line AC, making equal angles with the line BD; that is AB, GE, FH, successively increase in length and the perpendicular AB is shorter than all of them. For $\angle AEB < \pi/2 < \angle ABE$ making $\overline{AE} > \overline{AB}$ by *Elements* I, proposition 19.[146] Similarly $\angle AGE < \angle GAE$, so $\overline{GE} > \overline{AE}$, and hence $\overline{GE} > \overline{AB}$.

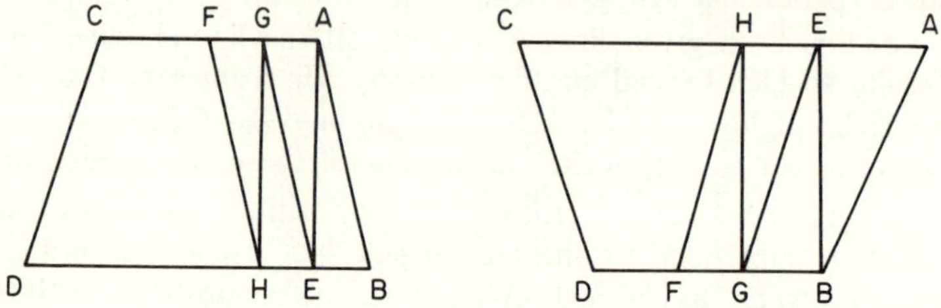

Figure 22

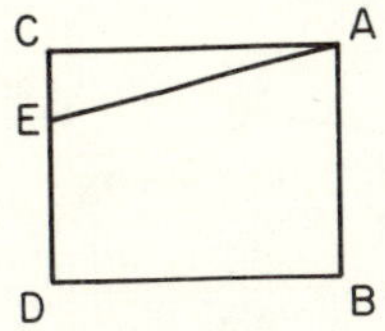

Figure 23

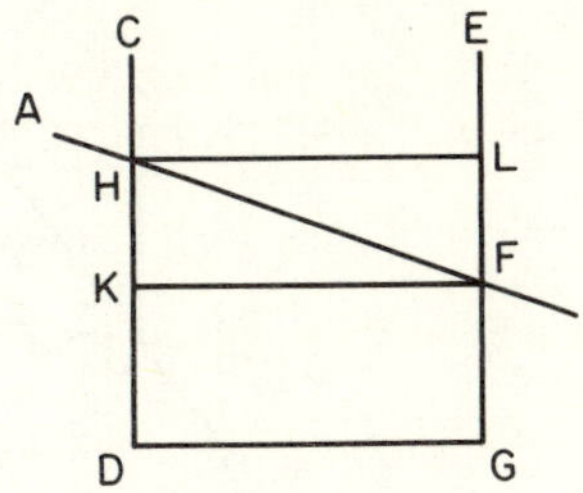

Figure 24

In the same manner one proves that $\overline{FH} > \overline{GE}$, and so on. Thus this sequence of perpendiculars from AC to BD forms an increasing sequence in the direction of C, and consequently lines AC and BD diverge in the direction of C and converge in the direction of A. (Here, al-Ṭūsī, unconsciously, applies his "implicit" proposition.) But $\langle$DCA is already obtuse, since it equals $\langle$BAC by the previous proposition. Thus, as before, he has proven that CA and DB diverge in the direction of C and converge in the direction of A. [But this is a contradiction; consequently, $\langle$BAC and $\langle$DCA are not obtuse].[147]

At this point, al-Ṭūsī goes on, by an analogous argument (second part of Figure 22) to deny the acute angle hypothesis. Consequently, we move on to his proposition 4: pairs of opposite sides of a right angled quadrilateral are equal (Figure 23). As the figure suggests al-Ṭūsī supposes that $\overline{AB} < \overline{CD}$ and marks off DE on CD with $\overline{DE} = \overline{AB}$, and applies the previous proposition. The proof that follows is exactly like al-Khayyāmī's. Since proposition 3 is the heart of al-Ṭūsī's proof, we merely sketch the remaining proposition and their accompanying drawings. In proposition 5, al-Ṭūsī follows al-Khayyāmī's lead, and proves *Elements* I, proposition 29. The figure (Figure 24) and proof differ only slightly from al-Khayyāmī's proposition 7. After this, al-Ṭūsī offers two versions of his parallel postulate proof. The first depends on his sixth proposition: if two distinct straight lines meet to form an angle, and a perpendicular is dropped from a point interior to the angle to one of the sides, its extension necessarily cuts the other side. The proof requires Figure 25, which at first glance, may suggest some elements in common with al-Abharī's, but the actual object is to construct the point of intersection of HG extended and AB. The Archimedian axiom is used in

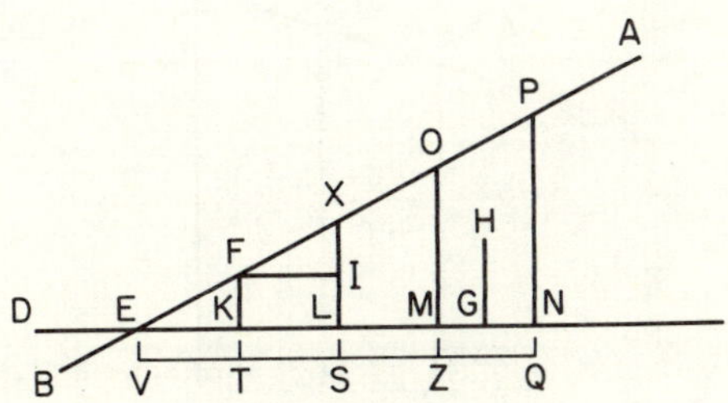

Figure 25

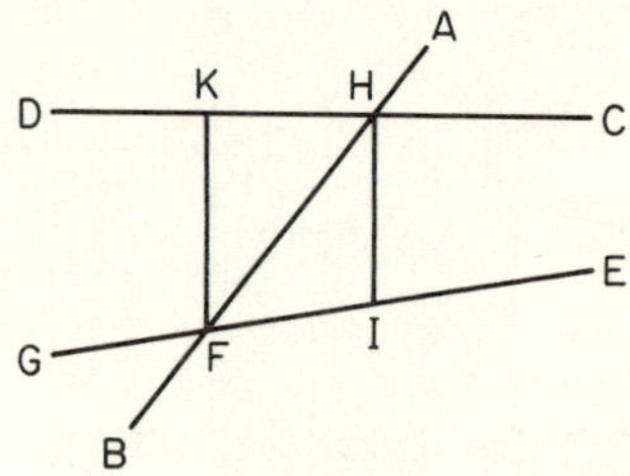

Figure 26

an essential way, and the proof resembles, in many respects, that of Aganis, underlining al-Ṭūsī's familiarity with the commentary of al-Nayrīzī.

Al-Ṭūsī then deduces the parallel postulate from this preliminary result, in proposition 7, which uses Figure 26, where line AB cuts lines CD and EG, forming $\angle$CHF and $\angle$EFH with sum less than π. Not content with this proof, al-Ṭusī generates another, based on an alternate version of proposition 6: through any point interior to an angle, one can draw a line which cuts both sides of the angle. In his alternate proposition 7, he uses this to give another elegant proof of the Fifth Postulate.

As Youschkevitch points out, al-Ṭūsī treated the subject again in the second version of his edition of Euclid.[148] This time, he replaced the postulate we might call "Simson's axiom", by two. One of these proposes that if two lines are such that perpendiculars to one of them from points of the other (Figure 27) strictly decrease in one direction (say from D to C), then the lines converge in that direction, and diverge in the other. Moreover the angles formed by the perpendiculars with AB, on the "converging" side are acute, and obtuse on the "diverging" side. The other begins with the angle condition, and concludes that the perpendiculars are "converging" or "diverging".[149] He tried to use these to establish the crucial proposition 2 above, but again succumbed to an unconscious use of an assumption equivalent to the Fifth Postulate. This is unfortunate, but al-Ṭūsī did go on, in his analysis, to show the connection between the fact that any triangle has angle sum π and the parallel postulate.[150]

We summarized the importance, and novelty of al-Ṭūsī's work in the introduction, so here we offer only a few statistical remarks. His recensions of

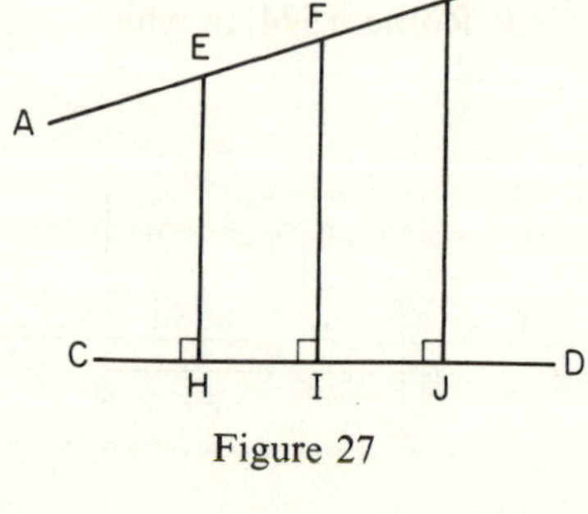

Figure 27

Euclid survive in over thirty manuscripts and five printed editions with glosses and commentary by al-Ǧurǧānī (1339–1413/14), Qāḍī Zādā al-Rūmī (*c.* 1346–*c.* 1436), al Hafrī (d. 1551), al-Husainī (fl. 1680), al- ᶜAlawī (d. 1651), al-Maibudī (d. 1466); while *Risāla fī Muṣādarāt Uqlīdis* and *ar-Risāla aš-šāfiya* together are extant in at least eight manuscripts, and one printed edition.[151] Whether this testifies to the popularity of al-Ṭūsī's geometrical work, or to luck, we leave it to the reader to conclude. By contrast, al-Khayyāmī, who exerted the most influence in shaping al-Ṭūsī's work, had his work on Euclidean geometry published, in the modern sense, for the first time in Teheran in 1936.[152]

Before summarizing the connections between the work of al-Ṭūsī, his predecessors and contemporaries, giving a sketch of Dehn's classification scheme, and remarking on Gray's thesis, we offer a brief section on the Near Eastern successors of al-Ṭūsī, and some European mathematicians whose work al-Ṭusi directly influenced.

V. Successors of al-Ṭūsî

The first of these, probably one of the first in Europe to be concerned with the foundations of geometry, as opposed to translation of geometry from Greek and Arabic into Latin and the vernacular, is Levi ben Gerson (1288–1344) of France.[153] His commentary on the first five books of the *Elements* uses Moses ibn Tibbon's (fl. 1270, Montpellier)[154] translation of Euclid into Hebrew; his other geometrical work *Ḥibbur ḥokmat ha-tishboret* is only known through fragments.[155] His attempt to prove the parallel postulate has been commented upon by Rosenfeld.[156] His approach is to try to prove that no quadrilateral can have all acute, or all obtuse angles.

One of the few Near Eastern mathematicians known to have commented on the foundations of geometry, as late as the fifteen century is Qāḍī Zādā al-Rūmī (*c.* 1364–*c.* 1436), who served as rector of Ulugh Beg's university in Samarkand.[157] He learned astronomy and geometry from Mullā Shams al-dīn Muḥammad al-Fanārī (1350–1431) and wrote a gloss,[158] *Sharḥ Ashkāl al-taʾsīs* (Fundamental Theorems) on al-Samarqandī's tract of 35 propositions of *Elements* I, and his proof of the parallel postulate, following al-Abharī.[159]

Finally Clavius (1537–1612), Cataldi (1552–1626), Wallis (1616–1703), Saccheri (1667–1733), and Lambert (1728–1777) show varying degrees of al-Ṭūsī's influence. Clavius, for example, did a Latin translation of the *Elements* (*Euclidis elementorum libri XV*, Rome, 1574) in which he criticises Proclus' proof, and then gives one of his own, using as a fundamental theorem: "the line equidistant from a straight line is a straight line".[160] Bonola claims that the ensuing proof has much in common with al-Ṭūsī's "Aganis-like" proof.[161] This could merely mean a familiarity with Gerard of Cremona's Latin translation of the al-Nayrīzī *Commentary* (12th century),[162] since al-Ṭūsī's *Euclidis elementorum* (spurious) was not published in Rome until 1594, while Clavius' dates from 1574. Cataldi, on the other hand, could have had more direct knowledge, since his geometrical works start to appear after 1603, and in one *Aggiunta all'operetta delle linec rette equidistanti et non equidistanti* (Bologna, 1604),[163] he supposes that "straight lines which are not equidistant converge in one direction but diverge in the other".[164] But this is only speculation.

For Wallis, Saccheri, and Lambert, the connection is more demonstrable. Wallis, for example, discussed al-Ṭūsī's proof, based on a Latin translation by Edward Pocock[165] of a copy of a *Taḥrīr* erroneously attributed to him.[166] Nonetheless, the proof of the parallel postulate in this translation is al-Ṭūsī's. Wallis then criticised the proof, and went on to give one of his own in a public lecture at Oxford, in 1663. This work is preserved in a Latin essay by Wallis "De Postulato quinto et definitions quinta". Wallis' proof is based on a postulate of his own: for any given triangle, there is a similar triangle of arbitrary size. As Torretti points out, Wallis begins and ends the lecture with a disclaimer of the necessity of proving the parallel postulate.[167]

Saccheri, in his justly famous work *Euclides ab omni naevo vindicatus: Sive conatus geometricus quo stabiliuntur prima ipsa universae geometriae principia*...(Milan, 1733), attempted to prove the parallel postulate, taking for his start the first 26 postulates of Euclid.[168] He employed a method outlined in his earlier work *Logica demonstrativa* (Pavia, 1701): assume that a hypothesis to be proved true is false, and thereby arrive at its proof.[169] His proof of the Fifth Postulate then devolves around the isosceles quadrilateral so familiar to the reader from Thābit, al-Haytham, al-Khayyāmī, and al-Ṭūsī. Yet his own work is a monumental step forward, since in the course of his deductions, be derives many of the elementary parts of non-Euclidean geometry.[170] However, the very strangeness of these results caused him to claim, for example, that the acute angle hypothesis was untenable, and to lamely conclude the truth of the right angle hypothesis. (He had disposed of the obtuse angle hypothesis previously, but with the restriction that lines are finite.)[171] None of his forerunners had pushed logic and geometry so far. But he, and certainly none of these forerunners, including al-Khayyāmī and al-Ṭūsī had any idea of the possibility of geometries other than Euclid's.

A note in *Euclides vindicatus* cites Wallis' postulate, and thereby provides a link between Saccheri and al-Ṭūsī. In the same way one can tie al-Ṭūsī to Lambert; for Lambert certainly knew and praised the doctoral dissertation of G. S. Klugel (1739–1812) (*Conatuum praecipuorum theoriam parallelarum*... Göttingen, 1763), which discusses Saccheri's work at length. Lambert went on to do a treatise on parallels himself: *Theorie der Parallelinien*, which was published posthumously.[172] But lest the reader begin to take such "links" too seriously, let us quote a line from Frankland to end this section: "If two explorers, strangers to each other, enter an unknown land beyond the mountains by the same pass, they may quite possibly be found to have chosen the same route for some distance into the interior."[173]

VI. Summary and conclusions

In this section we propose to summarize and highlight the connections between the various proofs of the parallel postulate that culminate in the work of al-Ṭūsī.

Of course, the picture we present can not possibly be complete. As a matter of fact, we offer two rather different pictures of the connections based on the same historical record! This comes about, essentially, because these particular records deal with geometrical systems which may be subjected to analysis based on the idea of logical equivalence, as well as on the more orthodox method of historical inference, which depends on internal evidence and "direct citation" for its conclusions. Specifically, when we consider only logical equivalence of the geometrical systems we are classifying them modulo a certain fundamental set of hypotheses. For example, most of the theories of parallels we have examined use, either explicitly or implicitly, some starting hypothesis. If these hypotheses are absolutely equivalent, in the sense that each implies the other without recourse to another hypothesis, we say that the parallel theories are absolutely equivalent. As a case in point, consider a theory which uses only the hypothesis: one and only one parallel to a given line can be drawn through an external point. Then consider another, which uses the one hypothesis: two straight lines parallel to a third are parallel to each other. These are evidently absolutely equivalent. More generally, suppose that a geometrical theory is based on a set of hypotheses $\{H_1, H_2,\ldots,H_n\}$ and H, H^* are new hypotheses such that $\{H_1,\ldots,H_n, H\}$ implies H^* while $\{H_1,\ldots,H_n, H^*\}$ implies H. Then we say that H and H^* are equivalent relative to $\{H_1,\ldots,H_n\}$, as are the resulting theories $\{H_1, H_2,\ldots,H_n, H\}$ and $\{H_1, H_2,\ldots,H_n, H^*\}$. Of course if H and H^* are absolutely equivalent, they are also equivalent with respect to $\{H_1,\ldots,H_n\}$.

Now we follow Dehn,[174] and take $H_1,\ldots,H_4$ to be respectively, Hilbert's axioms of incidence, order, congruence, and the Archimedian axiom,[175] and adopt $\{H_1,\ldots,H_3\}$ as our fundamental set. With respect to this fundamental set, one has the following hypotheses equivalent to each other and to the parallel postulate: H_5, two parallel straight lines are equidistant; H_6, if a straight line intersects one of two parallels, it intersects the other; H_7, a transversal of two parallel lines makes interior angles on the same side which are supplementary; H_8, through a point external to a straight line, one and only one parallel can be drawn; H_9, through a point interior to an angle, a line can be drawn which intersects both the sides; H_{10}, given a triangle, a similar one can be constructed of arbitrary size.[176] On the other hand, with respect to this fundamental set, the following postulates are equivalent, but none is equivalent to the parallel postulate: H_{11}, if a straight line L is perpendicular to a transversal, while another line M makes an acute angle with it, perpendiculars dropped from M to L are less than the segment of the transversal on the same side as the acute angle; H_{12}, the locus of points equidistant from a straight line is a straight line; H_{13}, the sum of the angles of a triangle is π. Finally, we remark that H_5 through H_{13} are all equivalent under the set $\{H_1,\ldots,H_4\}$. These results are all consequences of Max Dehn's paper of 1900.[177]

Now if we go back and re-examine the hypotheses underlying the various theories of parallels from Ptolemy to al-Samarqandī, we may construct a table similar to that of Figure 28. Subscripts refer to the order in which we have treated the theories in the text. The bracket around al-Jawharī's use of H_9 is meant not only to indicate that he essentially derives it from H_4 and H_5, but also to highlight

	Hypothesis assumed									
	4	5	6	7	8	9	10	11	12	13
Ptolemy				×	×					
Aganis	×				×				×	
Proclus		×	×							
al-Jawharī	×	×				[×]				
al-Nayrīzī	×				×				×	
Thābit_1		×								
Thābit_2	×								×	
al-Qass	[?]								[?]	
al-Haytham_1	×								×	
al-Haytham_2	×				×				×	
al-Khayyāmī	×	×								
al-Maghribī		×								
al-Ṭūsī_1	×	×								
al-Ṭūsī_2	×	×								
al-Ṭūsī_3	×							×		
al-Abharī						×				
al-Samarqandī						×				

Figure 28. Hypothesis assumed

the connection between his proof and al-Abharī's. (Of course, al-Qass' proof is unknown.)

If we now apply Dehn's scheme, we see that the various theories fall into two groups, with some overlap, since, for example Aganis uses hypotheses H_5 and H_{12}, which come from two inequivalent groups modulo $\{H_1, H_2, H_3\}$. This is summarized in Figure 29, in which the overlap is indicated by dotted lines.

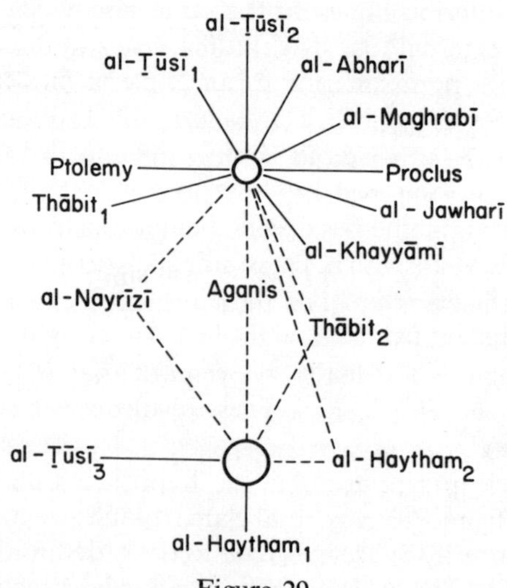

Figure 29

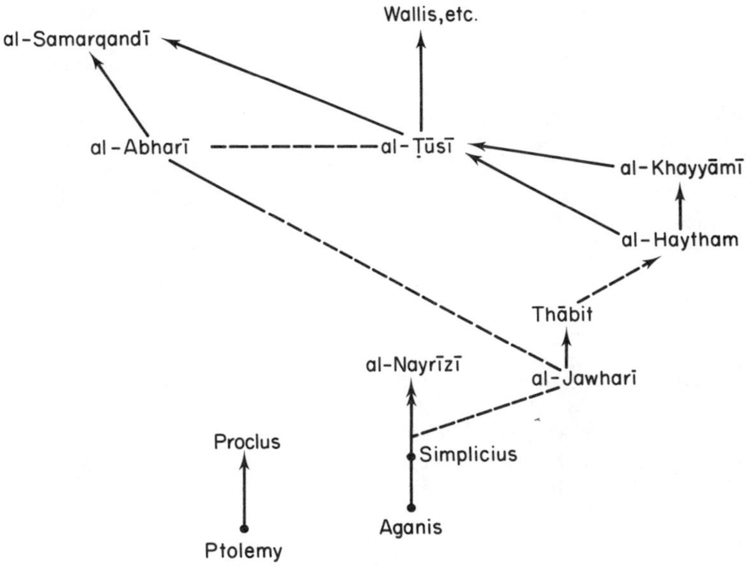

Figure 30

Next, on the basis of internal evidence and direct citation (as presented in our text), we have the connection exhibited in Figure 30.

The double arrow under al-Nayrīzī is meant to indicate a direct connection with all listed authors who come after al-Jawharī. The dotted lines indicate inferred relations, based on internal evidence, the solid lines, those based on citations.

Since we have already indicated in our introductory remarks, the reasons for singling out al-Ṭūsī's work on the Fifth Postulate, we need not repeat them here. We can only hope that the reader is satisfied with our demonstration of these remarks.

We conclude by calling attention to a historical phenomenon underlined in Jeremy Gray's recent paper on the history of non-Euclidean geometry: that the impetus behind the current development of a mathematical subject may be far removed from its classical foundations.[178] By the same token, a historian should see his subject with an eye to emerging concepts lest he misrepresent it. This might, equally well, serve as a warning to those who would try to read "non-Euclidean ideas" into works by al-Haytham, al-Khayyāmī, and al-Ṭūsī. The force behind the tremendous development in non-Euclidean geometry, namely differential geometry, is just too far removed from Euclid to make any such claim more than laughable.

Notes

1. The injunction is reproduced in Greek at the start of the paper: "This ought to be struck right out of the Postulates; for it is a theorem, . . ." (Thomas (89), vol. 2, 367. Of

course, Proclus was not part of the Arabic tradition, which relies mainly on Simplicius. (Murdoch (44), especially 458–9).

2. Youschkevitch (101), 122 gives essentially these same reasons. As he notes, the first three also apply, to some extent to al-Khayyāmī.
3. Our biographical coverage is purposefully brief. Among Arabic sources giving some general information on al-Ṭūsī are al-Daffa' (13), 192; al-Gasri (22), 200–2; Ḥājjī Khalifah (25), vol. 1, 139; Kahalah (33), 33; Mazhar (39), 283; al-Mūlla (43), 163; Toogan (90), 408–9. For al-Ṭūsī's theological work, see Nasr (45), 321–7, and Siddiqui (81), 564–80.
4. Nasr (46), 44–6. See also Nasr (44), *passim*; Sarton (74), 1001–13. Kamāl al-Dīn's work is treated in Sarton (74), 600. See also Cantor (10), Bd. I, 779–80.
5. Nasr (46), 508–9. Geometrical manuscripts of al-Ṭūsī are listed, with locations, by Sezgin (78), 113–14. His astronomical works are similarly listed in various places in Sezgin (79). See also Wiedemann (99), 701–28.
6. Youschkevitch (101), 120. See also Sezgin (78), 113–14.
7. Heath (29), vol. 1, 78; Suter (85), 146–53. Murdoch (44), 453–4; notes 5–6; Sabra (72), no. 11; Sabra (71), 18; Rosenfeld, *et al.* (56), 51–3.
8. Suter (85), 151 and Murdoch (44), 453, n. 5. The latter includes Lucknow 1873–4, Delhi 1873–4. Tehran 1881. See also Sezgin (78), 113.

The proofs arising from the two editions, actually come from two variant Arabic manuscripts which are precursors of the 15 book printed edition.

9. Nasr (46), 509, and Siddiqui (81), 565.
10. Bonola (7), 118–21, and Dehn, M. "Die Legendre'sche Sätze über die Winkelsumme in Dreiecke", *Math. Annalen* **53** (1900), 405–39. (102).
11. Gray (24), 253–4.
12. Heath (29), vol. 1, 190–4. This is our standard source for citations from the *Elements*.
13. Heath (29), vol. 1, 202.
14. Heath (29), vol. 1, 311–12.
15. Heath (29), vol. 1, 202. Aristotle (48) or (65), *Anal. prior.* II 16, 65a4.
16. Toth (92), 249–422.
17. The passage in question reads (e.g. Ross (65)): "It is impossible for instance, on a certain hypothesis that the triangle should have its angles equal to two right angles, and an another the diagonal is commensurable." Toth sees this as "Ist die Winkelsumme des Dreiecke nicht gleich π, so ist die Diagonale des Quadrants mit der Seite kommensurabel." (Toth (92), 317ff.)
18. Heath (29), vol. 1, 202–3. See also Wolfe (100), 19–20.

19–20. Heath (29), vol. 1, 202–3.

21. Heath (29), vol. 1, 190. The source is al-Nayrīzī's *Commentary*. See also Heath (27), (28) *passim*.

22–3. Heath (29), vol. 1, 191. This time the source is Proclus' *Commentary*.

24–6. Heath, *loc. cit.*

27. Sabra (69), 6.
28. Heath (29), vol. 1, 38–9. See also Dicks (15), 344–7.

29–30. Heath (29), vol. 1, 204–6.

31–2. Heath, *loc. cit.*, and Wolfe (100), 26–7.

33. Heath (29), vol. 1, 27, citing Tannery: *Bibliotheca mathematica* **3** (2) (1900), 9–11.
34. Sabra (69), 6.
35. Petrosian (50), 153ff.
36. Our account follows Heath (27), vol. 2, 228–30.
37. Playfair's *Elements of Geometry* (1795); see Frankland (20), 31–2. For the Proclus citation, see Heath (29), vol. 1, 220.
38. Heath (27), vol. 2, 230.

39–40. Heath (29), vol. 1, 207.
41. Our account follows the more detailed one Heath (29), vol. 1, 206–8. See also Frankland (20), 7ff.
42. Heath (29), vol. 1, 208.
43. Wolfe (100), 27.
44. Verbeke (96), 440.
45. Heath (29), vol. 1, 203. See also Suter; *Z. Math. Phys.* **37**, 194.
46–8. Sabra (68), 79–80.
49. Youschkevitch (101), 113. Our account of the proof follows Sabra (68), 79–80, and the Russian translation of it in (94).
50–1. Sabra (68), 79–80.
52. As we noted above (in the text), al-Nayrīzī's *Commentary* dates from around 897 (certainly before 922), while al-Jawharī's work appears to have been done around 70 years earlier.
53. Sabra (68), 79. See text, pp 5–7.
54. Heath (29), vol. 1, 213–19.
55–6. Sabra (68), 79. See also (94).
57–8. Sabra (69), 5–7.
59. Sezgin (78), 284–5.
60. Sabra (69), 7.
61. Rosenfeld (57), 288.
62–4. Youschkevitch (101), 114.
65. See text, preceding note 20.
66. Rosenfeld (57), 290. Also Youschkevitch (101), 114 and Rosenfeld (59), 587–92.
67. Youschkevitch (101), 115.
68. Heath (29), vol. 1, 226–7.
69–71. Rosenfeld (57), 290. Youschkevitch (101), 115. For note 69 (parallel translation), the reader might also compare al-Nayrīzī's argument for the existence of parallels, based on "the natural priority of equality over inequality". (See Sabra (69), 6.)
72. Youschkevitch (101), 118. See also Ahadova (1), which cites kinematic principles in Thābit, al-Haytham, and Ibn Sīnā.
73–75. Sezgin (78), 298.
76. See (94), 523–4.
77. Anawati (4), 494–8.
78–9. Ibn Sīnā (82), vol. 2, 94–7.
80. See Section IV of our paper on Ibn Sīnā for details.
81–84. Sabra (67).
85. Youschkevitch (101), 115–16.
86–7. Sabra (67), 201; 204ff.
88. Sabra (67), 201.
89–91. Heath (29), vol. 1, 195–6.
92–4. Youschkevitch (101), 116.
95. Scriba (75), 600.
96. Our account follows Sabra (67), 202, and Youschkevitch (101), 119.
97. See for examples Gans (21), Martin (38).
98. Youschkevitch (101), 117.
99. This is also one of Hilbert's "order axioms". See Wolfe (100), 9.
100. Sabra (67), 202.
101. Rosenfeld (62), 323.
102–7. Youschkevitch (101), 118.
108. Our account follows Rosenfeld (61), 329–30, Youschkevitch (101), 118–20, and the Rosenfeld and Youschkevitch edition of al-Khayyāmī's manuscript (63), 67–84.
109. Youschkevitch (101), 172, note 67.
110. The figures follow Rosenfeld and Youschkevitch (63).

111. Heath (29), vol. 1, 311.
112–13. Youschkevitch (101), 119.
114–15. Sarton (74), 623–4.
116. This is just lack of information on our part. According to Murdoch (44), 455, note 7, excerpts of this work appear in English translation in either Sabra (71) or Sabra (72). At the time of writing, these are still unavailable to us.
117. In this case, because the manuscript is, presumably, lost. See Sezgin (78), 111, no. 44. Also see Sabra (68), 80 (Sezgin (78), 394) for mention of another anonymous proof of postulate five in the "Simplicius group".
118. Sabra (68), 80, and Tekeli (88), 555.
119. Tekeli (88), 555.
120. Sabra (68), 80. See also Sabra (71)–(72). According to Murdoch (44), 455, note 7, one of these papers also contains excerpts from al-Maghribī's proof.
121–2. Sarton (74), 867. See also Sezgin (78), 324; Suter (85), 145, 227.
123. Our account follows al-Dahir (14), 60–5.
124–5. Dilgan (18), 91.
126. Youschkevitch (101), 123.
127. Youschkevitch (101), 172, note 70.
128. Sezgin (78), 114–15.
129. Our account follows Dilgan (16), 191–6.
130. Dilgan (16), 193.
131. That is, a straight line may be drawn between two points.
132. Dilgan (18), 193.
133. Youschkevitch (101), 120. See also al-Ṭūsī (94), 475ff.
134. See text, preceding note 56.
135. Sabra (67), 202.
136. Jaouiche (30), 65–6.
137. Youschkevitch (101), 120.
138. Youschkevitch (101), 172, note 68. See Frankland (20), 23–4 for a brief account of Simon's axiom.
139. Our account follows the Rosenfeld and Youschkevitch translation of al-Ṭūsī's manuscript (94), 475–532.
140. Heath (29), vol. 1, 281: "In any triangle two angles taken together in any manner are less than two right angles".
141. Heath (29), vol. 1, 311.
142. Heath (29), vol. 1, 255: "If in a triangle two angles be equal to one another, the sides which subtend the equal angles will also be equal to one another".
143. Heath (29), vol. 1, 251: "In isosceles triangles the angles at the base are equal to one another, and, if the equal straight lines be produced further, the angles under the base will be equal to one another".
144. Heath (29), vol. 1, 269: "To draw a straight line at right angles to a given straight line from a given point on it".
145. Heath (29), vol. 1, 279: "In any triangle, if one of the sides be produced, the exterior angle is greater than either of the interior and opposite angles".
146. Heath (29), vol. 1, 284: "In any triangle the greater angle is subtended by the greater side".
147. Rosenfeld and Youschkevitch (94), 514 fill in an obvious (scribal) omission.
148–50. Youschkevitch (101), 121.
151. Sezgin (78), 113–14.
152. Youschkevitch (101), 123. See also Rosenfeld (61), 331–3.
153–6. Samso (73), 279–80.
157–8. Dilgan (17), 227.
159. See note 127 above.

160–1. Bonola (7), 13.
162. See Clagett (12) and Lemay (35).
163–4. Carruccio (11), 128. See also Bonola (7), 13.
165. Wallis (97), 667.
166. See note 8 above.
167. Toretti (91), 44; Wallis (97), 674–8.
168. Bonola (7), 22.
169. Struik (77), 55.
170. Bonola (7), 22ff.
171. Struik (77), 56.
172. Bonola (7), 44ff; see also Scriba (75), 595–600.
173. Frankland (20), 24.
174. Dehn (102).
175. Dehn (102), 405.
176–7. Bonola (7), 119–20.
178. See note 11 above.

5 Ibn Sīnā as a Mathematician

I. Mathematical Education

Abū 'Alī al-Ḥusayn ibn 'Abdallāh ibn Sīnā[1] (Avicenna) [980–1037] is well known for his work in philosophy and medicine, but few studies detail his accomplishments in mathematics. Woepcke, Eneström, de Montferrier, and Tannery,[2] for example, each devoted a short article to ibn Sīnā's arithmetic or to his account of the method of casting out nines, while Lokotsch, in a 27 page Bonn University Dissertation, *Avicenna als Mathematiker*, considered only the summary of the planimetric books of Euclid, appearing in *Kitāb al-Shifā'*.[3] A further note on the contents page of this 1912 dissertation promises a German translation of the mathematical writings of Avicenna within two years, but World War I intervened, and while Lokotsch survived, the projected translation evidently never was printed.[4] Later writers usually pass over ibn Sīnā's mathematics with a flat statement of his divisions of mathematics in the Shifā' into four principal (theoretical) sciences and four subordinate (practical) ones: arithmetic ('ilm al-ḥisāb), geometry ('ilm al-handasa), geography and astronomy (ilm al-hay'a), music ('ilm al-mūsiqī), with respective practical sub-sciences: Hindu mathematics and algebra, mechanics, preparation of zījs and calendars, use of musical instruments.[5] Of course, compared to the brilliant work of his elder contemporary al-Bīrūnī (973–1048), with whom he competed, all these encyclopedic summaries appear flat and unexciting. Possibly this, or similar evidence, prompted Sarton to declare that "Ibn Sīnā's interest in mathematics was philosophical rather than technical,... [as to be expected] in a late Neoplatonist."[6] While the present study may not overturn this statement, it will establish that it is a simplification, and it will offer the reader ample mathematical citations from ibn Sīnā's work. Let us add that for want of time, we shall restrict our discussion of ibn Sīnā's mathematics to mathematics in the modern sense. Moreover, we shall call the reader's attention to some excellent, and rather recent Russian studies on certain aspects of ibn Sīnā's mathematics.

Before listing and examining these mathematical works, let us first establish what is known of ibn Sīnā's mathematical education, and his general attitude toward mathematics. The principal source of this material is an autobiographical fragment, edited and supplemented by his student 'Abd al-Wāḥid ibn Muḥammad al-Jūzajanī (11th century), and known as *Sīrat al-Shaykh al-Ra'īs* (Life of Ibn Sīnā).[7] In the fragment, ibn Sīnā recounts that he had completed his

Qurānic education by the age of ten,[8] and that his father's house in Bukhārā served as a centre for discussions of "philosophy, geometry and Indian calculations."[9] He adds that both he and his father studied the *Rasā'il Ikhwān al-Ṣafā'*, (Epistles of the Ikhwān al-Ṣafā'), an important point to consider, when we shortly examine the structure of ibn Sīnā's mathematical texts. His father then sent him to learn "Indian calculation" (al-ḥisāb al-Hindī) and algebra (al-jabr wa'l-muqabāla) from a vegetable seller known as al-Maḥmūd al-Massāḥī, and at the same time he read logic and geometry with the philosopher Abu 'Abd Allāh al-Nātilī. Ibn Sīnā refers to him, in fact, as one who "claimed to know philosophy,"[10] and had no knowledge of the deeper parts of logic. Together, they started the *Elements* of Euclid and got through the "first five or six figures"[11]—possibly the first eighth of Book I—when ibn Sīnā got impatient and "undertook the solution of the rest of the book in its entirety"[12] by himself. He then began Ptolemy's *Almagest*, but al-Nātilī begged off when ibn Sīnā completed the first nine sections of Book I, saying, "Take over reading and solving them [the geometrical figures] by yourself; then show them to me, so that I can explain to you what is right and what is wrong."[13] Given al-Nātilī's performance with the *Elements*, his attitude toward the *Almagest* is understandable; section 10 of Book I assumes the reader has ready familiarity with Euclid. The very first figure is a simultaneous construction of a regular decagon and pentagon requiring, in order: *Elements* II, 6; I, 47; VI, definition 3; XIII, 9; IV, 15 corollary; XIII, 10, as justifications to be supplied. In fact, al-Nātilī was of little help, and ibn Sīnā became the lecturer. Whether this continued through the whole of the *Almagest* is uncertain, for the text of *Sīrat al-Shaykh al-Ra'īs* simply reports that al-Nātilī left for Gurgānj, leaving ibn Sīnā to his own devices. These, according to the fragment, consisted of readings in natural sciences ('ilm ilāhī), metaphysics (al-ilāhiyyāt), and medicine. By the age of 16, ibn Sīnā claimed a mastery of medicine, as well as jurisprudence, and turned again to the serious study of philosophy for about a year and a half. In the text, he remarks, at this point, "Everything which I knew at that time is just as I know it now; I have not added anything to it to this day."[14] He then took up the study of Aristotle's *Metaphysics*, and claimed that he read it so often (40 times) without comprehension, until he acquired, by chance, al-Fārābī's commentary, possibly the *Kitāb al-ḥuruf*,[15] which made Aristotle's intention clear. At about this time, an invitation to share in the treatment of the ailing governor of Bukhāra (Nūḥ ibn Manṣūr) gave him access to the governor's library, where he read in all the sciences. Unfortunately, the fragment gives no specific authors or titles. By this time ibn Sīnā was 18; and the text reiterates his claim to have acquired no new knowledge since that time!

Thus, according to the Sīrat, ibn Sīnā's mathematical training consisted of (possibly) the encyclopedia articles of the Ikhwan al-Ṣafā', some algebra and Hindu arithmetic, Euclid's *Elements*, Ptolemy's *Syntaxis Mathematica* (*Almagest*), a classical compendium of applied mathematics, as well as of astronomy, and (probably) the assimilation of much of Aristotle's notions upon the foundations of mathematics.

At this point al-Jūzajanī continues the Sīrat on his own, by introducing himself as a student who read the *Almagest* and studied logic with ibn Sīnā. He then offers a bibliography of 46 works of his teacher, 5 of which appear to contain mathematical studies: the Shifā', Najāt, Ala'ī, Summary of Euclid, The Finite and Infinite. The order of composition is not particularly certain from al-Jūzajanī's account, although he does mention that ibn Sīnā had already written book one of the Qānūn before embarking on the Shifā', and that he would often write 50 pages a day, even outlining whole books without consulting references. Al-Jūzajanī, who is known for his own work on astronomy, *Kitāb Kaifīyat al-aflāk* which influenced the work of Quṭb al-Dīn al-Shīrāzī [1236–1311],[16] goes on to underline his master's competence in general with remarks such as: "he was not outclassed in any of the sciences", during the Friday night gatherings of the learned at the court of Amir 'Alā al-Dawla. As one would expect, al-Jūzajanī offers no particular opinions on his master's mathematics, except to note that ibn Sīnā "presented..., in the Arithmetic [of Shifa'] some excellent numerical properties",[17] and to describe some additions he made to the drafts of the summaries of the *Elements* and the *Almagest* for the Shifā'. As will soon become apparent, al-Jūzajanī does, however, offer valuable insight into ibn Sīnā's motivation for writing these summaries, when he introduces sections of similar material in the Dânesh Nâmeh. Another early biographer, Ibn Funduq al-Bayhaqī (d. 1170), in the *Tatimmat ṣiwān al-ḥikma* (Supplement to the "Repository of Wisdom"), however, pronounces a judgment which has all the elements of Sarton's. In fact, he says ibn Sīnā was not much of a mathematician simply because "a person who has tasted the sweetness of metaphysics is niggardly is spending his thoughts in mathematics, so he fancies it from time to time and then abandons it."[18]

Al-Jūzajanī then briefly takes up ibn Sīnā's involvement in an observational astronomy project for al-Dawla,[19] and in a philological argument with Abū Manṣūr al-Jabbān. The latter seems to have taken some three years to conclude—in ibn Sīnā's favour—and is of interest as a comment on ibn Sīnā's character. It also gives the lie to his constant claim of "learning nothing new from age 18". After al-Jūzajanī's account of his master's death, another (unknown) writer adds a bibliographic postscript. This lists 92 titles of ibn Sīnā, includes the mathematical ones given by al-Jūzajanī, with some changes; for example, "Finite and Infinite" becomes "Infinity", and "The Summary of Euclid" is recorded as (possibly) "for the Najāt". It adds two further items: a letter to Abū Sahl al-Masīḥi on the Angle, and a "Summary [of the position] that the Angle which is formed by the Circumference and the Tangent has no magnitude." Mahdavi and Anawati both identify these treatises on the angle as being the same.[20]

II. Mathematical Texts

Having briefly remarked on the mathematical texts of ibn Sīnā as reported in *Sīrat al-Shaykh al-Ra'īs* and its anonymous appended bibliography, let us list them in detail, with manuscript locations, conjectural dates of composition, and evidence of translation into Latin in the Middle Ages. This will both define the limits of our main area of discussion, and, later, allow us to trace the influence of ibn Sīnā's mathematical ideas among the Latin Schoolmen.

A title by title examination of Anawati's bibliography yields 8 works of ibn Sīnā which seem to have mathematical content. We consider them almost in the order he adopts. The first is given the number 13: *Danish nāmē 'Alā'ī*,[21] is in Persian and contains seven parts: logic, metaphysics, physics, geometry, astronomy, arithmetic, and music. Part is the work of ibn Sīnā, while the rest, primarily the mathematical section, is a translation of writings of ibn Sīnā into Persian by al-Jūzajanī after his master's death. In fact, in the preface to the section on geometry, al-Jūzajanī claims that the original section on mathematics was lost, and that he replaced it with another.[22] Now al-Jūzajanī also composed prefaces for the mathematical sections of the Najāt, and these differ little from those of the 'Alā'ī.[23] In Najāt he asserts that the mathematical sections are assembled from other writings of ibn Sīnā, and that the arithmetic section is extracted from the Shifā'.[24] These mathematical parts of Najāt do not appear in the Cairo edition of ibn Sīnā's work, but are preserved in manuscript, e.g. Ahmet III, no. 34481, Istanbul. Mahdavi has shown that the mathematical sections of 'Alā'ī are merely a Persian translation of the corresponding Arabic text of Najāt.[25] Nonetheless, al-Jūzajanī himself remarks that the Najāt consists of three books—probably the logic, physics and metaphysics—which Goichon[26] has shown are mostly extracts from the Shifā'. In fact, he states that the original 'Alā'ī consisted of "one volume".[27]

Al-Najāt, 23 in Anawati's bibliography,[28] exists in at least 28 manuscripts, and al-Jūzajanī tells us that ibn Sīnā composed it "en route", "in the year 'Alā al-Dawla attacked Sābūr Khwāst".[29] Gohlman discusses the problem of exact date, noting that al-Dawla attacked Sābūr several times, and decides in favour of 1030 or 1032.[30] Al-Jūzajanī adds that the 'Alā'ī was written in Isfahan,[31] a fact confirmed by the anonymous bibliography appended to Sīrat.[32] This would place its composition somewhere between 1024 and 1037. But since our interest is in the mathematical sections this has little relevance. Still, the preceding discussion does establish that the arithmetic section of 'Alā'ī derives from Najāt, which, in turn, has its source in the Shifā', and that the geometric sections of 'Alā'ī and Najāt are the same. In a later section, we shall compare their geometric content with that of the Shifā'.

Besides the Shifā' itself, Anawati lists 5 other works of possible mathematical content. His no. 75: al-nihāyawal-lānihāya (the finite and the infinite) is really a discussion of the question: does time have a beginning, and so, will be dismissed from further consideration.[33] Number 160: al-Zāwiya (the angle), or taḥqīq mabādī' al-handasa[34] was probably written in Jurgān between 1012–14,[35] and discusses the *κερατοειδης* or "hornlike" angle between a circle and its tangent. This was a problem mentioned by Aristotle[36] and settled completely in Euclid.[37] Number 161: Ḥall mushkilāt mu'ayyana (method of solving problems) [Persian] exists in 1 ms, Aya Sofya 4853*25. We can offer no information on its possible content. The other two items, nos. 169 and 170 are respectively, extracts from the Shifā': Mukhtaçar Oklidos and Mukhtaçar Kitāb al-arithmātīqī.[38]

The Shifā', the evident source of much of ibn Sīnā's "other" mathematical writing, is preserved in over 100 ms.[39] Al-Jūzajanī tells us, as noted earlier, that it was begun, possibly with the physics section, shortly after book I of al-Qānūn.[40] The anonymous bibliographer claims that both the physics and metaphysics sections were done in twenty days in Hamadhān.[41] This would be after 1015. The logic section was partly done there,[42] and later completed in Isfahan after 1024.[43] Sections on arithmetic, geometry and music seem to have been worked out before then,[44] while the section on astronomy may have taken shape when ibn Sīnā was in Jurgān (1012–14).[45] Al-Jūzajanī also seems to indicate that the mathematical parts of al-Shifā' were expanded and polished during ibn Sīnā's stay in Isfahan.[46] The text that comes down to us consists of four parts: logic, physics, mathematics, and metaphysics, in around 430 folios, with the subjects being allotted roughly 43, 32, 16, and 9 per cent of the text, respectively. The 70 folios devoted to mathematics are split between geometry, arithmetic, music, and astronomy, with roughly 30 per cent to geometry, 50 to astronomy and the remaining 20 distributed with slight favour toward music.[47] The astronomy section also contains material on ratios and on trigonometry.

In addition to these works of evident mathematical content, noted from surveying Anawati's bibliography, two others give further information: the *als-As'ilah wa 'l-Ajwibah* (questions and answers)[48] exchanged between ibn Sīnā and al-Bīrūnī, and al-Bīrūnī's *Kitāb Taḥdīd al-Amākin li-taṣhih masāfat al-masākin* (determination of the coordinates of cities).[49] The former offers some insight into ibn Sīnā's notion of infinity,[50] while the latter contains a criticism by al-Bīrūnī of an astronomical calculation in a lost work of ibn Sīnā.[51]

Perhaps we should mention two further manuscripts: the first is a poem in praise of Euclid attributed to ibn Sīnā by Ünver;[52] (A translation appears in the notes.) The second, al-nīrūziyya fī ḥurūf abjad (risālat al-ḥuruf) has some connection with a magic square of order three, but ibn Sīnā's interest is cosmological and not mathematical, so we need not discuss it.[53]

Of course, we cannot claim to have exhausted all possible references of ibn Sīnā to mathematics. For example, in his Aristotelian commentary al-Burhān,[54] he states that the solvability of $x^3 + y^3 = z^3$ in integers is still unresolved in his time!

Steinschneider's classic study of European translations of Arabic manuscripts indicates that although ibn Sīnā's Canon of Medicine (al-Qānūn fīl ṭibb) was well known to the Latin West in the late Middle Ages, the mathematical sections of the Shifā' were not.[55] In fact it is only the (pure) logic section that was fully translated.[56] This and parts of the physics had considerable impact on European science in the time of the Latin Schoolmen (13th–14th centuries). Haskins,[57] Rashdall[58], and Crombie[59] discuss this impact: the first, in terms of early translators; the second, in the setting of the Medieval university; and Crombie, in the works of various scientists, including Robert Grosseteste, Roger Bacon, Albertus Magnus, and Leonardo da Vinci. In addition, Alverny's extensive series "Avicenna Latinus",[60] adds much to Steinschneider's work, and is invaluable to further study of ibn Sīnā's influence in Medieval Europe. Finally, the studies of

Goichon and Madkour cast additional light on the impact of his works in pure logic.[61]

Although ibn Sīnā's writings in mathematics had no impact on the Latin West, and probably little influence among Near Eastern mathematicians, his mathematical writing, particularly from the Shifā', does receive citation in the Arabic literature of the Middle Ages. We have already mentioned the remarks of al-Jūzajānī and ibn Funduq. Later, we shall give citations (rather uncomplimentary) from al-Bīrūnī and al-Maghrabī. More typical, however, are those references made by ibn Khaldūn in the *Muqaddimah*. In sections 19–21, he notes ibn Sīnā's contributions in arithmetic, geometry and astronomy, neglecting only trigonometry (Menelaus' theorem).[62]

Having detailed ibn Sīnā's mathematical education, and listed his mathematical texts, let us consider them in the next four sections: arithmetic, geometry, trigonometry, and foundations. This last division, of course, is not a special heading of al-Shifā', 'Alā'ī, or any of the mathematical essays, but we include it as a catch-all for discussing ibn Sīnā's position on infinity, infinitesimals, limiting processes, and other issues in the foundations of mathematics.

III. Arithmetic

As we have noted already, ibn Sīnā's writing in arithmetic derives primarily from the Shifā', with a line of dependence reaching from the 'Alā'ī via the Najāt to the Shifā' itself. Our discussion of the arithmetic will concentrate first on the 'Alā'ī, and then shift to the Shifā', which has recently appeared in an Arabic edition of Lutfy.[63] Because of the intimate relation between 'Alā'ī and Najāt, we give no special consideration to the latter. As will become apparent, even a survey based on the 'Alā'ī itself gives ample evidence of the content of ibn Sīnā's arithmetic, and omits little of interest in the Shifā', apart from certain sections on "speculative arithmetic".[64]

In his introduction to the arithmetic of the 'Alā'ī, al-Jūzajanī remarks that he includes only that part of ibn Sīnā's writing essential to the study of music.[65] The section is in seven parts, and we now offer a summary of each. Our purpose is twofold: first, to give the reader a clear notion of the contents, and second, to determine the sources. We use formulas for brevity only; ibn Sīnā's arithmetic is purely rhetorical. The impatient reader may skip ahead to our conclusions, returning to the summary as needed.

1. *Characteristics and kinds of numbers*

The author begins by defining number as a collection of units, and separates them into two kinds: the even, which are divisible into two equal parts, and the odd, which are not. He then notes that every number except 1 is the mean between two numbers equally distant from it, e.g. 5 is the mean of 6, 4; 7, 3; 8, 2; 9, 1. He defines the median of the sequence $1, \ldots, n$ for n odd to be the average of 1 and n, and notes that it is the middle term, but for n even, recognizes two medians, e.g. 5 and 6 for 1,...,10, recognizing that $5 + 6 = 10 + 1$, and the numbers equally spaced about the median as 7 and 4 have the same property.

The formula

$$1 + \cdots + n = \frac{n(n+1)}{2}$$

appears as $1 + \cdots + n$ equals half the number of elements times the number of elements plus one. Then it is noted that $1 + \cdots + n + (n-1) + (n-2) + \cdots + 1 = n^2$, with $n = 10$, but with the implication that the result is perfectly general. Curiously enough, there is no correlation in the text between this and the statement a few paragraphs later remarking that the sum of the odd numbers 1, 3, 5,...,$2k + 1$ is a square.

The author then suggests another addition formula for $1 + \cdots + n$, by observing the pattern: $1 + 2 = 3 = 2 + \frac{1}{2}(2)$; $1 + 2 + 3 = 6 = 2(3)$; $1 + 2 + 3 + 4 = 10 = 2(4) + \frac{1}{2}(4)$; $1 + 2 + 3 + 4 + 5 = 15 = 3(5)$; $1 + \cdots + 6 = 21 = 3(6) + \frac{1}{2}(6)$, etc. The section closes with the observation that odd numbers are not products of even numbers, and not "even" sums of even numbers, and the statement concerning sums of odd numbers given above. Apart from squares there is no reference to the other Pythagorean numbers.

2. *Even numbers*

In the second part, devoted to even numbers, the author notes that $2 + 4 + \cdots + 2n = n^2 + n$, using $n = 3$ as the generalizable instance. He then lists three kinds of even numbers: (1) evenly even; (2) evenly odd; (3) oddly even. The first kind have only the divisor 2 (and its powers), such as 64, and hence no odd factors. If an evenly even number $n = q(r)$, then both q and r are also evenly even. Such numbers are generated as a sequence 1, 2, 4, 8,...,2^n,.... There is also a median property which is best illustrated by the following self-explanatory diagram. Terms equidistributed (geometrically) about the median (or medians) have the same product and are called by the author, complimentary. (See Figure 1.) The translation, as given by Achena and Massé, then asserts that $1 + 2 + 4 + \cdots + 2^n$

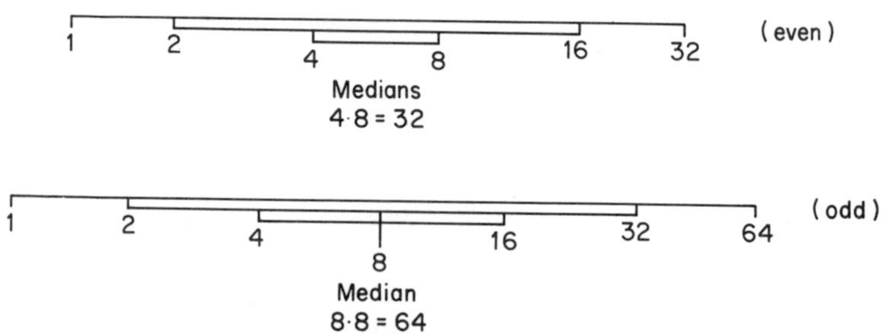

Figure 1

$< 2^{n+1}$, an unlikely reading[66] given that the correct formula $1 + 2 + \cdots + 2^n = 2^{n+1} - 1$ appears in Euclid IX, prop. 35,[67] and that the author shortly remarks that every evenly even number is deficient with deficiency 1![68] The author records other properties of these numbers: in the sequence $1,\ldots,2^n$, 2^k divides 2^{k+1}, that is, "successive members of the sequence measure each other" in the sense of Euclid; the squares of such numbers belong to the same class; $2^n - 2$ is evenly odd.[69] The latter term is yet to be defined in the text. Then comes the remark on the deficiency of 2^n; deficient, perfect and abundant numbers are defined later. This is followed with a paragraph which notes that successive evenly even numbers stand in geometric ratio, a more precise form of the "measure" statement.

Now there follows a rule for using these numbers to obtain perfect numbers: add $1 + \cdots + 2^n$ to a place where the sum $[2^{n+1} - 1]$ is prime, multiply this sum by 2^n. The product is then perfect. The rule is illustrated by 6: $1 + 2 = 3$, a prime and $2(3) = 6$ is perfect, and by 28: $1 + 2 + 4 = 7$, prime and $4(7) = 28$.

Evenly odd numbers are then defined as those having only a single factor of 2, e.g. 6, 10, and are obtained by multiplying successive odd numbers by 2. Successive members of evenly odd sequences thus differ by 4, and no such number can be a square or cube, because squares or cubes are either odd numbers divisible by odd numbers, or even numbers divisible by even numbers, the quotients being the same [in parity] as the divisor. Oddly even numbers have the divisor 2 more than once, "but don't end in one",[70] i.e. aren't successively divisible by 2 as are evenly even. The author conceives of them as between evenly even and evenly odd. They arise by multiplying odd numbers by evenly even numbers. He then shows by examples that oddly even numbers may be deficient, perfect, or abundant, e.g. $68 = 4(17) > 1 + 2 + 4 + 17 + 34$ (deficient) $28 = 4(7) = 1 + 2 + 4 + 7 + 14$ (perfect), $12 = 4(3) < 1 + 2 + 3 + 4 + 6 = 16$ (abundant).

3. *Odd numbers*

In the third section the author notes that there are three kinds of odd numbers: the primes (first non-composite); the second composite, which are products of two odd numbers (excluding 1 as a factor); the third composite. These are actually pairs of numbers of the second composite class that are relatively prime, e.g. 9 and 25. The primes have no factors but themselves and unity. Examples are 3, 5, 7, 11. The text then reads "All other numbers are composed of these prime numbers, whereas, they themselves are not reducible."[71] Evidently the author either intends "all other numbers" to be the odd ones since 2 is not a prime in his sense, or, more likely, that the evenly even numbers provide the factorizations by powers of 2. (See previous section.) There is no statement of the number of primes, and no explicit mention of "unique factorization". This part ends with a method for finding primes and second composite numbers: consider a square table of successive odd numbers, and "do as other scholars have shown".[72] Presumably this is a reference to a sieve-method.

4. [*Abundant, deficient, and perfect numbers*][73]

In section four the author states that abundant numbers are those whose proper divisors, in sum exceed the number. 12 is an example. Deficient numbers have the sum of their proper divisors less than the number e.g. 8; and perfect numbers, like 6, are equal to the sum of their proper divisors. He then remarks that it is necessary to know that a number can be perfect only if it is the product of an even and odd number.[74] Now among the first 9, there is but one, 6; among the first tens, only 28; between 100 and 1000 only 496, and between 1000 and 10,000 only 8128.[75] Next he gives a prescription for finding perfect numbers: multiply an evenly even number by a prime number, subject to the condition that the evenly even number exceeds by one half unit, half the prime. [That is $2^n - \frac{1}{2} = P/2$ or $P = 2^{n+1} - 1$, the condition of section 2.] If the evenly even number exceeds by more than one half unit, half the prime, the resulting number is abundant, and if it is less by one unit than half the prime, the resulting product is deficient.[76] Examples are 2(3) and 4(7) (perfect), 4(5) (abundant); 4(9) is cited as deficient [however, it is abundant. Moreover 9 is not prime and the rule does not apply as $4 + \frac{1}{2} = \frac{9}{2}$]. This is a curious oversight. However, the author does supply an example with 4(11).[77]

5. [*Ratio*]

In section five, ibn Sīnā notes that there are different meanings attached to the term ratio; for example, there are ratios which compare size of numbers [the sense adopted here], and ratios which compare qualities like kind and class. Having made this point, he states a trichotomy law, and an inversive principle for ratios: of two different numbers, one is the larger and the other the smaller; when one knows the ratio of the larger to the smaller, one knows, inversely that of the smaller to the larger.

Ratios are classified as simple or composite; simple, if the larger is twice, or some greater multiple of the smaller, or is a unit plus a part or many parts of the smaller;[78] composite, if the larger exceeds the smaller by twice and a part, etc. The first simple ratio is the double, in which the excess of the larger over the smaller is equal to the smaller, as 1 and 2. The next type, the triple, has the larger 3 times the smaller, as 1 and 3. Letting the smaller term 1 "run through the natural numbers," while the larger goes through their triples generates such ratios, e.g. $\frac{3}{1}, \frac{6}{2}, \frac{9}{3}, \frac{12}{4} \ldots$. The next type of ratio has the larger exceed the smaller by the mean of the smaller, e.g. 2, 3, since $\frac{3}{2} = 1 + \frac{1}{2}$. To generate other ratios of this type 3 goes through the triples of the natural numbers and 2 through the doubles: $\frac{3}{2}, \frac{6}{4}, \frac{9}{6}, \ldots$. The remainder of this part takes up, in order, analogous discussions for the ratio $1 + \frac{1}{3} : 1$, i.e. 3 and 4; $1 + \frac{2}{4}$, i.e. 4 and 6, etc.. The author wishes to indicate how to get the ratios in lowest terms, noting that ratios of the form one plus two or more parts may either be reduced or not.[79] As examples: $1 + \frac{2}{6}$, i.e. 6 and 8 are not, but $1 + \frac{2}{5}, 1 + \frac{3}{4}$, i.e. 5 and 7, 4 and 7 are. The general rule is: take a number equal to the *part* as the smaller number, and add it to the numerator e.g. $1 + \frac{1}{3}$; $3 + 1$, so 3 and 4, if there are 2 parts, add it to 2, e.g. $1 + \frac{2}{3}$; $2 + 3$, so series of 3, 5.

The author then takes up the rules for composites by example; thus $2 + \frac{1}{2}$, gives 2 and 5, $2 + \frac{1}{3}$, 3 and 7; $2 + \frac{1}{4}$; $2 + \frac{2}{3}$; $2 + \frac{3}{4}$; $2 + \frac{2}{5}$.[80] Thus to get the series for $2 + \frac{1}{4}$, he multiplies 4 and 2 and adds one, giving 4 and 9. This continues to $3 + \frac{1}{4}$, to get 4 and 13, and concludes with the observation that in the triples and a part, the small terms proceed by one, while the larger terms grow by three, that is $3 + \frac{1}{2}$, $3 + \frac{1}{3}$, $3 + \frac{1}{4}$, etc. yield 2, 7; 3, 10; 4, 13. The same analysis was applied earlier, in the text, to the unit and a part, a double and 2 parts.

6. [*Composition of ratios*]

In this section the composition of ratios is essentially their products, but such a statement does violence to the actual method of the author, since ratios are never thought of as fractions. This is best illustrated by ibn Sīnā's own examples. As usual he observes that each particular example is representative of a whole class, following this with an example to show that the ratio of $1 + \frac{1}{3}$ composed with $1 + \frac{1}{2}$ is 2: Let AB = 4, AC = 2, AD = 3 (see Figure 2). Then he notes that AB to AD = $1 + \frac{1}{3}$ to 1, AD to AC = $1 + \frac{1}{2}$ to 1 and so AB to AC, the "composition" is the "double". Moreover, he argues that this is perfectly general, in the sense that actual integral values of AB, AC, and AD are irrelevant; only the ratios have bearing. In fact he asserts that any double can be decomposed into a composition of a $1 + \frac{1}{3}$ to 1 and a $1 + \frac{1}{2}$ to 1. The simple argument establishing the latter, essentially reconstructs a figure similar to ABCD (above) using only ratios, identifies corresponding points, asserts that corresponding segments all lie in the same ratio to each other, and hence by equality and composition decomposes the "double" of the similar figure in the same manner as ABCD.[81]

There follow a number of such numerical examples. A few will suffice to give the flavour of the work.

(a) Let CD be a third of AC, and AC twice AB, then AD is the double of AB plus a third of AB. [$CD/AC = \frac{1}{3}$ and $AC/AB = 2$, $AD/AB = 2 + \frac{2}{3}$.] For the proof, divide AC into three parts, by M and N (see Figure 3). Then AM must be equal to CD, for the latter is also a third of AC, and AC is the double of AB. Now the mean of AM is a third of AB, consequently AM must be two thirds of AB. Thus $AD = AC + CD = 2AB + \frac{2}{3}AB$.[82]

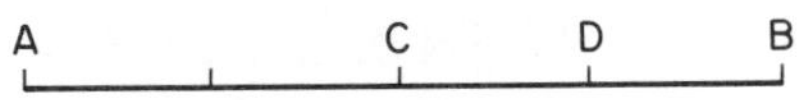

Figure 2

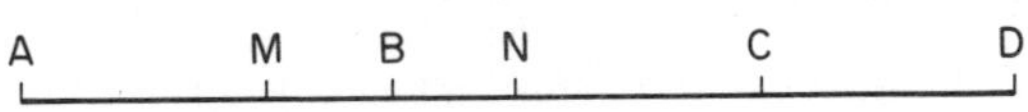

Figure 3

Figure 4

(b) Or again, if AC to AB is one plus one third, and BC to BD is one plus one half, then AB to BD is one to one half. For the proof, divide AB into three parts by the points M and N (see Figure 4). Then AM, MN, NB, and BC are equal, and the mean of each of these is an eighth part of AC, and must equal CD. Thus BD is three times CD, and AB is six times CD, which is the ratio of one to one half. Moreover, the ratio of BC to CD is the same as that of AB to BD, or equivalently BD to AB equals DC to BC, which is the ratio of the same to half the same.[83]

(c) Similar constructions are produced to show that if AC to AB is 5 to 4, and AC to CD is 6 to 5, AD to AB is 3 to 2; AC to AB 6 to 5 and AC to AD 7 to 6 imply AD to AB is 7 to 5. In addition, there are arguments to show that $\frac{7}{6}\cdot\frac{8}{7}=\frac{4}{3}$, $\frac{9}{8}\cdot\frac{10}{9}=\frac{5}{4}$, $\frac{10}{9}\cdot\frac{11}{10}=\frac{11}{9}$, $\frac{11}{10}\cdot\frac{12}{11}=\frac{6}{5}$, $\frac{15}{14}\cdot\frac{16}{15}=\frac{8}{7}$.[84] The last demonstration of this section should show that if AC to AB is 6 to 5 and AD to AC 5 to 4, then AD to AB is 3 to 2, but the text is garbled and the result is twice stated as 4 to 3, i.e. one and one third instead of one and one half. The author closes with the assurance that this suffices for the proofs of all assertions in the section on music.

7. [*Proportions*]

Al-Jūzajanī begins this section by remarking that he has omitted some notes of ibn Sīnā, namely the definition of integers, and their quantitative aspects. He then mentions offhandedly, that there are 20 kinds of proportions, 6 of which are useful to music theory.[85] In particular, there are proportions depending on the notion of surplus: one speaks of the surplus of a over b being equal to the surplus of c over d, i.e. the difference of a and b equal to the difference of c and d. Another sort of proportion he calls geometric; this involves a comparison of ratios, or in modern notation $a/b=c/d$. He notes that these are, in general, different; for example, 6 is to 4 as 9 is to 6, but $6-4\neq 9-6$; while an arithmetic proportion 4, 6, 8 where $6-4=8-6$ is not geometric, since $\frac{6}{4}\neq\frac{8}{6}$.

Next the author asks us to look at numbers with composition such that the ratios between them are successive, with the greatest ratios being those between the largest numbers, and the smallest ratios those between the smallest numbers. For example 6, 4, 3 gives $\frac{6}{4}=1+\frac{1}{2}$ and $\frac{4}{3}=1+\frac{1}{3}$. He calls these ratios of composition because of their use in musical composition, and, possibly, also because the ratio of the two extremes equals the ratio of the surpluses. [$\frac{6}{3}=\frac{6-4}{4-3}$; this actually defines the harmonic mean.]

The author next takes up the study of the geometric mean, whose square is the product of the extremes. Similarly, if there are two means, their product equals the product of the extremes

$$[a/m=m/b, a/m_1=m_2/b \quad \text{give} \quad m^2=ab, m_1m_2=ab.]$$

On the basis of this he notes that a three term geometric series is constituted by interposing consecutive rectangular numbers between consecutive squares. [The allusion is apparently of the sort: $a^2/x=y/b^2$ implies $xy=(ab)^2$ a rectangular number, "interposed" between squares, a^2 and b^2.] Also he states that between two squares there can be only one geometric mean, and between cubes there can be two.[86]

In the case of arithmetic proportions, the text continues, the numbers grow uniformly, be it by units, or by tens, etc.. They are continuous with respect to a single mean, but not continuous with respect to two. This mean is determined by the extremes. When two elements stand in arithmetic proportion, all is known. The mean is just the average of the extremes, and has the property that the product of the extremes is less than its squares. [Given a, $a + d$, $a + 2d$, $(a + d)^2 > a(a + 2d)$.] "And the measure of this difference is just the square of the excess of one extreme over the other."[87] [Or, so the text reads in Achena-Massé's version. Actually $(a + d)^2 = a(a + 2d) + d^2$, i.e. the measure of the difference is just the square of the excess of either extreme over the mean. That the translation is merely garbled here is indicated by ibn Sīnā's clear illustration: for 2 and 6 the mean is 4, and $4^2 - 2(6) = 4 = 2^2$.]

As to composite proportions [ratios of composition], the author observes that we know that they differ from arithmetic proportions, and how. To obtain the mean term between the extremes a and b (a less than b), take for it a times the excess of b over a, divided by the sum of a and b, and add this quantity to a. This is the mean; [i.e. $a(b - a)/(a + b) + a = 2ab/(a + b)$]. For example the mean between 6 and 3 is $3 + 3(6 - 3)/(6 + 3) = 4$. If the mean and the larger term are known, then the smaller term is $m - m(b - m)/(2b - m)$.[88] [The translation of Achena and Masse, however, gives a nonsense expression "obtenir la difference entre le puis divisor le resultat par le moyen terme *une seconde fois* et en deduire enfin le resultat du moyen terme; c'est alors le petit terme."[89]] When only the smallest term a and the mean are known, then the largest term is

$$\frac{m}{\dfrac{a}{m - a} - 1} + m.$$

The author then lists further properties of such composition proportions: if a, m, b are respectively the smallest, mean, and largest term, then $(a + b)m = 2ab; mb = 2ma = 2ab$. [The latter two equalities are, of course, nonsense, and may well be interpolations by another hand. The first $mb = 2ma$ may have arisen when someone noted that ibn Sīnā's example for this part 3, 4, 6 satisfies it, but this does not explain $2ma = 2ab$.][90] Al-Jūzajanī closes this section by announcing the omission of seven other kinds of proportions, since ibn Sīnā himself admitted that they have no application to music theory.

Having briefly sketched the contents of these seven sections of the arithmetic of the 'Alā'ī, let us note, at the outset, that the purported purpose of ibn Sīnā's tract, to expose the reader to the mathematical elements necessary for studying music is amply met. The musical part of the 'Alā'ī requires nothing more in the way of mathematical introduction. This leaves us with the problem of establishing the possible sources of the arithmetic itself. Such sources will, of course, not be Greek originals, since he himself tells us that he knew no Greek.[91]

Ibn Sīnā's mathematical education gives us strong hints: the Rasā'il of the Ikhwān al-Ṣafā' contains a section on arithmetic, and so does the *Elements*

(Books VII–IX). The former has a number of areas in common with ibn Sīnā's work: division of even numbers into evenly even, evenly odd, oddly even, and division of odd numbers into prime and composite odd, dropping the third class.[92] However, the remainder of the Rasā'il's arithmetic contains no theory of ratios,[93] except a cursory mention in the geometry,[94] and some use in the musical theory.[95] A later section on the relationship between arithmetic, geometry, and music[96] offers a few more details. But the Rasā'il provides a further possibility by referring to Nicomachus.[97] Much as the Rasā'il is unsuitable as the primary source of ibn Sīnā's study, so is the *Elements*, as a perusal of Book VII–IX readily shows. So we turn to Nicomachus of Gerasa, who lived around 100 A.D., probably in the Gerasa of Palestine.[98] His *Arithmetic* remained a popular mathematical handbook well into the sixteenth century in Europe, particularly among philosophers. An Arabic translation *Kitab al-Mudhal ilā 'ilm al-'adad*, attributed to Thābit ibn Qurra (836–901) survives.[99] Moreover the topics treated, as well as the order match ibn Sīnā's arithmetic very closely. Nicomachus' *Arithmetic*, after a six chapter prologue to book one, defines number, and even and odd (Chapter 7). Chapter 8 gives the median property, also treating one as exceptional; then lists the three classes of even numbers, and gives 64 as an example of an evenly even number. At this point ibn Sīnā's text interpolates the formula for $1+2+\cdots+n$ and $1+2+\cdots+n+(n-1)+\cdots+1$, as well as the property, of which $1+\cdots+6=21=3.6+\frac{1}{2}(6)$ is an example. None of these except the first (Book II, Chapter 9) occur in Nicomachus; nor does the formula for the sum of the first n odd integers. Then Nicomachus gives the median property for evenly even sequences, and the correct formula for $1+2+\cdots+2^n$, remarking on its usefulness for constructing perfect numbers. Chapter 9 follows with the properties of evenly odd numbers, while Chapter 10 is devoted to oddly even numbers. The next chapter details the three classes of odd numbers; he also excludes even numbers from the class of composites.[100] The 13th chapter gives the sieve of Eratosthenes to which ibn Sīnā only alludes, but there is also no reference to the number of primes. As one would expect, this is followed by a chapter on abundant, deficient, and perfect numbers. Even the set of perfect numbers 6, 28, 496 and 8128 appears, as well as the formula for perfect numbers $2^n(2^{n+1}-1)$ where the latter factor is a prime. However, ibn Sīnā's criteria for abundant and deficient numbers are not in Nicomachus,[101] but he does assert that all perfect numbers are even, and in fact, ending in 6 or 8.[102] Nicomachus' examples of deficient numbers are 8 and 14, not ibn Sīnā's 44. Chapter 18 of book one then introduces ratios, with a similar set of preliminary remarks, including trichotomy. Furthermore, ratios greater than one are divided into five groups: multiple, superparticular, superpartient, multiple superparticular, and multiple superpartient, where the last four refer to

$$1+\frac{1}{n},\quad 1+\frac{n}{m+m},\quad m+\frac{1}{n},\quad p+\frac{m}{m+n}.$$

Ibn Sīnā calls the group consisting of multiples, superparticulars, and superpartients simple, and the last two together complex. Nicomachus' discussion, paralleling ibn Sīnā's construction of series for the various ratios occupies Chapters 28–33 of book one.

Chapter 5 of book two of the *Arithmetic* handles the next section in ibn Sīnā—composition of ratios. The first example is, in fact, concerned with the observation that the composition of $\frac{3}{2}$ and $\frac{4}{3}$ is 2. But the approach is different. While ibn Sīnā proceeds geometrically, in a manner like that used in manuals like Al-Khwārīzmī's Algebra to solve inheritance problems,[103] Nicomachus uses the mnemonic devices inherent in his classification scheme; e.g.[104]

2 sesquialter 3 sesquitertian 4
double

6 double 12 sesquilater 18.
triple

Indeed, Nicomachus does not give any evidence of concern for geometric justification; as he puts it "double with sesquialter is triple, triple with sesquitertian is quadruple, quadruple with sesquiquartan is quintile, and as far as you wish to proceed, no contrary result will appear."[105] His Chapters 6–20 of book two are devoted to Pythagorean topics: the figurate numbers, omitted by ibn Sīnā. Finally, in Chapter 21, he introduces ibn Sīnā's last arithmetic topic, proportions, noting that six were known to the ancients, and ten in his own time.[106] In Chapters 22–25 Nicomachus considers, in order, the arithmetic, geometric, and harmonic (composition) proportions, noticing (Chapter 24) that geometric and harmonic proportions are different, but with numerical example: 2, 4, 8.[107] Nicomachus also refers, but less transparently, to the fact that rectangular numbers are the mean proportionals between consecutive squares,[108] and observes that squares have only one geometric mean, while cubes have two.[109] Indeed, he also mentions the characterization of the arithmetic mean by square differences and the inequality $(a+d)^2 > a(a+2d)$ given by ibn Sīnā.[110]

Nicomachus then devotes Chapter 26 to a digression on the harmonic (composition) proportion in music. In Chapter 29, he gives the same recipe for the harmonic mean as ibn Sīnā does, but neither offers formulae for the smallest term, given the largest and the mean, nor for the largest, given the mean and the smallest. Finally, the error in ibn Sīnā's text: $mb = 2ma$, has an analogue—though not an exact one—in Nicomachus![111]

Now, let us summarize the most striking differences between ibn Sīnā's work and that of his apparent model Nicomachus' *Arithmetic*, and try to account for them. It is evident that we should look first, for such variations in Thābit's *K. al-mudhal.*

These differences are primarily interpolations by ibn Sīnā, and consist of: (1) the formula $1+2+\cdots+n+(n-1)+(n-2)+\cdots+1=n^2$; (2) formulae like $1+\cdots+6=21=3(6)+\frac{1}{2}(6)$, etc.; (3) criteria for abundant and deficient numbers; (4) a geometric justification for composition of ratios; (5) formulae for any term in a harmonic proportion, given the other two. Curiously enough, they do not appear in Thābit's al-mudẖal, which, at least in the edition of Kutsch available to us, follows the Greek text of Nicomachus' *Arithmetic* quite closely, without additions or commentary.

Nonetheless, some of these items do appear in other Greek arithmetics: Iamblichus, for example, in his *On Nicomachus' Introduction to Arithmetic*, discusses the Pythagorean formula

$$\left.\begin{matrix} 1+2+3+\cdots+(n-1) \\ +1+2+3+\cdots+(n-1) \end{matrix}\right. + n = n^2,$$

explaining how the Pythagoreans looked upon the sum, n^2, as a "racecourse" (δίαυλος). Iamblichus even cites the same particular instance, $n = 10$.[112] It doubtless appears in other manuals and handbooks of mathematics—perhaps some of them Arabic. On the other hand, formulae (2) may be original with ibn Sīnā. The faulty criteria for abundant and deficient numbers seem, however, to have no counterpart in Greek literature. Thābit, of course, is known to have left criteria in *Maq̄ala fī stiẖrāǧ al-a'dād al-mutaḥābba bi-suhūlat al-maslak ilā ḏālika.*[113] He states, for example, that if $s = 1 + 2 + 2^2 + \cdots + 2^n$, then $2^n s$ is perfect, if s is prime, while $2^n p$ is deficient if p is a prime less than s, and abundant if p is a prime exceeding s. These are proven, and followed by additional criteria for abundant and deficient numbers.[114] The whole tract, as the title indicates, is concerned with establishing a criterion for "amicable" numbers.

As we remarked earlier, the geometric approach to the justification of composition of ratios is already present in such compendia as al-Khwārīzmi's *Algebra* (section on legacies, for example). This leaves: (5) formulae for one term in a harmonic proportion, given the other two. The geometrical versions of these are certainly preserved in Greek sources, e.g. Pappus' *Collection* III, propositions

١٠	٩	٨	٧	٦	٥	٤	٣	٢	١
٢٠	١٨	١٦	١٤	١٢	١٠	٨	٦	٤	٢
٣٠	٢٧	٢٤	٢١	١٨	١٥	١٢	٩	٦	٣
٤٠	٣٦	٣٢	٢٨	٢٤	٢٠	١٦	١٢	٨	٤
٥٠	٤٥	٤٠	٣٥	٣٠	٢٥	٢٠	١٥	١٠	٥
٦٠	٥٤	٤٨	٤٢	٣٦	٣٠	٢٤	١٨	١٢	٦
٧٠	٦٣	٥٦	٤٩	٤٢	٣٥	٢٨	٢١	١٤	٧
٨٠	٧٢	٦٤	٥٦	٤٨	٤٠	٣٢	٢٤	١٦	٨
٩٠	٨١	٧٢	٦٣	٥٤	٤٥	٣٦	٢٧	١٨	٩
١٠٠	٩٠	٨٠	٧٠	٦٠	٥٠	٤٠	٣٠	٢٠	١٠

Figure 5

9–11, which give an explicit construction of one term (as a line segment) in terms of the other two.[115] On the other hand, none of the Greek Arithmetics available to us—Pappus, Nicomachus, Theon of Smyrna, Iamblichus—contains a rhetorical solution of the problem. We comment on this again, momentarily.

At this point, to complete our picture of ibn Sīnā's work in arithmetic, we offer a sketch of those parts of the Shifā' which supplement the arithmetic of the 'Alā'i

1	2	3	4	5	6	7	8	9	10
2	4	6	8	10	12	14	16	18	20
3	6	9	12	15	18	21	24	27	30
4	8	12	16	20	24	28	32	36	40
5	10	15	20	25	30	35	40	45	50
6	12	18	24	30	36	42	48	54	60
7	14	21	28	35	42	49	56	63	70
8	16	24	32	40	48	56	64	72	80
9	18	27	36	45	54	63	72	81	90
10	20	30	40	50	60	70	80	90	100

Ratio atq₃ expositio digeste formule. Cap. 27.

S Igitur duo pma latera ppositę formulę q̄ faciūt āgulū: ab vno ad. 10. et. 10. pcedentia respiciāt: τ his subteriores ordies cōparent: qui scilicet a. 4. angulum incipientes: in rigenos terminū ponunt: duplex id est prima species multiplicitatis ostenditur: ita vt primus primū sola superet vnitate: vt duo vnū secūd' scdm binario supuadat: vt q̄ternari' binariū. terti' tertiū tribus: vt senari' ternariū. qrtus qrtū qternarij numerositate transcendat: vt 8 qternariū: τ p eādē cūcti sequētiā sese minoris pluralitate pretereant. Si vero terti' angulus aspiciat: q ab 9. inchoāis lōgitudinē latitudinēq₃ tricenis altrinsecus numer(extēdit: et hic cū pma latitudine et lōgitudine cōpararetur: triplex species multi-

MULTIPLICATION TABLE, 1488 BOETHIUS

Figure 6. Reproduced with permission from D. E. Smith, *Rara Arithmetica*, 4th edition Chelsea Publishing Company, N.Y., 1970, figure 12 p. 26

and Najāt. These are: (a) a short account of "practical" arithmetic, e.g. multiplication tables to 10×10 (Figure 5) arranged very much like those in Boethius' manual of arithmetic;[116] (Figure 6) [See Figures 5 and 6.]; (b) various rules of formal algebra; (c) a list of triangular, square,..., octagonal numbers, which coincides, up to scribal error, with Nicomachus II, Chapters VIII–XII;[117] (d) a fairly complete discussion of the 10 "proportions" of the Greeks; (e) a theorem in the Indian arithmetic about squares and cubes (casting-out nines); (f) two rules for summation of square and triangular arrays filled with ordered arrangements of successive odd numbers.

The first of these occupies the third section of the Cairo edition[118] and needs no further comment. The various formal rules of algebra are all rhetorical and involve factorizations, like: $A^2 - A = A(A-1)$, $A^3 - A = (A-1)(A)(A+1)$, $A^4 - A = A(A-1)(A^2 + A + 1)$, $(A+B)^2 = 2AB + B^2 + A^2$, or properties of exponents, such as $2^{n+1} - 2^n - 1 = 3(2)^{n-1} - 1$ and $(9)2^{n-1} - 2 = 3(2^n) - 1 + 3(2^{n-1}) - 1$. These are scattered throughout the text, and Lutfy attempts to summarize them following the preface to his edition of al-Ḥisab.[119]

As we have noted, ibn Sīnā's account of the various Pythagorean numbers follows Nicomachus and is the content of Section 3 of al-Ḥisāb (pp. 53–61).[120] The section on the ten Greek proportions is best summarized by table 1[121] (pp.

Table 1. [$a < b < c$, positive numbers]

1.	$\frac{a}{a} = \frac{c-b}{b-a}$; $b = \frac{a+c}{2}$,		
2.	$\frac{c-b}{ab} = \frac{b}{a}$; $b = \sqrt{ac}$,		
3.	$\frac{c-b}{b-a} = \frac{c}{a}$; $b = \frac{2ac}{a+c}$,		
4.	$\frac{c-b}{b-a} = \frac{a}{c}$; $b = \frac{a^2+c^2}{a+c}$;	example	(3, 5, 6)
5.	$\frac{c-b}{b-a} = \frac{a}{b}$; $b = \frac{c-a}{2} + \sqrt{\frac{(c-a)^2}{4} + a^2}$,		(2, 4, 5)
6.	$\frac{c-b}{b-a} = \frac{b}{c}$; $b = -\frac{c-a}{2} + \sqrt{\frac{(c-a)^2}{4} + c^2}$,		(1, 4, 6)
7.	$\frac{c-a}{b-a} = \frac{c}{a}$; $b = \frac{2ac - a^2}{c}$,		(6, 8, 9)
8.	$\frac{c-a}{c-b} = \frac{c}{a}$; $b = \frac{c^2 + a^2 - ac}{c}$,		(6, 7, 9)
9.	$\frac{c-a}{b-a} = \frac{b}{a}$; $b = \frac{a}{2} + \sqrt{\frac{4ac - 3a^2}{2}}$,		(4, 6, 7)
10.	$\frac{c-b}{b-a} = \frac{a}{b}$; $b = c - a$,		(5, 3, 2)

65–9 in text). This is in the last (4th) section. (Section one is devoted to properties of natural numbers). It is worth noting that all the examples, except that for (10) are the same as those given by Nicomachus. (2) and (10) are in error, perhaps owing to mistakes in transcription from the manuscript; the former should read $c/b = b/a$, and the latter $b/a = (c - a)/(c - b)$ (or its inversion), with example (3, 5, 8). Finally we call attention to the fact that (4)–(6) are stated in the form given by Theon of Smyrna rather than that of Nicomachus, i.e. as "inversions".[122]

In the section of al-Ḥisab devoted to the Indian arithmetic, ibn Sīnā, without elaboration gives two rules. These are: (1) "If a given number, which is divided by 9 gives a remainder of 1 or 8, then the square of that number, on division by 9, gives a remainder 1. If the number, divided by 9, gives remainder 2 or 7, then the square of that number, on division by 9, gives remainder 4. If the number, divided by 9, leaves a remainder of 4 or 5, then its square, divided by 9, gives a remainder of 7. Finally, if the number, divided by 9, gives a remainder of 3, 6 or 0, then its square, on division by 9, leaves remainder zero." (2) "If a number, divided by 9, leaves a remainder of 1, 4 or 7, then its cube, on division by 9, gives a remainder of 1; if the number, divided by 9, gives a remainder of 2, 5 or 8, then its cube, divided by 9, leaves 8; and if the number, divided by 9, leaves remainders 3, 6 or 0, then its cube, on division by 9, leaves remainder zero." Woepcke[123] appears to have been the first to call attention to these rules, having found them in a Leiden manuscript.[124] Cantor[125] cites Maximus Planudes as offering support for ibn Sīnā's own claim that the rules are Indian in origin, while Tannery argues that "proofs by 9" may very well have evolved independently from the Apollonian notion of "pythmene".[126] However, this may be, the notion of "proof by 9" is certainly in evidence in Arabic arithmetics before the time of ibn Sīnā.[127] On the other hand, it is worth pointing out that ibn Sīnā offers his rules without application, and that, although they contain the germ of an idea for a "proof by casting 9's", they are merely statements about congruence classes of squares and cubes modulo nine.

Of much more interest are ibn Sīnā's two rules for summations of square and triangular arrays of odd numbers. These appear in the first section of al-Ḥisāb. The first reads: "If the successive odd numbers are placed in a square table, the

9	7	5	3	1
19	17	15	13	11
29	27	25	23	21
39	37	35	33	31
49	47	45	43	41

Figure 7

3	1
7	5

5	3	1
11	9	7
17	15	13

Figure 8

sum of the numbers lying on the diagonal will be equal to the cube of the side; the sum of the numbers filling the square will be the fourth power of the side."[128]

Ibn Sīnā illustrates the rule by first giving a 5×5 square (Figure 7), whose diagonal sums $9 + 17 + 25 + 33 + 41 = 125 = 5^3 = 1 + 13 + 25 + 37 + 49$, and whose total sum is $625 = 5^4$. In addition, he illustrates the way the squares are filled by using 2×2 and 3×3 examples (see Figure 8).[129]

Later in the same section,[130] he discusses square arrays of the "doubles", giving as an example that shown in Figure 9. He notes that the sum of each diagonal is equal to the other.

The second rule, which follows immediately after the first is:[131] "If the successive odd numbers are placed in a triangle, the sum of the numbers taken from one row equals the cube of the [row] number." Ibn Sīnā illustrates this with one figure (Figure 10). As the reader may verify $1 = 1^3, 3 + 5 = 8 = 2^3, \ldots, 21 + 23 + 25 + 27 + 29 = 125 = 5^3, \ldots$.[132]

These two rules are quite striking, and for the modern reader, seem like sparkling gems in ibn Sīnā's otherwise, rather ordinary, but thorough account of the elements of Pythagorean and elementary Indian arithmetic. The second is easily traced; Nicomachus gives it in a rather less eye catching form in the *Arithmetic*:[133] "For when the successive odd numbers are set forth indefinitely beginning with 1, observe this: The first one makes the potential cube; the next two added together, the second; the next three, the third. . . ." The first is a different matter, however. Since our inspection of Arabic manuals of arithmetic like that of al-Uqlīdisī (952/3),[134] or the later summary of the Hindu arithmetic of Khūshar ibn Labbān (1000),[135] fail to turn up anything like it, we can only speculate on ibn

22	18	14	10	6	2
46	42	38	34	30	26
70	66	62	58	54	50
94	90	86	82	78	74
118	114	110	106	102	98
142	138	134	130	126	122

Figure 9

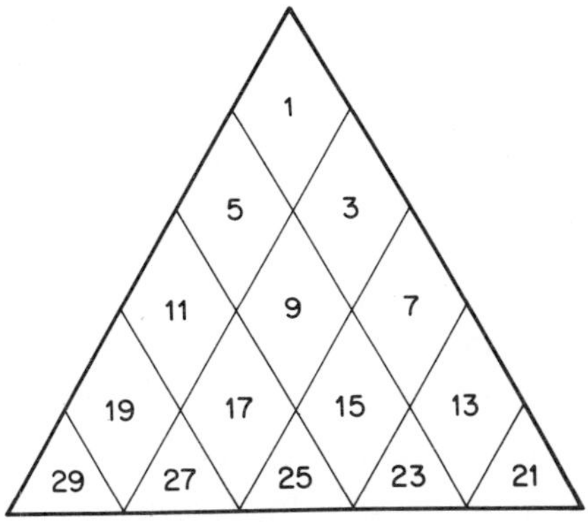

Figure 10

Sīna's source. Perhaps it was the same as that for his rule of nines, an Indian method, or, perhaps, it was his very own discovery.

Finally, let us return to our consideration of possible sources for ibn Sīnā's computation of one term of the harmonic proportion in terms of the other two. As our table of the 10 Greek proportions found in al-Ḥisab indicates, ibn Sīnā also solved for the middle term as a function of the other two, indicating familiarity with a (rhetorical) quadratic formula. Perhaps, in the case of the 'Alā'ī as well as the Shifā', this work is an exercise the author introduces to show his familiarity with methods for solving linear and quadratic equations.

Curiously enough, the author himself is rather misleading about his actual sources. As a matter of fact, in the introduction to the arithmetic section,[136] he says that he relies on the *Kitāb al-Usṭuqusāt* for his number theory, and *Kitāb al-Katēguryas* for his basic ideas about numbers. The former is, of course, Euclid's *Elements*,[137] and the latter is probably Aristotle's *Categories* (which does contain some primitive ideas about numbers), or some commentary on it.[138]

IV. Geometry

The geometric treatise of ibn Sīnā in the 'Alā'ī is, as al-Jūzajanī tells us in his introduction, a resumé of the *Elements* of Euclid, in particular the parts necessary for the study of the *Syntaxis Mathematica* (*Almagest*) of Ptolemy.[139] It consists of 13 sections, does not list definitions, postulates and axioms, and contains no formal proofs. In what follows we offer a brief summary of these sections, test the statement that they offer sufficient geometric preparation for the *Almagest*, and examine the possibility of a more immediate source than the *Elements*.

For the reader's convenience we list those propositions of Euclid used implicitly in the *Almagest* (Table 2).

In Table 3 is a list of the Euclidean propositions in order by book. This allows a verification of the fact that only the first six books of the *Elements* play an essential role in the *Almagest* (with the exception of a proposition from Book VIII and another from Book XIII).

Table 2. Euclidean propositions implicit in the *Almagest*

Location in *Almagest*	Proposition of *Elements*
Book I, section 10	I, 11,10; II, 6; I, 47; VI (def. 3) XIII, 9; IV, 15 (Cor.); XIII, 10 III, 21; VI, 4, 16; II, 1; III, 27; I, 26 VI, 8; III, 26, 29; VI, 3; V, 8; VI, 1 III, 3; I, 21; VI, 33
Book I, section 13	VI, 4, 2
Book III, section 3	I, 16; III, 7, 3; III, 16, Por; VI, 8 VI, 7
Book III, section 5	III, 31
Book IV, section 6	III, 36
Book IX, section 10	I, 32.[140]

Table 3.

Book of *Elements*	Proposition
I	10, 11, 16, 21, 26, 32, 47
II	1, 6
III	3, 7, 16 Por, 21, 26, 27, 31, 36
IV	15 (Cor.)
V	8
VI	(def. 3), 1, 2, 3, 4, 7, 8, 16, 33
VIII	9
XIII	10

The first three sections of ibn Sīnā's geometry are concerned with the properties of intersecting straight lines, parallel lines, and notions related to angles and triangles. Ibn Sīnā begins the first section by observing that two lines intersect at right angles or obliquely, and in either case the sum of the angles on the same side is two right angles (*Elements* I, 13).[141] In addition the opposite angles are equal (I, 15). His definition of perpendicular lines: one "does not incline toward one side or the other of the second line", stands in contrast to Euclid's definition (I, definition 10): "making equal adjacent angles". Lines and planes are undefined. Ibn Sīnā then closes this section with the following propositions: a line perpendicular to two intersecting lines is perpendicular to every line in their plane and passing through the intersection point (XI, 2, 4). He adds that such a line is said to be perpendicular to the plane (XI, def. 3), and states that lines perpendicular to the same plane are parallel (XI, 6), offering a heuristic argument based on the notion of "inclination".

In section 2, the author notes that line segments, when extended, either intersect, or are parallel. They intersect when one is "inclined toward the other", the intersection being along the direction of inclination, but not in the opposite. Moreover, "if they always remain separated in such a way that the distance between the ‘extremities’ are equal, and the perpendicular to one is also perpendicular to the other...then they are parallel.” This is essentially Posidonius’ definition of parallels, which entered the Arabic tradition through Al-Nayrīzī’s commentaries on Simplicius’ Euclid, and lies within the “equidistance”, “convergence–divergence” theory of parallels which extends from Thābit through al-Ṭūsī.[142] Ibn Sīnā then continues with a circular proposition: “each time that a line cuts two parallels without inclining toward one or the other, it is perpendicular to both.” Next, he makes an assertion equivalent to *Elements* I, 29, and follows this with: “if a line cuts two other lines so that the two interior angles on the same side equal two right angles, then the two lines are parallel.” (I, 28). He argues that if the two lines were not, then all three lines would form a triangle with the sum of its angles being greater than two right angles, and having an exterior angle equal to one of its interior angles. Of course, the argument implicitly assumes *Elements* I, 32 to reach a contradiction. Ibn Sīnā then adds that if the sum of the two angles is less than two right angles, the lines actually intersect. He follows this with a construction of a line through a given point and parallel to a given line (I, 31). However, in contrast to Euclid, ibn Sīnā uses *Element* I, 11 implicitly in his construction instead of I, 23, and evidently replaces Euclid’s use of I, 27 with I, 28. The second section closes with a statement of I, 30: lines parallel to the same line are parallel to each other.

He begins section 3 with a partial statement of I, 32: the sum of the angles of a triangle is two right angles. The proof offered is accompanied by a diagram (Figure 11). The angle HCD is constructed equal to $\angle$ ABC to make HC parallel to AB, making implicit use of I, 23 instead of simply citing the previous section. The proof then proceeds as the diagram suggests by reference to ibn Sīnā’s version of I, 29. Next, he enunciates one case of congruent triangles, namely that of equal angles included between two equal sides (I, 4); he later makes implicit reference to three sides equal (I, 8), but omits the case of one side and two angles (I, 26) entirely. His proof of congruence is by superposition, phrased as “apply angle B [in one triangle] to the corresponding angle H by an act of imagination”, which seems to indicate an opposition to the introduction of kinematics to geometry, a principal favored by Thābit. Ibn Sīnā next gives a proof of the fact that triangles with two equal sides have two equal angles, with equal supplementary exterior angles (I, 5). It depends on the following figure in which Z bisects AC (I, 10), consequently

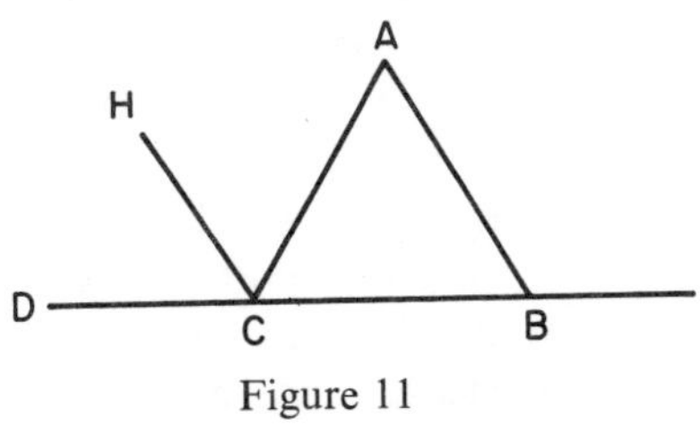

Figure 11

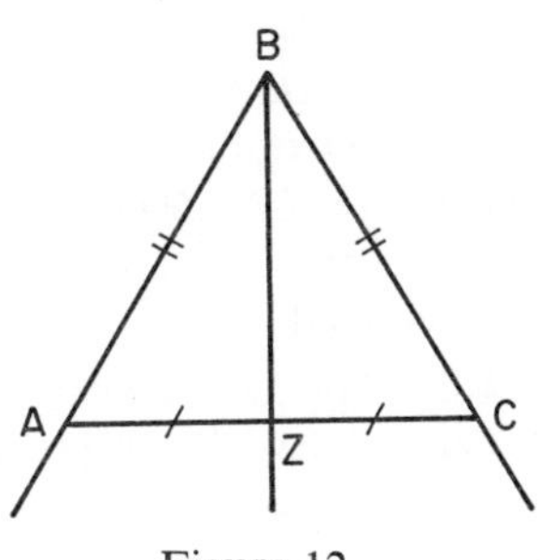

Figure 12

making $\triangle ABZ = \triangle BCZ$ (I, 8). (See Figure 12.) This, in combination with implicit use of I, 13 completes his proof. It is, of course, circular since the I, 10 requires I, 9, which in turn depends on I, 8 and I, 8 invokes I, 7 which depends on I, 5 itself. No mention is made of Euclid's partial converse: I, 6, which is considered in totality in al-Nayrīzī's commentary. Ibn Sīnā then observes that if a $\triangle ABC$ has base AB increasing, the $\langle ACB$ will also. This leads him to assert that the greatest side of a triangle is opposite the greatest angle, and conversely (I, 18; I, 19).

In the fourth section, ibn Sīnā discusses three ruler and compass constructions: perpendicular to a line at a point on the line (I, 11), perpendicular to a line from an external point (I, 12), bisector of a given angle (I, 9). The following section is devoted to parallelograms, triangles, and their areas. It begins with a definition of parallelogram, followed by the assertion that a diagonal separates the parallelogram into two identical triangles (I, 34 approximately). Then there appear two propositions: parallelograms (triangles) with the *same* base, lying between parallel lines (i.e. with the same height) are equal in area. The discussion then "generalizes" this to include parallelograms (triangles) with equal bases (I, 35, 36, 37, 38). The final proposition seems to be in reference to a figure that does not appear in the manuscript of 'Alā'ī; it asserts that the parallelograms formed by completing the triangles, on either side of the diagonal of a given parallelogram, to parallelograms, gives rise to equal parallelograms.

In section 6, ibn Sīnā takes up the "multiplication of lines". First he shows how to "multiply" a line segment by itself, that is, how to construct a square with side equal to the given segment (I, 46). He then offers sketches of proofs of the Pythagorean theorem (I, 47) and its converse (I, 48), following Euclid in spirit. Yet, he makes no mention of Thābit's alternate proof of I, 47 (which is noted in al-Nayrīzī's commentary), and Thābit striking generalization to arbitrary triangles is also omitted.[143] Ibn Sīnā's next proposition in geometric algebra is: if a line segment be divided into two parts, the square of the whole segment equals the sum of the squares on the two subsegments plus twice the product of these two segments $[x^2 + (l-x)^2 + 2(l-x)x = l^2]$. This is *Elements* II, 4 and the treatment again follows Euclid. In an analogous fashion, ibn Sīnā ends the section with *Elements* II, 6, which asserts in modern notation that

$$(l+m)m + \left(\frac{l}{2}\right)^2 = \left(\frac{l}{2} + m\right)^2.$$

Section 7 of the 'Alā'ī treats circles. Since the reader already has a fair idea of ibn Sīnā's method, our account will be very much abbreviated. In order, ibn Sīnā gives *Elements* III, 26–29 (approximately), that is, propositions concerned with equal circles, chords, central angles, and arcs. Then he establishes III, 3, part of III, 18, but skips III, 16—Euclid's account of the "horned angle", a topic treated in the manuscript "al-Zawiya".[144] After that he takes up III, 19, 20, 21, 31 (partially), and 36, with the proof in two parts as it appears in the *Elements*.

The next three sections (8–10) deal with proportion, with the last of these devoted to geometric figures. The language is clearer and the wording more concise than in the corresponding section of the *Arithmetic* of the 'Alā'ī. For example, ibn Sīnā explains the concepts of proportion and composition of ratios straight away, then takes up "inverses", i.e. $A/B = C/D$ then $B/A = D/C$. The geometric material consists of 11 propositions from *Elements* VI: 1, 2, 4 (implicitly), 5, 8, 13 (construction of mean proportional), 11 (third proportional), 14 (reciprocally proportional parallelograms), 16, 19, 20. These, for the most part relate ratios of areas of similar polygons to ratios of squares of lengths of corresponding sides.

In sections 11 and 12, ibn Sīnā again returns to the topic of circles. Section 11 actually begins by repeating certain parts of section 7, then sketches proofs of *Elements* XII, 1 and 2, while section 12 is concerned with chords of circles, and begins with the construction of a regular hexagon (IV, 15). However ibn Sīnā's concern is not with the construction itself but rather the relationship between the side of the hexagon and the diameter of the circle ($s_6 = r$). This is borne out by the fact that he dispenses with the construction of an equilateral triangle altogether, and just assumes it as given (inscribed in a circle), when arguing that $s_3^2 = r \cdot 3r$. Then he discusses *Elements* XIII, 9 which asserts that the side of a hexagon inscribed in a circle is the mean proportional between the side of a decagon inscribed in the same circle and sum of the sides of the hexagon and decagon, and *Elements* XIII, 10($s_5^2 = s_6^2 + s_{10}^2$).

The last section concerns itself with some notions of solid geometry and the properties of spheres. Solids are defined in the Euclidean manner, and spheres, cubes, and prisms are given as examples. Then ibn Sīnā remarks that solids will be considered to be equal if their extremeties (surfaces) are equal; he follows this with definitions of cones and prisms. The former includes Euclid's separate class of pyramids, while the latter is more restrictive than the Euclidean class, admitting only prisms with triangular bases and tops, and sides that are parallelograms. This set of definitions is succeeded by a loose discussion of the content of XI, 25, 29–31. After this, ibn Sīnā devotes considerable attention to pyramids with triangular base, starting with Euclid's dissection (XII, 3) of such a pyramid into two similar, equal pyramids and two equal prisms, each greater than half the original pyramid. Ibn Sīnā's restriction of the class of prisms causes no difficulty, since those referred to in Euclid are of the same type. He follows this with expositions of XII, 4 and 5, but omits Euclid's extension of XII, 5 to pyramids with polygonal bases, preferring instead to illustrate XII, 7: the decomposition of a triangular based prism into three equal triangular based pyramids. Next ibn Sīnā gives *Elements* XII, 16 and 17: the construction of polygons between distinct concentric circles, not touching the smaller circle, and polyhedral solids between concentric spheres, not touching the smaller sphere. After mentioning that spheres stand to one another in volume as the ratio of the cubes of their diameters (XII, 18), ibn Sīnā ends his account of geometry with the assurance that anyone who understood it would be prepared to understand the *Almagest* under the guidance of a good teacher.

Table 4 summarizes the specific Euclidean propositions that appear in the 'Alā'ī geometry by book as uncovered by our examination. A comparison with

Table 4

Book of *Elements*	Propositions
I	4, 5, 8, 9, 10, 11, 12, 13, 15, 18, 19, 23, 28, 29, 30, 31, 32, 34, 35, 36, 37, 38, 46, 47, 48
II	4, 6
III	3, 18, 19, 20, 21, 26, 27, 28, 29, 31, 36
IV	15
VI	1, 2, 4, 5, 8, 11, 13, 14, 16, 19, 20
XI	2, 4, dfn. 3, 6, 25, 29, 30, 31
XII	1, 2, 3, 4, 5, 7, 16, 17
XIII	9, 10

the needs of the *Almagest* reader indicate that ibn Sīnā omits I, 16, 21, 26; II, 1; II, 7, 16 Por; VI, 3, 7, 33; VIII, 9. However I, 32 implies I, 16, while III, 16 Por. and VI, 33 are implicit in the 'Alā'ī. So one must indeed admit that ibn Sīnā provides his reader with a good summary of geometry prerequisite for the *Almagest*.[145]

If ibn Sīnā used an existing summary (of the *Elements*) by another author, as he did when writing of arithmetic, it may never be known. In fact, it would even be a considerable achievement to determine whether ibn Sīnā worked with a copy of the *Elements* based on the translation of al-Ḥajjāj ibn Yūsuf ibn Maṭar (fl. 786–833), or the later translation by Isḥāq, improved by Thābit ibn Qurra. The present writers are of the opinion that the summary is probably ibn Sīna's own, perhaps based on the translation of al-Ḥajjāj. He certainly would have had to prepare one in his initial, almost single-handed study of the *Almagest*, and prior to that he says, in the autobiographical fragment, that he studied Euclid, making no mention of summaries, as he specifically does when telling how a gloss, or summary of Aristotle by al-Fārābī helped him to understand the intent of the original. Moreover, as al-Jūzajanī tells us, ibn Sīnā revised at least one of his geometrical summaries to include additional figures, a further suggestion that a manuscript of the *Elements* may have been the primary source. A careful examination of the figures themselves provides little evidence for identification of the source, since they differ from both the al-Ḥajjāj and Isḥāq–Thābit versions. It is also known that Abū 'Uthman al-Dimashqī (fl. *c.* 908–932) translated parts of the *Elements*, but only the part dealing with Pappus' commentary on Book X survives, a fact which does not help in deciding the tradition for the 'Alā'ī text, since it does not contain propositions from the tenth book of the *Elements*.[146]

As Murdoch points out, summaries (mukhtaṣar) were not unknown, even in the Arabic, by ibn Sīnā's time. In fact Ḥajjāj favoured his second version of the *Elements* for its shorter length.[147] Moreover, those who would argue for a more immediate source of ibn Sīnā's summary than any translation of the *Elements*, will certainly point out that ibn Sīnā, when preparing the Shifā', Najāt, and 'Alā'ī, was writing in a well established classical tradition: the encyclopedia format, which has roots in the Greek and Latin worlds of the second century A.D..[148] One of the earliest of these encyclopedias, Theon of Smyrna's *περὶ τῶν κατὰ τὸ μαθηματικόν χρησιμων εἰς τὴν Πλατωνος ἀνάγνωσιν* (*On the Mathematical Knowledge which is Needed to Read Plato*) contained five parts: arithmetic, geometry, stereometry, astronomy, and music.[149] Only the first part, with an appendix on music (from a Byzantine edition), and the part on astronomy survive in full, although some are of the opinion that Theon simply never fulfilled the promise of his introduction, to write detailed sections on geometry and stereometry.[150] Instead he limited his remarks in these areas to "geometrical numbers", e.g. pyramidal numbers, spherical numbers, etc.. Heath, in fact, gives some striking evidence to support this opinion.[151] There is, of course, no evidence that Theon's "vade-mecum for philosophy students"[152] entered the Arabic tradition. The same seems to be true of the work of Iamblichus of Chalcis [in Syria] (*c.* 250–330) who wrote an encyclopedia of Pythagoreanism, which contained a summary of Nicomachus' *Arithmetic*, as well as books on gematria, geometry, and music.[153]

On the other hand, various compendia of the sciences in Arabic, such as the *Mafātiḥ al-'ulūm* (Keys of the Sciences) of Abū 'Abd Allah Muḥammad ibn Aḥmad ibn Yūsuf Al-Khuwārizmī (fl. *c.* 975 in Khuwarizm),[154] may well have been known to ibn Sīnā. The Mafātiḥ relies on Nicomachus as well as on the encyclopedia (Rasā'il, literally "epistles") of the Ikhwan al-Ṣafā', but appears to be more a catalogue of sciences and terms, with emphasis on their etymologies, rather than a purposeful summary of the fundamentals of each science. Wiedemann devoted his Beiträge XIV to a translation of its mathematical sections, and mentions therein other tracts in the scientific encyclopedia tradition in Arabic.[155] The Mafātiḥ itself seems to have been written as a manual to assist the reader in partaking of scholarly conversations, or at least to allow him to glean something from listening to them. The author dedicates his work to the vizier of the Samanid ruler Nūḥ II (976–997), Abu'l-Ḥasan al-Utbī, much as ibn Sīnā dedicated his Danesh-nameh to 'Alā' al-Dawla. Yet none of these compendia mentioned, nor those others studied by Wiedemann have any claim for being a source for ibn Sīnā's geometry, coming between him and Euclid.

Now let us consider ibn Sīnā's much fuller treatment of the *Elements*, the geometry section of the Shifā'. This has recently been edited (in Arabic) by Sabra,[156] and it is his edition to which we shall refer. Unlike the geometry of the Alā'ī, this work is a summary of the *Elements* I–XIII, as well as of the post-Euclidean Books XIV–XV. It has no general introduction, preface, or statement of purpose, but each book contains some of the customary Euclidean definitions. The first also enumerates the common notions ('ilm gami'), but omits the postulates, except for the fifth.[157]

Ibn Sīnā's treatment of the Euclidean propositions roughly resembles the *construction* which *follows* each proposition in the *Elements*. In other words, ibn Sīnā rarely states the proposition itself, but allows it to be inferred from the geometric construction, which acts, in turn, as a guide to the proof. Moreover, he follows Euclid's lead in rarely citing propositions upon which the proof depends, rather than emulating Ḥajjāj, who sprinkled his account of the *Elements* with phrases like "by proposition so and so", etc..[158]

Table 5

Book	Contents (number of propositions)	
	Euclid	Ibn Sīnā
1	48	53 (but no. 42 is missing), accounting for 46 Euclidean propositions, and omitting, I, 40, 45. [53 (no. 42 missing); 46E; omits I 40, 45.]
2	14	18 (but several numerals repeated or added by editor); 14E; omits none.
3	37	36 (no. 24 is missing and might corresponded to III 26); 35E; omits III, 26, 29.
4	16	18; 16E; omits none
5	25	25; 25E; omits V 21, 22, 23
6	33	31 (also one un-numbered); 27E; omits VI 7, 9, 12, 16, 27, 29
7	39	41; 39E; omits none
8	27	25; 24E; omits VIII 11, 17, 20
9	36	36; 36E; omits none
10	115	108; 100E; omits X 7, 8, 10, 16, 19, 27, 29, 30, 31, 32, 83, 84, 112, 113, 114
11	39	39; 37E; omits XI 23, 34
12	18	16; 15E; omits XII 6, 13, 14
13	18	17; 15E; omits XIII 4, 5, 7
	(465)	(463) (425E)
14	(9)[a]	11; 6
15	(5)[b]	5; 5
	479	479 436

[a](see Appendix and Heath (46), III, 512–19)
[b](see Appendix, and Heiberg and Menge (48), V, 40–67)

Table 6. Order of propositions in the Shifā'

Shifā' Book	
1.	Same as *Elements* I (Heath edition throughout)
2.	Same as *Elements* II
3.	Same as *Elements* III
4.	Same as *Elements* IV
5.	Interchanges V, 12, 13
6.	VI, 1–8, 13, 11, 14, 15, 10, 17, 19, 20, 18, 21, 22, 24, 26, 25, 23, 28, 30, 32, 31, 33
7.	VII, 1–20, 22, 21, 23–28, 31, 32, 29, 30, 33–39
8.	Same as *Elements* VIII
9.	IX (27, 25, 28)
10.	X (15, 11, 17, 14, 18); (21, 13, 22); (74, 76, 75, 77)
11.	Same as *Elements* XI
12.	XII (7, 9, 8, 10, 12, 11, 15)
13.	XIII (6, 12, 9, 10, 8, 11, 13, 15, 14, 16)
14.	Same as Hypsicles[159]
15.	(3, 2, 1, 4, 5)[160]

Tables 5 and 6 are extracted from Appendix I of this paper and together summarize the contents of ibn Sīnā's geometry, and the order of the propositions. These are extremely important pieces of information for determining the Arabic edition of the *Elements* used in the Shifā'.

Curiously enough, although ibn Sīnā does omit certain propositions from his summary of the *Elements*, there are only two additions: Book 11, 32 and 33, which are introductory to Book 11, 34 (XI, 33). As the reader will immediately observe from Table 5, a mere count of propositions tells us very little about the actual content of the Shifā's geometry. For example, the Heath–Euclid and the Shifā' contain 479 propositions, but, in the latter case, these actually collapse to 436 Euclidean propositions, the remainder being accounted for by errors in numbering, case-wise treatment, or addition of propositions.

As Sabra notes in his introduction,[161] it is very difficult, linguistically to decide whether ibn Sīnā based the Shifā' on the al-Ḥajjāj or the Isḥāq–Thābit edition of the *Elements*. He does point out, for example, that both ibn Sīnā and al-Ḥajjāj (2nd version) use 'ilm gami' for "common notion", while Thābit prefers 'ilm 'am mutafq alaih. Of course this is far from conclusive, and Sabra goes on to observe that the figures in the Shifā' differ from both those in al-Ḥajjāj and Thābit.

If we assume either a Ḥajjāj or īsḥāq–Thabit precursor, at least three types of evidence seem to be available for making a choice: figures, linguistic analysis, order of propositions, including omissions and additions. In the case of al-Ḥajjāj, none of these are entirely compelling since his translation is really an integral part of al-Nayrīzī's commentary, and may, consequently, have been extensively reworked. Moreover, any judgment based on the geometric figures tends to be extremely subjective. For example, one of the authors of this paper compared the figures in Book I of al-Hajjāj with those in ibn Sīnā, and felt that 16 figures agree, 23 show partial agreement and 7 show marked discrepancy. In fact, al-Jūzajanī seems to imply, in the biography, that ibn Sīnā made some entirely new figures for the *Elements*.

Since the present writers had only Book 1, 5–6 of Codex Leidensis 399, 1 (al-Ḥajjāj) and fragments of Nicoll 279 (Isḥāq–Thābit) at their disposal while preparing this paper, no additional linguistic analysis of the three versions of Euclid is offered here. However, to give the reader some idea of the complexity of the problem in this area, we offer some excerpts (V, 7, 15) from ibn Sīnā (Sabra's edition), al-Ḥajjāj (Codex Leidensis) and Isḥāq–Thābit (Copenhagen 81, studied by Klamroth), in Appendix II. As the reader will note, from the short, though typical selection, the principal difficulty is caused by the brevity of ibn Sīnā's summaries. We hope to return to this intriguing identification problem, from the linguistic point of view, in a later paper. However, we can contribute something by turning to the third type of evidence: the ordering of propositions. From this point of view, the evidence seems to be slightly in favor of an al-Ḥajjāj precursor over Isḥāq–Thābit. Let us review this evidence, from Appendix I, by book. Again, the evidence is not decisive, because there is really not much variation in order, omission, and addition.

In book one al-Ḥajjāj, Isḥāq–Thābit and ibn Sīnā follow the Euclidean order.

The first gives 47 propositions, the second 48, and the last 46. Both al-Ḥajjāj and ibn Sīnā however omit proposition 45. In fact, ibn Sīnā also omits proposition 40, both in the Sabra edition, and in the Leiden manuscript studied by Lokotsch. In book two, all three editions are identical with each other, in order, and indeed agree with Euclid. Nonetheless, ibn Sīnā seems to lean slightly toward al-Ḥajjāj by giving II, 14 for triangles, instead of arbitrary rectilineal figures.

In book three, all versions are in the Euclidean order, with ibn Sīnā only, omitting III, 26, 29. He also agrees with both al-Ḥajjāj and Isḥāq–Thābit by treating III, 11 and 12 as a single proposition. The omission of part two of III, 15 which occurs in the Leiden manuscript (Lokotsch) is amended in the Cairo edition. In the fourth book, the only distinguishing characteristic, two different proofs of a single proposition in Isḥāq–Thābit is not shared by either al-Ḥajjāj or ibn Sīnā.

Book 5 in the Cairo edition differs somewhat from the Leiden manuscript. While the former omits V, 21, 22, 23 (possibly a last leaf?) and inverts V 12, 13, the latter omits only V, 6 and inverts V, 10, 11. Consequently the Cairo edition agrees in its ordering with both Ḥajjāj and Isḥāq–Thābit, although the latter editions are both complete.

The peculiar ordering of the last planimetric book, six: 1–8, 13, 11, 12, 9, 10, 14–17, 19, 20, 18, 21–22, 24, 26, 23, 25, 27–30, 32, 31, 33 *should* be decisive. Unfortunately, both al-Ḥajjāj and Isḥāq–Thābit share it. Ibn Sīnā, in the Cairo edition presents the propositions as: 1–8, 13, 11, 14–15, 10, 17, 19–20, 18, 21–22, 24, 26, 25, 23, 28, 30, 32, 31, 33, omitting 7, 9, 12, 16, 27, 29. The Leiden manuscript (Lokotsch) nearly agrees: 1–8, 13, 9, 14–15, 10, 17, 19–20, 18, 21–22, 24, 26, 23, 27–30, 32, 31, 33, omitting 11–12, 16.

Since the al-Ḥajjāj edition exists only in six books, with fragments of seven, it is impossible to carry this triple comparison further. In summary, we see that ibn Sīnā seems to favour al-Ḥajjāj slightly in books one and two. In the other four books, all editions agree remarkably well.

If one carries this analysis further, comparing only Isḥāq–Thābit and ibn Sīnā (Cairo edition, Leiden ms. has only I–VI), one finds extremely good agreement. For example, both preserve the curious orders of books seven, twelve, thirteen, namely VII, 1–20, 22, 21, 23–28, 31–32, 29, 30, 33–39; XII, 1–5, 7, 9, 8, 10, 12, 11, 15–18; XIII, 1–3, 5, 4, 6, 7, 12, 9, 10, 8, 11, 13, 15, 14, 16–18.[162] Each has Euclidean order in book eight, although ibn Sīnā omits Isḥāq–Thābit non-Euclidean propositions 24–25. Moreover, since ibn Sīnā omits XI, 33, 34, which are reversed in Isḥāq–Thābit, there is again agreement order-wise, although ibn Sīnā fails to divide XI, 31 into two parts, a feature of Isḥāq–Thābit. This leaves only IX and X of the Euclidean books; here there are some discrepancies, as the reader may observe from consulting Appendix I.

V. Trigonometry

Ibn Sīnā devotes a section of the Shifā' (*Dzawamī' Kitāb Bitlimus al-Kabir al-ma'mul Madjisti wa 'ilm al-haya*) to Ptolemy's astronomical treatise, the

Almagest. A part of this, in turn, contains his work on trigonometry, which is unique to the Shifā', in the sense that it is not summarized in the Najāt or in the 'Alā'ī. Since the Cairo edition of this part of the Shifā' has not yet appeared, we can only report on its contents and sources by referring to a Russian analysis given by Hairetdinova.[163] That writer has proven that ibn Sīnā trigonometric materials follow those of Abu Nasr al-Fārābī's (870–950) *Sharh al-Madjisṭī* (Commentary on the Almagest),[164] almost verbatim. Both works treat the transversal theorem of Menelaus' (*c*. 100 A.D.) *Sphaerica*,[165] following Menelaus' own account more closely than the later adaptation by Claudius Ptolemy. They also diverge in some noteworthy aspects from both of these authors, for example, "sine" is defined, possibly (in al-Fārābī's case) for the first time in an Arabic commentary on Ptolemy.[166] It is given as "half the chord of double the arc."[167] In addition, al-Fārābī and ibn Sīnā present a result equivalent to the planar law of sines for triangles inscribed in a circle, in the form, "If the angles are known then we determine the ratios of their sides. If the angles are inscribed in a circle and the arcs of each angle are known to me, then the [ratio of sides equals] the ratio of the corresponding chords to the diameter of the circle; moreover if the angle is right, then chord and diameter coincide. Therefore, if one of the angles is known, or one of the arcs of a side is known and its ratio with the chord of a right angle, this is sufficient [information] for us to determine the arc of that corresponding angle."[168] They also take special care in treating certain cases of Menelaus' theorem missing from both *Sphaerica* and the *Almagest*, as well as in justifying Ptolemy's handling of composite and "inverse" ratios.[169] Their treatment, in the latter, is strictly Euclidean.

Hairetdinova goes on to show that al-Fārābī's work is quite different from Thābit ibn Qurra's tract *Risala fi shakl al-kita*' (Treatise on the transversal theorem).[170] In fact, the differences extend even to the terminology: al-Fārābī, for example, uses "nayf" rather than "fadl" for "difference", and "nisba al-murakkaba" rather than "nisba al-mu'allafa" (composite ratio). From this Hairetdinova concludes that al-Fārābī may well have translated the *Sphaerica* directly from the Greek, coining his own terms as he worked.

It is also worth noting that this section of the Shifā' received some attention by other Arab astronomers, e.g. al-Nasāwī (11th century) cites it in his *Kitāb al-ishba' fi shark ash-shakl al-kita'*.[171]

VI. Foundations

When one recalls the appraisal of a thirteenth century editor of Euclid, probably Muḥyi 'l-Din al-Maghrabī,[172] of ibn Sīnā's work in geometry, one may question the usefulness of looking for statements on the foundations of mathematics in his writings. The editor wrote, rather bluntly, "Ibn Sīnā left out the postulates and many definitions, and had nothing to do with resolving obscurities and difficulties!"[173] On the face of it, this shows a profound disregard on ibn Sīnā's part for the underlying logical structure of mathematical argument. However, as Lokotsch aptly notes, ibn Sīnā's purpose in summarizing the geometry in the

Shifā' is to provide quick access to mathematical knowledge.[174] Similarly, his purpose in producing a geometry section in the 'Alā'ī is simply to introduce material necessary for studying the *Almagest*.[175] In fact, one has only to look at ibn Sīnā's logical analysis of Euclid's first proposition in the section of the 'Alā'ī on syllogism to see that he shows ample regard for the relations between logic and mathematics.[176]

Before pursuing ibn Sīnā's views further, let us briefly survey the opinions of his time on mathematics. Medieval Arab philosophers tended to take the Aristotelian view of mathematics, regarding "mathematical objects... as existing by abstraction,"[177] and taking the proper domain of mathematics to be "only the quantitative and continuous",[178] not matter, and the qualities of matter like weight, hardness, heat, etc.. Ibn Rushd (1126–98), in his *Commentary on the Metaphysics of Aristotle* (Tafsīr Mā Ba'd at-Tabi' at), is a prime exponent of this view, and as Pines[179] points out, it is probably for similar reasons that al-Kindī, the Ikhwan al-Safa, Maimonides and ibn Rushd regarded mathematics as providing an inferior model of the real world than that of Aristotelian physics.

Ibn Sīnā, on the other hand, appears to take a different view, especially, if one recalls his classification of the sciences, which puts mathematics ahead of physics, but just behind metaphysics.[180] Yet, his extended writings on the subject put him firmly in the Aristotelian camp. For example, in the *Kitāb al-Najāt*, he notes that the objects of mathematics are "things conceived by the estimative faculty [al-wahmiyyāt]", as opposed to things which (physically exist).[181] Al-Ṭūsī, in his commentary on ibn Sīnā's *Kitāb al-Ishārāt wa'l-Tanbīhat*, goes on to assert that it is just this faculty (al-wahm) which accounts for the uniformity of opinion on what constitutes mathematics.[182] Moreover, somewhat later in the Najāt, ibn Sīnā says that "the mathematical sciences are not [qualified] to investigate either the natures of physical substances or quantities inherent in these substances",[183] echoing a passage in al-Kindī's *Rasa'il al-Falsafiyya*.[184] Ibn Sīnā's firm commitment to the Aristotelian view of mathematics and physics is further in evidence in his famous exchange of questions and answers with al-Bīrunī, the *als-As 'ilah wa 'l-Ajwibah*, in which the former plays Aristotle to the latter's Zeno.[185]

Even ibn Sīnā's view of the foundations of arithmetic shows an Aristotelian outlook, or at least a Pythagoreanism channeled through Aristotle, and modified by the Ikhwan al-Ṣafā and Nicomachus. His position in analysis, that is, his view of infinity and infinitesimals, also shows the influence of Aristotle, albeit with some modifications. We close this section with a detailed examination of this position.

As does Aristotle,[186] ibn Sīnā argues against the actual infinite. The argument appears in the Najāt, as well as in the section on metaphysics of the 'Alā'ī, in a discussion of the question of efficient causes ('illat-hā-i fā'il) (see Figure 13).[187]

"Let there be an infinite line AB. Let us indicate on it the point J and consider

Figure 13

the line JD a finite measure or a finite quantity. If the line DB is finite, and we augment it by JD, then JB will also be finite. If the line DB is infinite, let us allow DB to correspond to JB until they are at the same place. If DB should correspond to JB, then a lesser and a greater would correspond, which is impossible, DB being the lesser and JB being the greater. If DB stands still while JB continues, then the endpoint B will be limited, and JB will be greater than DB by a measure equal to JD which is finite. Thus, JB is also finite. It becomes evident, then, that such a number and such a measure are not infinite."

Al-Shahrastani (d. 1153) reproduced this argument of ibn Sīnā in a slightly clearer form in the *Kitāb al-Milal wal-Niḥal*,[188] and through al-Tabrīzī (2nd half, 13th century),[189] it finds its way into Hasdai Crescas' (1340–1410) *Critique of Aristotle*.[190] It also appears in the 1593 Rome dual edition of the Canon and al-Najāt.[191]

However, as Marmura observes, ibn Sīnā was obliged to modify his stance in his treatise *Risāla Aḍḥawiyya fī Amr al-Ma'ād*.[192] In effect, ibn Sīnā admits the existence of infinitely many *immaterial* bodies, in this case souls, arguing that since they are immaterial, they are also unordered. The implication is that his previous argument on infinity then does not apply. Evidently, Al-Ghazali (1058–1111), in *Tahāfut al-Falāsifa*,[193] rejects this argument, asserting that the immaterial bodies (souls) may be ordered just as the sequence of days.

Interestingly enough, none of these reworkings of Aristotle's argument ever arrives at the notion set forth in Adam of Balsham's (Parvipontanus) *Ars Disserendi*, that a proper subset of an (infinite) set may be equinumerous with it,[194] much less Galileo's discovery that the natural numbers and their squares are in one-to-one correspondence.[195]

Specific instances of ibn Sīnā's position on infinitesimals and limits are somewhat more difficult to give. As noted in section two, the treatise "al-zawiya" on the "hornlike" angle may contain some relevant passages; however, we have been unable to acquire a copy of the manuscript.[196] This leaves *als-As 'ilah wa 'l-Ajwibah*; a few of its exchanges are sufficient to give the general direction of ibn Sīnā's thoughts on the infinitesimal. Al-Bīrūnī asks: "If a square is divided into four equal squares, how are the surfaces of the squares at opposite corners contiguous with each other?"[197] Ibn Sina's answer is built on two Aristotelian precepts: "a point cannot be contiguous with a point",[198] and "a line is not made up of points".[199] In a second passage (answer to al-Bīrūnī's question 4),[200] ibn Sīnā bases his reply on Aristotle's position that infinite subidivision exists only potentially,[201] while al-Bīrūnī counters that such an answer implies that the side of a square and its diagonal are of equal length because they are, potentially, equinumerous.[202] Of course, to do justice to both ibn Sīnā and al-Bīrūnī we must note that these arguments are more about the classical Atomistic theory than about mathematics. Similar discussions appear in the works of al-Aš'arī (873/4–935/6) and al-Baqilanī (d. 1013), who seem to espouse the view that the side of a square and its diagonal are commensurable.[203]

The other notion of infinitesimal, based on the principle of Eudoxus (axiom of Archimedes),[204] and used so effectively by Archimedes in his treatise: Quadrature of the Parabola, and On the Sphere and Cylinder, and emulated by Thābit ibn Qurra, alKūhī and ibn al-Hāyatham[205] appears in ibn Sīnā's writings only peripherally. For example, he includes Euclid's version of it in the geometric section of the Shifā' (Elements V, dfn 5).

VII. Summary

As we remarked in Section I, Sarton's off hand dismissal of ibn Sīnā as a mathematician gave partial impetus to this study. Recall those words "Ibn Sīnā's interest in mathematics was philosophical rather than technical,... [as to be expected] in a late Neoplatonist". We are inclined to feel that this opinion is rather unfounded. Certainly ibn Sīnā's work on mathematics goes much beyond the traditional philosophical manuals, such as Theon of Smyrna's *Mathematics Useful for Understanding Plato*, occasionally improves on its sources, as is in evidence by his treatment of topics in common with Nicomachus, and often has a fixed, practical purpose: the arithmetic of the 'Alā'ī serves to introduce musical theory, and its geometry prepares the reader for a study of the *Almagest.*

Of course, others have criticized ibn Sīnā's style of presenting his material, as well as his philosophical bias. Al-Bīrūnī, in a treatise on chords,[206] notes that ibn Sīnā regarded any collection of assorted proofs of the same theorem "fūḍhūl" [redundant],[207] while the writer of the preface to a 13th century edition of the *Elements* (Nicoll, 280), accuses him of leaving out postulates and axioms and neglecting to address the difficulties of the subject.[208] We have already discussed the latter accusation. As for al-Bīrūnī's comment, suffice it to note that lemmas, porisms, and alternate proofs were regarded as "fūḍhūl" in most Arabic mathematical writings, notably in the versions of the *Elements.*[209]

On the other hand, one cannot altogether dismiss ibn Funduq's observation that a person like ibn Sīnā wasn't much of a mathematician because he'd tasted the sweetness of metaphysics and fancied mathematics from time to time, only to abandon it.[210] Ibn Sīnā's mathematics certainly has none of the originality of Thābit's, nor even the care for detail of Isḥāq. The geometry probably derives from the translation of al-Ḥajjāj, the arithmetic from Nicomachus and certain Indian sources, the trigonometry from al-Fārābī, the philosophy of mathematics and notions on its foundations from Aristotle, possibly through al-Kindī and al-Fārābī. Still, the arrangements, and many of the details are ibn Sīnā's own. Moreover, as al-Jūzajanī points out, mathematics, a subject in which ibn Sīnā was mostly self-taught, and only one of many subjects that occupied him, did so for his entire life.

Notes to Text

1. Recent biographical accounts appear in the DSB articles of Anawati (9) and Iskandar (53). For other accounts, see S. H. Nasr (81), and his text *Science and Civilization in Islam* (82); also the biographies by B. Carra de Vaux (19) and S. M. Afnan (1).
2. Woepcke (122), Eneström (32), de Montferrier (76). Cantor (18) covers the same ground.
3. Lokotsch (67). The paper of Ahadova (4) essentially does this work over, using the geometric part of the Dânesh Nâmeh 'Alā'ī instead of the Shifā'.

4. Says Lokotsch, "Die deutsche Übersetzung der mathematischen Schriften Avicennas soll nebst sachlichen Anmerkungen in ungefähr zwei Jahren in Buchform erscheinen." The *Jahrbuch über der Fortschritte der Mathematik*, which carries an entry for the dissertation, does not carry one for the promised translation, and a search by the Prussian State Library (Berlin) [1980] failed to turn up any record of one. A photocopy of the dissertation in the possession of Prof. Dr. F. Sezgin indicates that Lokotsch donated the dissertation to the Carl v. Kothschildsche Bibliothek in Frankfurt in 1929.
5. See Anawati (9), for example.
6. Sarton (94), I, 710.
7. We refer to the edition and commentaries of Gohlman (38), in what follows.
8. Tradition has it that ibn al-'Arabi and 'Umar ibn Aḥmad mastered the Qurān by age 9, while al-Ṭabari seems to hold a record for mastery by age 7. See for example Totah, K. A., *Contribution of the Arabs to Education*, N.Y., 1926, p. 49, or his references: Al-Maqqari, vol. I, 343; Yaqūt; Dictionary of Learned Men, VI, 30, 36–7.

9–14. See Gohlman (38), pages 19, 21, 23, 25, 31 respectively.

15. Gohlman (38), note 30, page 122.
16. Sezgin (98), vol. VI, 280–1. More information on al-Shīrāzī appears in S. H. Nasr's biographical sketch, DSB XI, 247–253. Al-Shīrāzī was, most notably, a pupil of Naṣīr al-Dīn al-Ṭūsī. He refers to al-Jūzajanī's work while discussing his own model for the motion of Mercury.
17. Gohlman (38), 65–6.
18. Gohlman (38), 122, note 27. See also ibn Funduq: *Tatimmat Ṣiwan al-ḥikma*. Muḥammad Shafī', ed. Lahore, 1935, volume 1, p. 42. This supplement continues the *Ṣiwān al-ḥikma* of Abū Sulaymān al-Sijistani (d. 985), so the mathematical opinion is ibn Funduq's own. Apparently, he was primarily a biographer and historian. See, for example, Brockelmann's GAL, I (supplement), 557–8.
19. See also Wiedemann, E. and Juynboll, T. W., "Avicennas Schrift über ein von ihm ersonnenes Beobachtungsinstrument", *Acta Orientalia* **5** (1927), 81–167, and Wiedemann's "Uber ein von ibn Sīnā (Avicenna) hergestelltes Beobachtungsinstrument", *Zeitschrift für Instrumentenkunde* **45** (1925), 269–75. These are translations and commentaries on the manuscript (Leiden, Or. 184/8): *Maqāla fi l-Ālāt arraṣadiya.*
20. Gohlman (38), 148 and 152. The references are to Y. Mahdavī: *Fihrist-i muṣannafāt-i Ibn-i Sīnā*, Publications of the University of Tehran, no. 206, Tehran, 1954, and G. C. Anawati: *Mu'allafāt Ibn Sīnā*, Cairo, 1950. See also Osman Ergin: *İbni Sina Bibliografyasi*, İstanbul, Osman Yalçin Matbbasi, 1951, XXXI, 168p (İstanbul Üniversitesi Tip Fakültesi yayinlarindan, 20). There is a critical review by Rudolf Sellheim in *Oriens* **11** (1958), 231–9.
21. Ergin 74. See Anawati (note 20) for actual locations of manuscripts. Carra de Vaux (Avicenna, 147, n. 3) identified *Danesh name*'-*i 'Alā'ī* with Anawati 11: *Ḥikma 'alā' iyya.*
22. *Le Livre des Science* II, 91.
23. *Le Livre des Sciences* II, 6.
24. *Ibid.*, 6.
25. Mahdavī: *Fihrist-i muṣannafāt-i Ibn-i Sīnā*, 101 and 226.
26. Anawati (8), 419, or A. M. Goichon: *La Distinction de l'essence et de l'existence d'apres Ibn Sīnā* (*Avicenne*), Paris, 1937; 449–503.
27. Gohlman (38), 47.
28. See also no. 24 in Anawati's list: al-Hidāya, ms. Aya Sofya 2475*1, also in 3 parts—logic, physics, and metaphysics.
29. Gohlman (38), 67.
30. Gohlman (38), 132–3, no. 89.

31–2. Gohlman (38), 69 and 97.

33. Anawati (8), 424. Ibn Sīnā argues that time has no beginning.
34. Anawati (8), 431.
35. Gohlman (38), 99. The author of the anonymous bibliography records this as "a letter to Abū Sahl al-Masīḥī" written in Jurjān. But see also Gohlman (38), 139, no. 5, and 124–5, no. 43.
36–37 See Heath (46) I, 177–8; Aristotle, *Analytica Priora* I 24; 41b, 13–22 (Oxford ed. vol. I) and *Elements* III, 16.
38. Anawati (8), 432.
39. Anawati (8), 417.
40. Gohlman (38), 55.
41. Gohlman (38), 92–3.
42. Gohlman (38), 61.
43. Gohlman (38), 65.
44. Gohlman (38), 65.
45. Gohlman (38), 45.
46. Gohlman (38), 65–6.
47. These figures are derived from Anawati's description of the al-Azhar ms. of the Shifā'; Anawati (8), 418.
48. S. H. Nasr and M. Mohaghegh: *Al-Bīrūnī and Ibn Sīnā: als-As 'ilah wa 'l-Ajwibah*, Tehran, 1973. There is also a Russian edition: I. M. Mumizov, ed., *al-Bīrūnī and Ibn Sīnā Correspondence* (Russ.), Izdat. "Fan," Uzbek SSR, Tashkent, 1973. This is a translation of ms. 2385, Arch. Inst. Eastern Studies, Uzbek Acad. Sci. Parts of these "questions and answers" are translated into English by Nasr in (82), 133–8. For additional manuscripts, see Anawati (8), 415, no. 1–2; 422, no. 54.
49. There is a Russian edition by P. G. Bulgakov (with commentary): *Abu Reihan, Izbrannye Proizvedeniya III* (*Geodeziya*), Akad. Nauk. Uzbek. SSR, Tashkent, 1966. The English translation by J. Ali (5), is supplemented by Kennedy's commentary (57).
50. Certain passages of the metaphysics of al-Najāt are also relevant. See §VI of text.
51. Kennedy (57), 124–5; al-Bīrūnī (5), §65.
52. Unver (116), 199. His free rendering reads: "Euclid in science is so great that his name is synonymous with events happening in heaven and on the horizons. He is like a stairway whose steps are his figures, on which ascend passengers to endless heights. A person who makes his acquaintance is sure to rise to a high level. That level is ever so pleasant and the person who reaches it is ever so happy." However, see Murdoch (78), 453, note 4.
53. Anawati (9), p. 422, no. 49. This is part of a nine essay collection: *Tis ra sā'il*, Cairo, Hindīyah Press, 1908. For partial translation into English, with commentary, see Nasr (81), 209–12.
54. See Anbouba (10), 92 and *al-Burhān*, ed. Badawī (Cairo, 1954), 135.
55. Steinschneider (104), II, 92, etc. Even the printed editions of the Shifā' done in the early 16th century are merely selections. For example, *Avicene... philosophi: ac medicorum facile primi opera in lucē redacta...*, Venice, 1508 contains the Logic (Book 1, fen 1, makālah 1, 2 translated by Gundisalvus); Physica (Book 2, fen 1, makālah 1, 2 and faṡl 1, 2 of Makālah 3, translated by Gundisalvus); De Anima (kitāb al-nafs) (Book 2, fen 6 translated by Hispalensis and Gundisalvus); De animalibus (al-hayawān) (Book 2, fen 8 translated by Michael Scott); Philosophia prima (Metaphysics, i.e. Book 4, as well as sections titled "De intelligentiis. Alpharabius de intelligentiis," and "De celo et mundo," which the descriptive catalogue (B. M. Arabic Books I, 664) lists as a collection based on Aristotle, thereby distinguishing it from al-samā' wal 'ālam (Shifā', Physica, fen 2).

    Apparently al-Najāt also enjoyed some currency; for example, the 1593 Rome edition (Arabic) of *al-Qānūm fi al-Ṭibb* contains it as an appendix. (See B. M. Arabic Books I, 670, 683).

56. Anawati (8), 417.
57. Haskins (42), 13, 245, 360.
58. Rashdall (86); i, 236n., 353, 365, 436; ii, 33n, 121n, 127; iii, 245n.
59. Crombie (24), 26, 72, 90, 92–3, 100, 370; and Crombie (26), 79–81, 324.
60. Alverny, M.-Th. d', "Avicenna latinus" in *Arch. hist. doct. et litt du moyen âge*, I, t. 28 (1961), 281–316; II, t. 29(1962), 217–333; III, t. 30(1963), 221–72; IV, t. 31(1964), 271–86; V, t. 32(1965), 257–302; VI, t. 33(1966), 305–27; VII, t. 34(1967), 315–43; VIII, t. 35(1968), 301–35. See also d'Alverney's "Les traditions manuscrites de l'Avicenne latin", in *Mélanges Taha Hussein*, Cairo, Dār al-Ma'āref, 1962, 67–75.
61. For example: A. M. Goichon: *Lexique de la langse Philosophique d'Ibn Sīnā*, Paris, 1938, etc., and I. Madkour: *L'Organon d'Aristotle dans le monde arabe*, Paris, 1934.
62. Rosenthal (59), III, 118–34, or Woepcke (123).
63. *The Arithmetic* and *Geometry* appeared in 1970 and 1975 respectively, see (101) and (102).
64. See also: Eneström (32), Montferrier (76), Tannery (110), and Woepcke (122).
65. Achena, Massé edition (110), 190.
66. *Ibid.*, 194.
67. Shifā' (Cairo edition of *Geometry*), 293–294, no. 35.
68. *Ibid.*, 194.
69. *Loc. cit.*
70. *Ibid.*, 195.
71. *Ibid.*, 197
72. *Ibid.*, 198.
73. Title added by translators.
74. *Ibid.*, 198.
75. *Ibid.*, 199.
76–7. *Ibid.*, 199.
78. *Ibid.*, 199.
79. *Ibid.* 202.
80. *Ibid.*, 203–6.
81–3. *Ibid.*, 206–9.
84. *Ibid.*, 210.
85. *Ibid.*, 211. Actually there are 10.
86. *Ibid.*, 214.
87. *Ibid.*, 215. "Son caractère propre consiste en ce que le produit de la multiplication des deux extrêmes est moindre que le carré du moyen, et cela à la mesure du carré de l'exces de l'un des deux extrêmes sur l'autre."
88–9. *Ibid.*, 215–16.
90. *Ibid.*, 216.
91. Afnan (2), 190. Ibn Sīnā makes this admission in his commentary on Aristotle's *Poetics*, among other places.
92. Dieterici (29), 11; see also Goldstein (40).
93. Dieterici (29), *passim.*
94. *Ibid.*, especially 33–4.
95. *Ibid.*, 117ff.
96. *Ibid.*, 154ff.
97. *Ibid.*, 2.
98. Tarán (111), 112.
99. Our references will be to the edition of Kutsch (64).
100. D'Ooge (83), 202–3.
101. D'Ooge (83), *passim.*
102. D'Ooge (83), 210. As the editor notes, this remark also appears in Boethius, *De Institutione Arithmetica* I. 20 [Edition Friedlein, Leipzig, 1867].

103. See, for example, Rosen (87), 91ff.
104. D'Ooge (83), 234–5.
105. *Ibid.*, 236.
106. *Ibid.*, 266ff.
107. *Ibid.*, 270.
108. *Ibid.*, 272.
109. *Ibid.*, 273.
110. *Ibid.*, 279.
111. *Ibid.*, 282, no. 2.
112. Heath (44), I, 114. However, as Zeuthen points out ("Sur l'arithmétique géométrique des Grecs et des Indiens", *Bibliotheca mathematica* (ser. 3) **5** (1904), 97–112), Nicomachus' statement that a square is composed of two consecutive triangular numbers (*Arithmetic*, Book II, chapter 12) is equivalent, since this assertion is just $(1+2+\cdots+n)+(1+2+\cdots+n-1)=n^2$.
113. Sezgin (98), V, 270, no. 13. See also F. Woepcke, "Notice sur une théorie ajoutée par Thābit ben Korrah à l'arithmétique spéculative des Grecs", *J. Asiatique* **20** (1852), 420–9.
114. See Woepcke cited in 103; also Dickson (28), 5.
115. Hultsch (51), I, 75ff.
116. Ibn Sīnā, *Al-Shifā': al-Ḥisāb* (101), 39. Compare with the table in Boethius, *Arithmetica Boetii*, Augsburg 1488, reproduced in Smith (103), 26, and given in our text. For an interesting study of multiplication tables (sexagesimal) used by Medieval Islamic astronomers, see King (60)–(61).
117. D'Ooge (83), especially, 248; *al-Ḥisāb* (Cairo, 1975), 13.
118. *al-Ḥisāb* (Cairo, 1975), especially 39.
119. *al-Ḥisāb* (Cairo, 1975), 11–14.
120. I.e., in Lutfy's edition.
121. *Ibid.*, 13–14.
122. See D'Ooge (83), 281–4 for tables giving respectively the Nicomachus and the Theon forms. Note, however, that D'Ooge reverses our a and c making $a>b>c$. Some comments on Theon's *Mathematics Useful*... are in order. His treatment of ratio and proportion are relegated to the section of music; the section on arithmetic has the following order:

    1. The order in which mathematics should be studied
    2. One and the monad
    3. Even and odd numbers
    4. Prime numbers
    5. Composite numbers
    6. The various kinds of even numbers
    7. Square, unequilateral and parallelogramic numbers
    8. Promecic numbers
    9. Triangular numbers, the way they are obtained and the polygonal numbers.
    10. Pyramidal numbers, lateral and diagonal numbers
    11. Perfect, abundant and deficient numbers

    Chapter 11 amply illustrates the quality of the work. The translation is Lawlor's: "Furthermore, among numbers some are called perfect, others abundant, and others deficient. Those are called perfect which are equal to (the sum of) their aliquot parts, such as 6. The parts of 6 are in fact its half, 3, its third, 2, and its sixth, 1, which when added together make 6."

    "This is how the perfect numbers are created: If we arrange the numbers in a progression by doubles, starting from unity, and we add them until we obtain a prime, incomposite number, and if we multiply this sum by the last term added, the product will be a perfect number. Let us then arrange the numbers in a progression by doubles, 1, 2, 4, 8, 16. Let us add 1 and 2; the sum is 3; if we multiply this by the last number added, which is 2, we will have 6 which is the first perfect number (since $1+2+3=6$). If we now add the three successive doubles, 1, 2, and 4, the sum is $7\ldots, 28$ perfect." [The section concludes with a cosmological discourse on the tetraktes!]

123. Woepcke (122), 502–4.
124. Cantor (18), I, 756.
125. *Ibid.*, 756–7.
126. Tannery (110), 142.
127. Saidan (92), and his article "Al-Uqlīdisī," DSB XIII, 544.
128. Cairo ed., 25.
129. *Ibid.*, 25–6.
130. *Ibid.*, 30.
131. *Ibid.*, 26.
132. In their interesting paper on this material from al-Ḥisāb, Kahharov and Hodžiev (55), 122, point out that from explicit formulae for the line sums with $n=2k$ and $2k+1$, it also follows that "the sum of the numbers situated in successive lines, beginning with the first equals the square of the triangular number which corresponds to the number of the following line. For example, the sum of the numbers filling lines 1 and 2 is $1+3+5=9=3^2$, of 1, 2, and 3 is $1+3+5+7+9+11=36=6^2$."
133. D'Ooge (83), 263–4.
134. Saidan (92), *passim.*
135. *Kitāb fi usūl ḥisāb al-hind*; see tr. of M. Levey and M. Petruck (63).
136. Cairo ed., 17.
137. See for example, Sezgin VI, p. 103 where one is led to speculate on the origin of the title al-Usṭuqusāt. It is a corruption of the Greek *στοιχείων*, or from the Arabic usṭuqsā? The second author wishes to thank A. S. Ibrahim and Simon Samoeil of the University of Petroleum and Minerals for some interesting conversations on this point.
138. Perhaps he is referring to al-Fārābī's, or even his own.
139. Alā'ī (100), 91.
140. See *Syntaxis Mathematica*, ed. J. L. Heiberg. Teubner, 1899–1907.
141. Proposition numbers follow the edition of Heath (46).
142. See our essay, "Naṣir al-Dīn al-Ṭusī's Attempt to Prove the Parallel Postulate of Euclid."
143. Sayili (95), 35–7. See our essay, "The Pythagorean Theorem and Mumpsimus."
144. See section below on the "Foundations".
145. One might profitably compare this with al-Battānī's introduction to the *Almagest* (Nallino (80), 6ff.) which starts with an account of the sexagesimal arithmetic. Also, as the reader may discover for himself from perusing the appendix (K. al-Shifā'-Geometry), all the propositions that ibn Sīnā omits in the Alā'ī, with the exception of III, 16 (por), and VI, 7 do appear in the Shifā'. The omission of VI, 7 is itself a curiosity, since it appears in the "Arabian Euclids". See Heath (46), I, 80–1. However the omission of the porisms is a common feature of all the Arabic editions. Of course, there is also the fact that ibn Sīnā includes some 40 propositions not essential to the *Almagest*. On the whole, however, his inclusion of these propositions lends some continuity to his account of geometry, and gives us some feeling for his own geometric preference, viz., problems of dissection and construction.

146–7. Murdoch (78), 439.

148. Stock (107), 7–8. See also Collison (23) for more details on this tradition, especially 21–81.
149. Huxley (52), 325–6; Heath (44), I, 112–13; Sarton (94), I, 272.
150. Sarton (94), I, 272.
151. Heath (44), I, 112–13.

152. Huxley (52), 325.
153. Lloyd (66), 1; Heath (44), I, 113–17.
154. Vernet (118), 357–8.
155. Wiedemann (120), I, 400–63.
156. Sabra (102).
157. However, postulate 3 of *Elements* I is interpolated between ibn Sīnā's treatment of I, 2 and I, 3.
158. See Heath (46), I, 83.
159. Heath (46), III, 512–19.
160. See reference to Book 15 in table 1.
161. Sabra (102), 11.
162. This is nearly the case, since ibn Sīnā omits XIII, 4, 5, 7.
163. Hairetdinova (41).
164. A Russian edition of approximately one half of al-Fārābī's commentary [*Kommentarii k Almagisti Ptolemeya*] was prepared by A. Kubesov, *et al.*, Izdat. "Nauka" Kazah. SSR, Alma-Ata, 1975. His supplement to the Almagest, al-Lawāḥiq, is included (pp. 53–90) of B. A. Rosenfeld's *The Mathematical Treatises of al-Fārābī*, Alma-Ata, 1972. For reviews of each, see MR 57, no. 5845 and MR 55, no. 7711.
165. For a statement of the planar version of Menelaus' theorem and its extension to the sphere—along with some good historical perspective—see Kennedy (58), 342–5, and Heath (44), II, 261–74.
166. However, see Kennedy (58), 346 for a 4th or 5th century use of the sine in a Sanskrit astronomical work, *Surya Siddhanta.*
167. Hairetdinova (41), 30 gives exact reference to the Arabic ms. in note 9.
168. For example, if a, b, c are sides and A, B, C are opposite angles $a/b = (a/2r)/(b/2r) = \sin A/\sin B$, etc., where r is the radius of the circle.
169. See the section on arithmetic above.
170. See A. Björnbo, "Thabits werke über den Transversalensatz", *Abh. z. Geschichte d. Naturwiss. u. Med. Heft* VII, 1924.
171. Hairetdinova (41), 29.
172. Heath (46), I, 77 Conjectures the author to be al-Ṭūsī. Bodelian MS. 280, dating from 1260–1261. Murdoch (8), 453, however, identifies it as a copy of the Taḥrīr of al-Maghribī.
173. See Heiberg (47), 16, where Nicoll's Latin translation of the preface is reproduced in its entirety. The section we translate reads "Etenim doctor primarius (Avicenna) postulata et definitiones multas resecuit, difficilium quoque et obscurorum resolutionem detrectavit." Cf. Heath (46), I, 77.
174. Lokotsch (67), 25, "Avicenna dagegen kam es mehr darauf an, eine Zusammenfassung der Elemente zu schaffen, die sich in Verbindung mit mundlicher Unterweisung zur schnellen Aneignung mathematischer Kenntnisse eignete."
175. See section on "Geometry" above, in particular al-Jūzajanī's introduction to the 'Alā'ī's resumé of Euclid.
176. Zabeeh (126), 36–7.
177. Aristotle, *De Caelo* III. 1, 299a 15ff.; see Heath (45), 224–6.
178. Aristotle, *Metaphysics* X, 31, 1061a 28ff.
179. Pines (85), 79.
180. Pines (85), 79, or ibn Sīnā: Aqsām al-'Ulūm, Tis' Rasā'il, Cairo, 1908, 105ff.
181. p. 158, Cairo edition (1331 AH).
182. Pines (85), 83, note 16. Actually al-Ṭūsī is speaking only of geometry.
183. p. 339, Cairo edition (1331 AH).
184. See Pines (85), 80, or M. Abu Rida, ed. *Rasā'sil al-Kindī al-Falsafiyya*, Cairo (1950), 111. See also ibn Sīnā's *Al-Burhān* III, 196–201, Cairo ed. 1956.
185. For selections, see Heinen (49), *passim*; Nasr (81), 170–3; Nasr (82), 133–8. The

complete text was edited by S. H. Nasr and M. Mohaghegh: *Al-Bīrūnī and Ibn Sīnā, als-As 'ilah wa 'l-Ajwibah*, Tehran, 1973.

186. Aristotle, *Physics*, iii, 5, 204a, 20–29; *Metaphysics* V, ch 11, 1073a 25. See also Maziarz and Greenwood (73), 173–83; Heath (45), 102–13.
187. Morewedge (77), 44. *al-Najāt* (Cairo ed., 1938), 124–5. Also *al-Ishārāt wa l-Tanbīhāt* (J. Forget ed., Leiden, 1892), 94–5. See also ibn Rushd: *Tahāfut al-Tahāfut* (Bouyges ed., Beirut, 1930), 27–8.
188. (*Book of Religious and Philosophical Sects*), ed. W. Cureton, London, 1846, p. 403.
189. See Wolfson's edition of Crescas' *Critique of Aristotle* (124).
190. *Ibid.*, see his *Index of Citations.*
191. *Ibid.*
192. Marmura (71), especially 233–4. The *Risāla Aḍhawiyya fī Amr al-Ma'ād* was edited by Dunya, Cairo, 1949.
193. For specific citations to *Tahafut al-Falāsifa* (M. Bouyges, ed., Beirut, 1927), see Marmura (71), 236. We restrict our discussion to the mathematical aspects of al-Ghazāli's argument; the interested reader will find that the principle issue, is, however, a religious one.
194. Thomas (113), 133–4, and I. L. Minio-Paluello's *Adam Balsamiensis Parvipontani Ars Disserendi*, Rome, 1956. See also Brown (15).
195. Struik (108), 198–207; especially 205, and Drake (30), 28–47.
196. Patna [Bankipore] Oriental Public Library; Patna II, 456, 2361, 6. [Brockelmann I, p. 596 no. 79r].
197. Heinen (49), 53–4.
198. *De generatione et corruptione* I.2. 317^{a} 11: *οὖ γάρ εστιν ἐχόμενον σημεῖον σημείου 'ἡ στιγμὴ στιγμῆς.*
199. *Physika* IV. 8. 215^{b} 19: *οὐ σὖγκειται ἐκ στιγμῶν.*
200. Nasr (82), 134.
201. *Ibid.*, 135; *Physika* III. 6. 206^{a} 9–^{b}27; see Heath (45), 105–7.
202. *Ibid.*, 135–6, following S. H. Nasr's translation.
203 Juschkewitsch (54), 204–5.
204. *Physika* III. 7. 207^{a} 33–^{b}34; see Heath (45), 110–12.
205. For an excellent survey of this, see Juschkewitsch (54), 288–95.
206–7. *Istikhrāj al-Awtār fī al-Dāirah bi-khwāṣṣ al-khaṭṭ al-Munhanī fīhā.* See Saidan (93), 683.
208. See note 173 above.
209. Heath (46), I, 81.
210. See note 18 above.

Appendix I(A)

K. al-Shifā' —*Geometry*

(Preliminary identification of contents)

Divisions of text into books

1. Definitions of triangle and parallelogram (12)[1]
2. The straight line, segments, and applications thereof (5)
3. Circles (10)
4. Applications of triangles and circles (5)

5. Ratios (6)
6. Similar figures (7)
7. Common factors, inequalities, and their connections (10)
8. Proportions (5)
9. Proportions and related notions, factors, etc. (6)
10. [Same as 7] (17)
11. Solid geometry (6)
12. Polyhedrons (3)
13. Mean proportionals and regular polygons (4)
14. Mean proportionals and regular polyhedrons (3)
15. Inscribing regular polyhedrons inside each other (1)

The text of the geometric section then follows, in 433 pages in the Sabra edition. There is no introduction, preface, or statement of purpose, in contrast to al-handasa section of the 'Alā,ī.

Book I of al-handasa begins with the Euclidean definitions of point, line; some common notions ('ilm gami') but no postulates other than postulate 5 are initially listed.

Correspondence

Shifā'[2]	*Elements*
20, 1	I, 1
21, 2	I, 2
22, 3	
23, 4	I, postulate 3
24, 5	
25, 6	I, 3
26, 7	I, 4
27, 8	I, 5
28, 9	I, 6
29, 10[3]	I, 7
30, 11[4]	—
31, 12	I, 8
32, 13[5]	—
33, 14	I, 9
34, 15	I, 10
34–35, 16	I, 11
35, 17	I, 12
36, 18	I, 13
37, 19[6]	I, 14
38, 20	I, 15
38, 21[7]	—
39, 22[8]	I, 16
40, 23	I, 17
41, 24	I, 18
41, 25	I, 19
41–42, 26[9]	I, 20
42, 27	I, 21
43, 28[10]	I, 22
44, 29	I, 23
45, 30	
46, 31	I, 24[11]–I, 25
47–48, 32	

Shifā'	*Elements*
48, 33	I, 26
49, 34	I, 27
50, 35	I, 28
51, 36	I, 29
51, 37	I, 30
52, 38[12]	I, 31
52, 39	I, 32
53, 40	I, 33
53–5, 41[13]	I, 34–I, 35
56, 43	I, 36
56, 44	I, 37
57, 45	I, 38
58, 46	I, 39
58–59, 47	I, 41
59, 48	I, 42
60, 49	I, 43
61, 50	I, 44
62, 51	I, 46[14]
63–4, 52[15]	I, 47
65, 53	I, 48

Shifā' (Book 2)[16]	*Elements*
69, 1	II, 1
70, 2	II, 2
71, 3	II, 3
71–72, 4	II, 4
73, 5	II, 5
74, 6	II, 6
75, 7	II, 7
75–76, 8	II, 8
77, 9 }	II, 9
78, 9 }	
79, 9 }	
79, (10) }	II, 10
80, 12 }	
81, 13 }	II, 11
81, 14 }	
82, (12)[17] }	II, 12
83, (15) }	
83, (13)[18] }	II, 13
84, 16 }	
85, 17 }	
85, 14[19] }	II, 14
86, 18 }	

Shifā' (Book 3)[20]	*Elements*
90, 1	III, 1
91, 2	III, 2
92, 3	III, 3

Shifā' (Book 3)	*Elements*
93, 4	III, 4
93–94, 5	III, 5
94, 6	III, 6
95, 7	III, 7
96, 8	III, 8
98, 9	III, 9
98, 10[21]	III, 10
100, 11	III, 11–12
101, 12[22]	III, 13
102, 13	III, 14
103, 14	III, 15
104, 15[23]	III, 16
105, 16	III, 17
106, 17[24]	III, 18–III, 19
107–108, 18[25]	III, 20
109, 19	III, 21
109, 20	III, 22
110, 21[26]	III, 23–III, 24
111, 22[27]	—
111, 23	III, 25
113, 25[28]	III, 27
114, 26[29]	—
114–115, 27	III, 28
115, 28	III, 30
116, 29	III, 31
117–119, 30	III, 32
119–120, 31[30]	III, 33
121, 32	III, 34
122–6, 33–4[31]	III, 35
127, 35[32]	III, 36
129, 36	III, 37

Shifā' (Book 4)[33]	*Elements*
133, 1	IV, 1
134, 2	IV, 2
134–5, 3	IV, 3
136, 4	IV, 4
137, 5[34]	—
138, 6	—
139, 7	IV, 5
140, 8	IV, 6
140–141, 9	IV, 7
141, 10	IV, 8
142, 11	IV, 9
142–143, 12[35]	IV, 10
143–144, 13	IV, 11
144–145, 14	IV, 12
146, 15	IV, 13
147, 16	IV, 14
148, 17	IV, 15
148–150, 18	IV, 16

Shifā' (Book 5)	*Elements*
155, 1	V, 1
156, 2	V, 2
157, 3	V, 3
158, 4	V, 4
159, 5	V, 5
159–160, 6	V, 6
160–161, 7	V, 7
161–162, 8	V, 8
163, 9	V, 9
163, 10[36]	V, 10
164, 11	V, 11
164–5, 12	V, 13
165–6, 13[37]	V, 12
166, 14	V, 14
166–7, 15	V, 15
167, 16	V, 16
168, 17[38]	—
169–70, 18	V, 17
171, 19	V, 18
172, 20	V, 19
173, 21[39]	—
174, 22	V, 20
175, 23	V, 24
175–6, 24[40]	—
176, 25	V, 25

Shifā' (Book 6)[41]	*Elements*
179, 1[42]	VI, 1
181, 2	VI, 1
181–2, 3	VI, 2
182–3, 4	VI, 3
184, 5	VI, 4
185, 6	VI, 5
186, 7	VI, 6
187, 8	VI, 8
188, 9[43]	—
188–9, 10	—
189, 11	VI, 13
190, 12	VI, 11
190, un-numbered	VI, 14
191, 13	VI, 15
191–192, 14	VI, 10
192–193, 15	VI, 17
193, 16	VI, 19
194, 17[44]	—
195, 18	VI, 20
195–6, 19	VI, 18
196–7, 20	VI, 21
197, 21	VI, 22
198, 22	VI, 24

Shifā' (Book 6)	*Elements*
198–9, 23	VI, 26
199–200, 24[45]	VI, 25
200–1, 25	VI, 23
201–2, 26	VI, 28
202–3, 27	VI, 28
204, 28	VI, 30
204–5, 29	VI, 32
205–6, 30	VI, 31
206–7, 31	VI, 33

Shifā' (Book 7)[46]	*Elements*
213, 1	VII, 1
214–15, 2	VII, 2
215, 3	VII, 3
216, 4	VII, 4
217, 5	VII, 5
217–18, 6	VII, 6
218, 7	VII, 7
219, 8	VII, 8
220, 9	VII, 9
221, 10	VII, 10
222, 11	VII, 11
222–3, 12	VII, 12
223, 13	VII, 13
223–4, 14	VII, 14
224–5, 15[47]	VII, 15
225, 16[48]	VII, 16
226, 17[49]	VII, 17
226, 18[50]	VII, 18
227, 19	VII, 19
227–8, 20	VII, 20
228–9, 21	VII, 22
229, 22	VII, 21
229, 23	VII, 23
229–30, 24	VII, 24
230–1, 25	VII, 25
231, 26	VII, 26
231–2, 27	VII, 27
232, 28	VII, 28
233, 29	VII, 31
233, 30[51]	VII, 32
233, 31	VII, 29
234, 32	VII, 30
235, 33	VII, 33
236, 34	VII, 34
237, 35[52]	—
238, 36	VII, 35
238, 37	VII, 36
239, 38[53]	—
239–240, 39	VII, 37
240, 40	VII, 38
240–241, 41	VII, 39

Shifā' (Book 8)	*Elements*
245, 1	VIII, 1
246, 2	VIII, 2
247, 3	VIII, 3
248, 4	VIII, 4
250, 5	VIII, 5
251, 6	VIII, 6
252, 7	VIII, 7
252–3, 8	VIII, 8
253–4, 9[54]	VIII, 9
254–5, 10	VIII, 10
256, 11	VIII, 12
257, 12	VIII, 13
258, 13	VIII, 14
258–9, 14	VIII, 15
259, 15	VIII, 16
260, 16 } 260–1, 17 }	VIII, 18
261–2, 18	VIII, 19
263, 19	VIII, 21
264, 20	VIII, 22
264–5, 21	VIII, 23
265, 22	VIII, 24
265, 23	VIII, 25
265–6, 24	VIII, 26
266–7, 25	VIII, 27

Shifā' (Book 9)[55]	*Elements*
271, 1	IX, 1
271–2, 2	IX, 2
272, 3	IX, 3
273, 4	IX, 4
273, 5	IX, 5
274, 6	IX, 6
274, 7	IX, 7
275, 8	IX, 8
276, 9	IX, 9
277, 10	IX, 10
277–8, 11	IX, 11
278, 12	IX, 12
279, 13	IX, 13
280–1, 14	IX, 14
281–2, 15	IX, 15
283, 16	IX, 16
283–4, 17	IX, 17
284–5, 18	IX, 18
285, 19	IX, 19
286, 20	IX, 20
287, 21	IX, 21
287, 22	IX, 22
288, 23	IX, 23

Shifā' (Book 9)	*Elements*
288, 24	IX, 24
289, 25	IX, 26
289, 26	IX, 27
289–90, 27	IX, 25
290, 28	IX, 28
290–1, 29	IX, 29
291, 30	IX, 30
291–2, 31	IX, 31
292, 32	IX, 32
293, 33	IX, 33
293, 34	IX, 34
293–4, 35	IX, 35
294, 36	IX, 36

Shifā' (Book 10)[56]	*Elements*
299–300, 1[57]	X, 1
301, 2	X, 2
302, 3	X, 3
303, 4	X, 4
303–4, 5	X, 5
304–5, 6	X, 6
305–6, 7 306, 8	X, 9
306–7, 9	X, 12
307, 10	X, 15
307, 11	X, 11
308, 12	X, 17
309, 13	X, 14
310, 14 311, 15	X, 18
312, 16	X, 20
312, 17	X, 21
313, 18 313–4, 19	Lemma between X, 13 and X, 14
315, 20	X, 13
315–16, 21	X, 22
316–17, 22	X, 23
318, 23	X, 24
319, 24[58]	X, 25
320–1, 25	X, 26
322, 26	X, 28
323, 27	X, 33
323–324, 28	X, 34
324, 29	X, 35
325, 30[59]	—
325–6, 31	X, 36
326, 32	X, 37
326–7, 33	X, 38
327–8, 34	X, 39
328, 35	X, 40

Shifā' (Book 10)	*Elements*
328–9, 36	X, 41
329, 37	X, 42
329–30, 38	X, 43
330–1, 39	X, 44
331, 40	X, 45
331, 41	X, 46
331–2, 42	X, 47
333, 43	X, 48
333, 44	X, 49
333, 45	X, 50
334, 46	X, 51
334–5, 47	X, 52
335, 48	X, 53
335–6, 49	X, 54
337–8, 50	X, 55
338, 51	X, 56
338–9, 52	X, 57
339, 53	X, 58
339, 54	X, 59
339–40, 55	X, 60
340–2, 56	
343, 57	X, 61
343, 58	X, 62
344, 59	X, 63
344, 60	X, 64
344, 61	X, 65
345, 62	X, 66
345–6, 63	X, 67
346, 64	X, 68
347, 65	X, 69
347, 66	X, 70
347–8, 67	X, 71
349, 68	X, 72
349–50, 69	X, 73
350, 70	X, 74
350–1, 71	X, 76
352, 72	X, 75
353, 73	X, 77
353, 74	X, 78
353, 75	X, 79
353, 76	X, 80
354–5, 77	X, 81
355, 78	
355, 79	X, 82
355–6, 80	
356, 81	X, 85
357, 82	X, 86
358–9, 83	X, 87
358, 84	X, 88
358, 85	X, 89
358, 86	X, 90

Shifā' (Book 10)	*Elements*
359–60, 87	X, 91
361, 88	X, 92
361, 89	X, 93
361–2, 90	X, 94
362, 91	X, 95
362, 92	X, 96
362–3, 93	X, 97
364, 94	X, 98
364, 95	X, 99
364–5, 96	X, 100
365, 97	X, 101
365, 98	X, 102
365–6, 99	X, 103
366, 100	X, 104
366–7, 101	X, 105
367, 102	X, 106
367, 103	X, 107
367–8, 104	X, 108
368, 105	X, 109
368–9, 106	X, 110
369–70, 107	X, 111
370–1, 108	X, 115

Shifā' (Book 11)[60]	*Elements*
376, 1 (f. 330)	XI, 1
377, 2 (f. 331)	XI, 2
377, 3 (f. 332)	XI, 3
378, 4 (f. 333)	XI, 4
379, 5 (f. 334)	XI, 5
380, 6 (f. 335)	XI, 6
380, 7 (f. 336)	XI, 7
381, 8 (f. 337)[61]	XI, 8
382, 9 (f. 338)	XI, 9
382, 10 (f. 339)	XI, 10
382, 11 (f. 340)	XI, 11
383, 12 (f. 341)	XI, 12
383, 13 (f. 342)	XI, 13
384, 14 (f. 343)	XI, 14
384, 15 (f. 344)	XI, 15
385, 16 (f. 345)	XI, 16
385, 17 (f. 346)	XI, 17
386, 18 (f. 347)	XI, 18
386, 19 (f. 348)	XI, 19
387, 20 (f. 349)	XI, 20
388, 21 (f. 350)	XI, 21
388, 22 (f. 351)	XI, 22
389, 23 (f. 352)	XI, 24
389, 24 (f. 353)	XI, 25
390, 25 (f. 354)	XI, 26
390, 26 (f. 355)	XI, 27

Shifā' (Book 11)	*Elements*
391, 27 (f. 356)	XI, 28
391, 28 (f. 357)	XI, 29
392, 29 (f. 358)	XI, 30
392, 30 (f. 359)[62]	XI, 31
393, 31 (f. 360)	XI, 32
393, 32 (f. 361)[63]	—
394, 33 (f. 362)[64]	—
395, 34 (f. 363)[65]	XI, 33
395, 35 (f. 364)	XI, 35
396, 36 (f. 365)	XI, 36
397, 37 (f. 366)[66]	XI, 37
398, 38 (f. 367)[67]	XI, 38
398, 39 (f. 368)	XI, 39

Shifā' (Book 12)[68]	*Elements*
401, 1 (f. 369)	XII, 1
402, 2 (f. 370)[69]	—
402, 3 (f. 371)	XII, 2
403, 4 (f. 372)	XII, 3
404, 5 (f. 373)	XII, 4
405, 6 (f. 374)[70]	XII, 5
405, 7 (f. 375)	XII, 7
406, 8 (f. 376)[71]	XII, 9
406, 9 (f. 377)[72]	XII, 8
407, 10 (f. 378)[73]	XII, 10
408, 11 (f. 379)[74]	XII, 12
409, 12 (f. 380)	XII, 11
409, 13 (f. 381)	XII, 15
410, 14 (f. 382)	XII, 16
411, 15 (f. 383)	XII, 17
412, 16 (f. 384)	XII, 18

Shifā' (Book 13)[75]	*Elements*
415, 1 (f. 385)	XIII, 1, 2
416, 2 (f. 386)	XIII, 3
416, 3 (f. 388)[76]	—
417, 4 (f. 389)	XIII, 6
417, 5 (f. 390)[77]	—
418, 6 (f. 391)	XIII, 12
418, 7 (f. 392)	XIII, 9
419, 8 (f. 393)	XIII, 10
420, 9 (f. 394)	XIII, 8
420, 10 (f. 395)	XIII, 11
421, 11 (f. 396)	XIII, 13
422, 12 (f. 397)	XIII, 15
423, 13 (f. 398)	XIII, 14
424, 14 (f. 399)	XIII, 16
425, 15 (f. 400)	XIII, 17
426, 16 (f. 401) } 427, 17 (f. 402) }	XIII, 18

Shifā' (Book 14)[78]	*Hypsicles*
433, 1 (f. 403)[79]	XIV, 1[80]
434, 2 (f. 404)	
435, 3 (f. 405)	Lemma (following prop. 2)[81]
435, 4 (f. 406)[82]	XIV, 6
435, 5 (f. 407)	
436, 6 (f. 408)	XIV, 6[83]
436, 7 (f. 409)	XIV, 6[84]
438, 8 (f. 410)	XIV, 7
438, 9 (f. 411)	XIV, 8
439, 10 (f. 412)	XIV, 8[85]
440, 11 (f. 413)	XIV, 8[86]
Shifā' (Book 15)	**[Isidorus and Eutocius?][87]**
445, 1 (f. 414)[88]	XV, 3[89]
445, 2 (f. 415)	XV, 2[90]
446, 3 (f. 416)	XV, 1[91]
447, 4 (f. 417)	XV, 4[92]
448, 5 (f. 418)	XV, 5[93]

Appendix I(B)
Isḥāq–Thābit *Elements*

I 48 propositions
II 14 propositions
III 36 propositions III, 11, 12 as one prop.
IV 16 propositions (But text has 17 prop.; still count accurate since IV 15, 16 are proofs of same result)
V 25 propositions
order V 13; V 12
VI 33 propositions
order: 1–8, 13, 11, 12, 9, 10, 14–17, 19, 20, 18, 21, 22, 24, 26, 23, 25, 27–30, 32, 31, 33.
VII 39 propositions [omits Vulgo VII, 20, 22 of Gregory][94]
VIII 27 propositions [Isḥāq's VIII, 24, 25 are not Euclidean] omits VIII, 16, 17
IX 38 propositions
order: 1–13, 20, 14–19, 21–5, 27, 26, 28–36 [with 2 new props. before 30]
X 109 propositions. Omits X 7, 8, 13, 16, 24, 112, 113, 114 and Vulgo X, 13 (X, 117 of Gregory);[95] X 31, 32 divided into 2 each
order: 1–6, 9–12, 15, 14, 17–23, 26–8, 25, 29–111, 115
XI 41 propositions
omits XI, 38 of Gregory;[96] XI 31, 34 divided into two each
order: 1–32, 34, 33, 35–9
XII 15 propositions. Omits XII, 6, 13, 14
order: 1–5, 7, 9, 8, 10, 12, 11, 15, 16–18
VIII 21 propositions. XIII 1–3 each divided into 2 propositions
order: 1–3, 5, 4, 6, 7, 12, 9, 10, 8, 11, 13, 15, 14, 16–18
XIV [10]* propositions (but text has 12 props.)
XV [6]* propositions

478

Note: This material follows Heath's Euclid I, 79–83, and Klamroth (62).

* Heath's estimates

Appendix I(C)

al-Hajjāj *Elements*

(Codex Leidensis 399, I)

I 47 propositions missing I, 45
(1–44, 46–7)
II 14 propositions
III 36 propositions
IV 16 propositions
V 25 propositions
(1–11, 13, 12, 14–25)
VI 33 propositions
(1–8, 13, 11, 12, 9, 10, 14–17, 19, 20, 18, 21–2, 24, 26, 23, 25, 27–30, 32, 31, 33)
[See (11) for further details]

Notes to Appendix I

1. Numbers in parenthesis in 'Divisions...' 1–15 give the approximate percentage of the text of al-Handasa devoted to each division. (Sabra ed.)
2. Numbers under Shifā' refer to page and proposition numbers in the Sabra edition; those under *Elements* are the corresponding book and proposition number in Heath's edition. However, see note 60, below.
3. However, the illustration is for the second case of the proposition, where D lies in the △ABC, i.e. the case treated by Proclus.
4. The text is jumbled, and makes little sense.
5. This is a preliminary to I, 9. The figure is Proclus' second case objection to I, 9.
6. Ibn Sīnā's figure is at slight variance with the usual Euclidean figure, i.e. Heath's. Similar comments later in these notes are made with the same frame of reference.
7. This is a 'converse' to I, 15, namely, "if the vertical angles are equal, the lines are straight, etc."
8. Ibn Sīnā's figure shows a right angled triangle, but his statement of the proposition is not restricted.
9. The figure numbered 25 on p. 42 is really for 26.
10. The Euclidean caveat—it is necessary that two of the straight lines taken together in any manner should be greater than the remaining one—is included in ibn Sīnā's version.
11. The figure for I, 24 resembles Proclus' first of two cases.
12. The figure seems to suggest that ibn Sīnā is taking a perpendicular transversal, but the text shows no such restriction.
13. There is no. 42.
14. I, 45 is also missing from the al-Ḥajjāj ibn Yūsūf translation. See Heath (46), I, 76, and Besthorn (11), Partis I, fasc. I–II.
15. The figure is nearly Euclid's. There is no mention of al-Nayrīzī, or of Thābit's generalization of the Pythagorean theorem. (See Sayili (95)).
16. Again, ibn Sīnā begins with definitions.
17. 82, (12) is an introduction to 83, (15). It treats the (impossible) case of the perpendicular lying "outside".
18. 83, (13) plays the same preliminary role for 84, 16.
19. Ibn Sīnā restricts himself to triangles (rectilineal figures).
20. The book is introduced by the Euclidean definitions from *Elements*, III.
21. Ibn Sīnā's figure suggests a symmetry not in Euclid; but again the treatment is general.
22. III, 12 is not in the al-Ḥajjāj translation either; see Heath (46), I, 80, or Besthorn (11), Partis II.

23. It is of interest to compare this proposition with ibn Sīnā's manuscript "al-zāwiya". See note 196 of the text of this essay.
24. The figure and statements for III, 19 appear after 106, 17 without any separate numbering in Sabra's edition.
25. The three figures in ibn Sīnā represent decompositions of the usual Euclidean figure.
26. 110, 21 is followed, without separate numbering, by a figure and statement of III, 24.
27. Part of III, 24.
28. There is no 24 in Sabra's edition.
29. Part of III, 27.
30. Ibn Sīnā considers the three cases: right, obtuse, and actute angle.
31. Ibn Sīnā gives four figures; compare with Campanus' four cases and Euclid's two.
32. There is no suggestion that ibn Sīnā sees the tangent case as the limit of the secant case in his treatment of III, 36.
33. No definitions introduce this book.
34. 137, 5 and 138, 6 serve to introduce 139, 7.
35. Ibn Sīnā's figure departs somewhat from the usual Euclidean one.
36. Ibn Sīnā appears to have two figures (in Sabra's edition) for 163, 10. The first of these, however, is for 163, 9.
37. The single figure serves both 164–5, 12 and 165–6, 13.
38. This serves to introduce 169–70, 18.
39. This introduces 174, 22.
40. Introduction to 176, 25.
41. The book begins with definitions of similar and congruent figures.
42. This is VI, 1 modified to "triangles and squares...."
43. 188, 9 and 188–9, 10 are cases of 187, 8(= VI, 8).
44. Introduces 195, 18.
45. This appears to be part of VI, 28.
46. This book is prefaced by Euclidean definitions.
47.–9. In each case ibn Sīnā's figure differs from the Euclidean one.
50. No figure.
51. The identification is only probable since the text is unclear.
52. Part of 236, 34 (VII, 34).
53. Part of 238, 37 (VII, 36).
54. Ibn Sīnā's figure differs from the Euclidean one.
55. As in the *Elements*, no definitions introduce this book.
56. Ibn Sīnā gives the four Euclidean definitions.
57. There is a difference between ibn Sīnā's figure and the Euclidean one.
58. Apparently ibn Sīnā replaces rectangles by squares in his enunciation of 319, 24.
59. Introduces 325–6, 31.
60. Here we must change our manner of identifying the propositions in the Shifā, since neither ibn Sīnā, nor his editor supply numbers. Thus 376, 1 (f. 330) is a proposition on p. 376 with figure 330. Our label 1 is an interpolation which we change consecutively, e.g. the next proposition is 377, 2 (f. 331), etc.. The book begins with Euclid's definitions from solid geometry, as *Elements* XI.
61. This figure as well as figure 337 show a total lack of perspective, an element worth considering in analysis of Arabic mss. of Euclid.
62. Figure differs from Euclid.
63. 393, 32 is essentially the planar version of XI, 32; see VI, 1, etc. It and 394, 33 can be viewed as an introduction to 395, 34 (XI, 33). Literally 393, 32 is: "parallelograms of the same height have proportional bases".
64. See n. 63. Literally 394, 33 is: "equal parallelograms of equal base have equal height".
65. Figure is different from Euclid.
66. The figure is that from Heiberg's ms b; see Heath (46), III, 359.

67. Figure differs from Euclid.
68. Following *Elements* XIII, there are no introductory definitions.
69. 402, 2 serves to introduce 402, 3.
70–4. Figure differs slightly from Euclid. Cone is confused with pyramid in the text of 406, 9.
75. Again, following *Elements* XIII, there are no introductory definitions.
76. There is no Figure 387. Figure 386 serves to introduce 417, 4.
77. This introduces 418, 6.
78. Book 13 concludes the *Elements* proper; however, ibn Sīnā goes on to include the so-called *Elements* XIV of Hypsicles, followed by *Elements* XV (possibly by Isidorus of Miletus and Eutocius). See Heath (46), III, 519–20.
79. 433, 1 introduces 434, 2.
80. Designation follows Heath (46), III, 512ff.
81. "If an equilateral and equiangular pentagon be inscribed in a circle, the sum of the squares on the straight line subtending two sides and on the side of the pentagon is five times the square on the radius." [Heath (46), III, 519].
82. This introduces 435, 5.
83. Actually this is Hypsicles second proof of XIV, 6 including the "preliminary", see Heath (46), III, 516–17.
84. This completes the second proof.
85. This is another version of XIV, 8.
86. Actually, this is the second part of the statement of XIV, 8 involving the "content" ratio.
87. See note 78 above.
88. Since Heath's description of *Elements* XV [Heath (46), III, 519–20] is much abbreviated, our identifying reference will be to the Heiberg and Menge edition (48), V, 40–67.
89. Heiberg and Menge (48), V, 43; to inscribe an octahedron in a cube.
90. Heiberg and Menge (48), V, 41; to inscribe an octahedron in a pyramid.
91. *Ibid.*, 41; to inscribe a pyramid in a cube.
92. *Ibid.*, 45; to inscribe a cube in an octahedron.
93. *Ibid.*, 47; to inscribe a dodecahedron in an icosahedron.
94. Vulgo VII, 20 (Heiberg's Euclid, vol. II, 428–31): "Si tres numeri proportionales sunt, productum extremorum aequale est quadrato medii et si productum extremorum aequale est quadrato medii, tres numerii illi proportionales sunt." Vulgo VII, 22 (Heiberg's Euclid, vol. II, 431–3): "Si tres numeri dati sunt et alii iis multitudine aequales, duo simul coniuncti et in eadem ratione, et proportio eorum perturbata est, etiam ex aequo in eadem ratione erunt."
95. Authors were unable to check this reference.
96. Vulgo XI, 38 (Heiberg's Euclid, vol. IV, 354–5): "Si planum ad planum perpendiculare est, et a puncto aliquo alterius plani ad alterum planum perpendicularis ducitur, perpendicularis ducta in communem planorum sectionem cadet."

Appendix II

Elements V, 7, 15 from al-Shifā' (Sabra's edition 160–1, 166–7)

أ مثل ب ، فلنسبتها الى حـ واحدة ، ونسبة حـ اليهما واحدة .

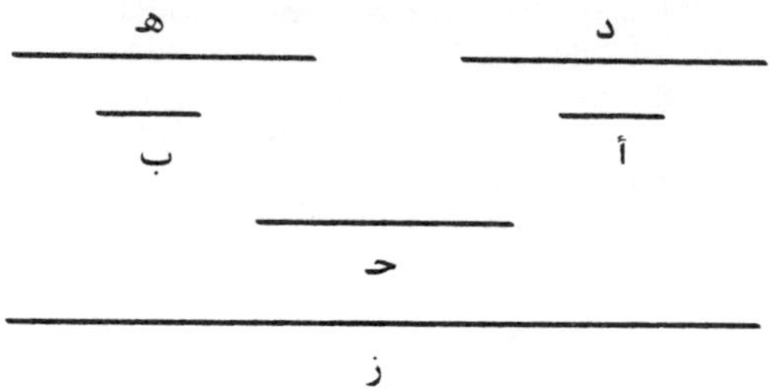

رسم رقم ١٣٤

فنأخذ د ، هـ أضعافا متساوية لهما ، و ز ل حـ كيف ما أتفق .

فـ ء مثل هـ ، فنقصانهما وزيادتهما ومساواتهما لـ ز واحدة ، وهما أضعاف متساوية للاول والثالث ، فنسبة أ ، ب الى حـ واحدة ، وكذلك نسبة حـ اليهما واحدة ، وبالعكس اذا كانت النسب واحدة فهى متساوية .

(١٥)

أ ب فيه من حـ ، ما فى د هـ من ز ، فنسبة أ ب الى د هـ كـ حـ الى ز . ونقسم أ ب ب ح ، ط على حـ . د هـ ب ل ، م على ز .

فنسبة أ ح الى د ل وكذلك البواقى واحدة ، فالمقدمات كلها ،

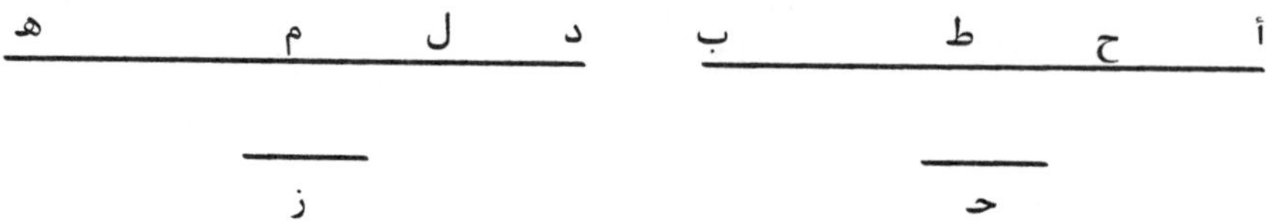

رسم رقم ١٤١

أعنى أ ب ، الى التوالى كلها ، أعنى د هـ كـ أ ح الى د ل أعنى حـ ، ز .

Elements V, 7, 15 from Ishāq-Thābit © Franz Steiner Verlag GMBH. Reproduced with permission from M. Klamroth, “Uber den arabischen Euklid”, *Zeitschrift der Deutschen Morgenlandischen Gesellschaft* **35** (1881), Arabic text, pp. 312–14.)

الشكل ز

الاقدار المتساويه نسبتها الى آخر واحدة ونسبته هو أيضا اليها واحدة . مثاله أنّ قدرى أ ب متساويان وقدر ج قدر آخر فأقول أنّ نسبة أ الى ج كنسبة ب الى ج ونسبة ج أيضا الى أ كنسبته الى ب .

بحين ذلك أن نأخذ لقدرى أ ب أضعافا متساوية وهى د ه ولقدر ج اضعافا ما كانت وهى ز واضعاف د لقدر أ كأضعاف ه لقدر ب وامثل ب قدر مثل ه وقدر آخر ما كان مقدرا د ه ان متساويان معا لقدر ز واما زائدان معا عليه واما ناقصان معا عنه وهما اضعاف متساوية لقدرى أ ب و ز أضعاف لقدر ج فنسبة أ الى ج كنسبة ب الى ج . واقول أن نسبة ج الى أ كنسبة الى ب لان تديرهما واحد وقد تبين أن د ه متساويان وان ز اما متساو لهما معا (واما زائد عليهما معا) واما ناقص عنهما معا و ز أضعاف لقدر ج د ه أضعاف متساوية لقدرى أ ب فنسبة ج الى أ كنسبة الى ب وذلك ما أردنا أن نبين :

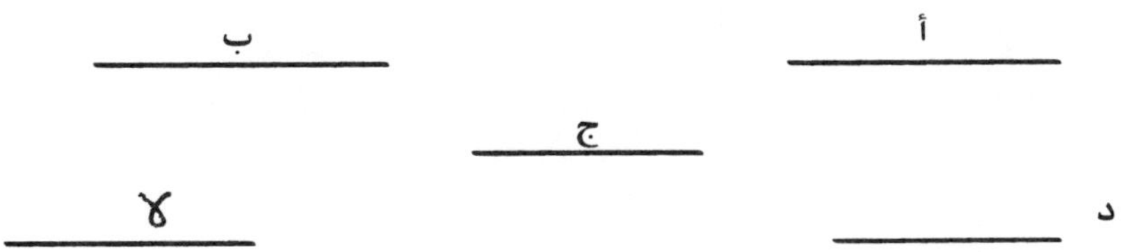

الشكـل يــــه

الاجزاء التي اضعافها متساوية فان نسبة بعضها الى بعض كنسبة اضعافها بعضها الـى بعض ٠

مثاله أن أضعاف أ ب لقدر ج كاضعاف د ꝿ لقدر ز فاقول أن نسبة أ ب الى د ꝿ كنسبة ج الى ز ٠

برهانه أن نقسم أ ب بقدر ج فأقسامه أ ح حط طب وهي متساوية ونقسم د ꝿ بقدر ز فأقسامه د ل لم مه وهي متساوية وعدة أ ح حط طب بعد (؟) (كعدة *vielleicht*) د ل لم مه فنسبة أ ح الى د ل كنسبة حط الـى لم وكنسبة طب الى مه ونسبته الواحد الى قرينه كنسبة الجميع الى الجميع فنسبة أ ح الى د ل كنسبة أ ب الى د ꝿ و أ ح مثل ج و د ل مثل ز فنسبته أ ب الى د ꝿ كنسبته ج الى ز وذلك ما أردنا أن نبين ٠

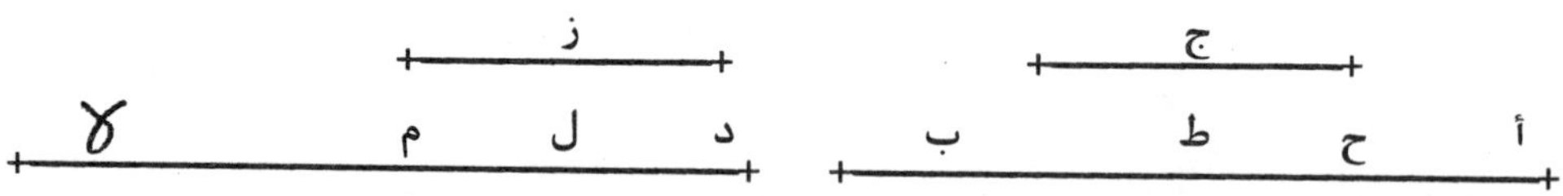

Elements V, 7 from al-Ḥajjāj (Besthorn and Heiberg (11), Partis III, Fasc. II, 42 and 44.)
Reproduced with permission from Gyldendalske Boghandel Nordisk Forlag Aktieselskab

الشكل السابع من المقالة الخامسة

المقادير المتساوية نسبتها الى مقدار آخر نسبة واحدة واذا نسب اليها ذلك المقدار فان نسبته
ايضا اليها واحدة مثاله ان مقدارى أ ب متساوين ومقدار جـ مقدار آخر فأقول أن نسبة أ الى
جـ كنسبة ب الى جـ ونسبة جـ ايضا الى أ كنسبة الى ب برهانة أنا نأخذ لمقدارى أ ب
اضعاف متساوية وننزل انها مقدارا د ه ونأخذ لمقدار جـ اضعافا أخر أى الاضعاف كانت
وننزل انها مقدار ز فمن أجل أن ما في د من أضعاف أ فرض مساويا لما في ه من أضعاف
ب و أ فرض مساويا لمقدار ب واذا كانت مقادير متساوية واخذت لها أضعاف متساوية فان
المقادير التي هي أضعاف المقادير المتساوية هي أيضا متساوية فمقدار د مساو لمقدار ه
ومقدار آخر مفروض وهو مقدار ز فظاهر أن مقدار د ان كان زائدا على مقدار ز فان مقدار
ه ايضا زائد على ز وان كان د ناقصا عن ز فان ه ناقص عن ز وان كان مساويا له فهو
مساو له فمقدار د ه أما أن يكونا زائدين معا على ز واما ناقصين معا عنه واما مساويين
معا له وهما أضعاف متساوية لقدرى أ ب ومقدار ز اضعاف لمقدار جـ فنسبة أ الى جـ
كنسبة ب الى جـ * * وأقول أيضا أن نسبة جـ الى أ كنسبة الى ب ونبين أيضا كما
بينا ان د ه متساويان وان ز ان كان زائدا على د فهو أيضا زائد على ه وان كان
ناقصا عن د فانه أيضا ناقص عن ه وان كان مساويا له فهو أيضا مساو له فمقدار ز اما
ان يكون زائد على د ه معا واما ناقصا منهما معا واما مساويا لهما معا ومقدار ز فرض
اضعافا لمقدار جـ ومقدارى د ه فرضا اضعافا متساوية لمقدارى أ ب فنسبة جـ اذن الى
أ كنسبته الى ب وذلك ما أردنا أن نبين :

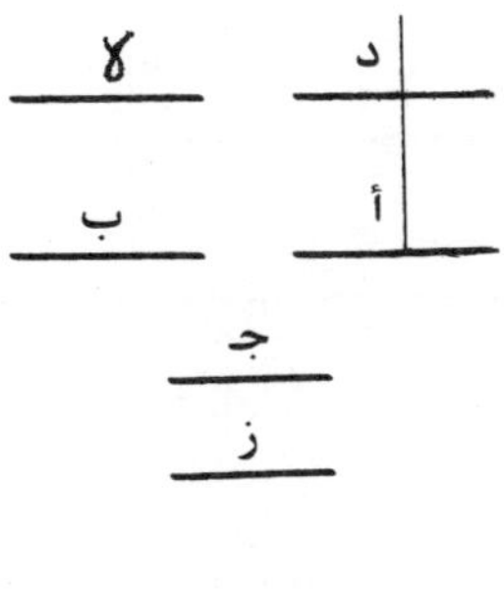

Elements V, 15 from al-Ḥajjāj (Besthorn and Heiberg (11), Paris III, Fase. II, 60 and 62.)
Reproduced with permission from Gyldendalske Boghandel Nordisk Forlag Akiteselskab

الشكل الخامس عشر من المقالة الخامسة

اذا كانت مقادير وأخذ لها أضعاف متساوية فان نسبة المقادير بعضها الى بعض كنسبة اضعافها بعضها الى بعض مثاله ان مقداري جـ ز مفروضان وقد أخذ لهما أضعاف متساوية وهي أ ب و د ه وفي أ ب من أضعاف جـ مثل ما في د ه من أضعاف ز فأقول ان نسبة جـ الى ز كنسبة أ ب الى د ه برهانه أنا نفصل ما في أ ب و د ه من أضعاف جـ ز وهي أ ح ح ط ط ب و د ل ل م م ه فمقادير أ ح ح ط ط ب متساوية لان كل منها مثل جـ(١) وأيضا مقادير د ل ل م م ه متساوية لان كل واحد منها مثل ز(١) وعدة أ ح و ح ط و ط ب مساوية لعدة د ل و ل م و م ه فظاهر أن نسبة أ ح الى د ل كنسبة ح ط الى ل م وكنسبة ط ب الى م ه فبين من برهان ١٣ من ه أن نسبة أ ح الى د ل كنسبة أ ب الى د ه و أ ح مساو لمقدار جـ و د ل مساو لمقدار ز فنسبة جـ الى ز كنسبة أ ب الى د ه وذلك ما أردنا أن نبين :

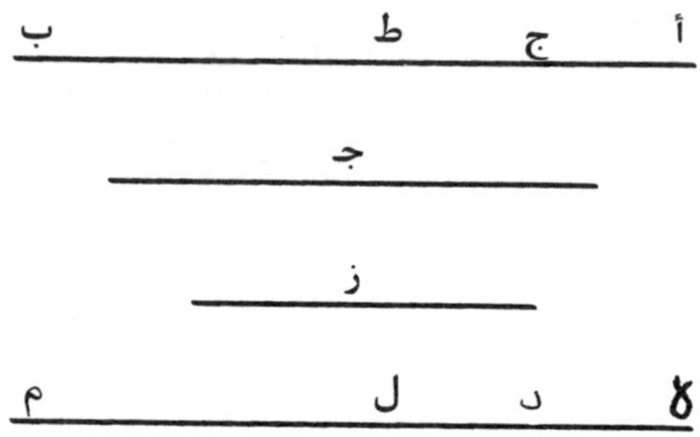

6 Numerical Analysis in the Middle East: Ninth Through Fifteenth Century

I. Introduction

The phrase "numerical analysis" as a name for a distinct mathematical discipline came into general use about 1947, after the establishment of the Institute of Numerical Analysis at UCLA.[1] However, numerical analysis, either as a part of astronomy, or of an elementary theory of equations, or of mensuration has been with us at least since Babylonian times. Neugebauer [1] even goes so far as to distinguish two types in late Babylonian and pre-Ptolemaic ephemerides, one based on "linear methods", that is, first order difference schemes, and a second on numerical procedures suggested by geometric models.[2] Needham [2] and Libbrecht [3] have discussed related developments in China, while Bose, Sen and Subbarayappa [4] have written on India. Van der Waerden's *Science Awakening* [5] and Neugebauer's monumental *A History of Ancient Mathematical Astronomy* [6] go a long way toward forming our current view of the subject's early history.[3]

A recent study by Goldstine [7] traces the development of numerical analysis, as a part of the rekindling of interest in science in Europe, beginning in the sixteenth century, and ending with the masterful work of Gauss, Jacobi, Cauchy, and Hermite. Since this work makes only passing reference to numerical methods in the Middle East, e.g. al-Kāshī's iterative scheme for solving a cubic, and al-Bīrūnī's attempt to use quadratic interpolation, it seems fitting to bridge the apparent gap between Babylonian and Ptolemaic numerical analysis, and that described by Goldstine.[4] Fortunately, this can be easily done, at least for the Middle East, owing to a series of studies by Suter [8–10], Luckey [11–12], Kennedy [13–19], Youschkevitch [20], Saidan [21–22], Rashed [23–24], and others.

The present authors attempt this, offer a comparison of the "state of the numerical art" in the Middle East and in Europe in the fifteenth century, and conclude with some comments on the failure of the work of Middle Eastern numerical analysts to have a direct effect on the development of numerical analysis in Europe.

Owing to the diverse views held even today about numerical analysis—

ranging from "data processing" to mathematical "drudgery"—it may help, for our purposes, to adopt Henrici's definition "a [collection] of procedures[s] which permit us to obtain the solution of a mathematical problem with arbitrary precision in a finite number of steps, that can be performed rationally, [i.e. by means of an algorithm]."[5] The modern subject then can be divided into three main areas: solution of equations, which includes iterative schemes for single equations and systems, linear difference equations, the Bernoulli method, and quotient-difference algorithms; interpolation and approximation, which treats interpolation polynomials, numerical differentiation and integration; computation, which subsumes the number system and the propagation of round-off error.[6] This may then be contrasted with the subject as it was known in the early sixteenth and seventeenth centuries in Europe: the works of Napier, Briggs, and Bürgi on logarithms, the development of interpolation schemes, and Vièta's trigonometric reduction of certain algebraic equations.[7] By the age of Newton, most of the first main area of the subject had been investigated, while Euler and Lagrange later contributed to interpolation theory and difference equations, leaving numerical integration and a theory of rounding-errors to Gauss.[8] The nineteenth century saw Gauss' work supplemented by Jacobi, and the numerical treatment of differential equations.[9]

Against this background, how does one answer the question: "What was 9th–15th century Middle Eastern numerical analysis?" The authors have tried to do this by matching achievements with the three main areas of Henrici's scheme, using a chronological arrangement within the scheme. The result is a sketch of what a numerical analysis text might have looked like over a span of 700 years of evolution. Roughly the effect is that of comparing the arithmetics of al-Uqlīdīsī (Kitāb al-Fuṣūl fī al-Ḥisāb al-Hindī, 952/3 A.D.) [25] and Kūshyār ibn Labbān (Kitāb fī uṣūl ḥisāb al-hind, *c.* 1000 A.D.) [26], with al-Kāshī's fifteenth century work *The Key to Reckoning* (Miftāḥ al-Ḥisāb, before 1429 A.D.). This analogy is, of course, misleading since the subject did not develop so smoothly, and, in fact, also evolved outside these texts, and in the domain of astronomy proper. For the latter reason much of our sketch may change as more astronomical works and ephemerides are studied. We begin with the third of Henrici's divisions: computation, the number system, and round-off errors, since this section is the shortest and least technical.

II. Computation

Middle Eastern arithmeticians inherited the rudiments of a number system from various sources: sexagesimals from Babylonia,[10] base ten schemes both from Babylonia and India,[11] finger reckoning,[12] some formalized arithmetic, including the rational field axioms of Euclid's elements and Eudoxus' theory of proportions from the Greeks.[13] The latter theory, which is expounded in Book V of the *Elements* is so geometrical, however, that it could not serve as a basis for the real number system. For example, it inherits the Pythagorean abhorrence of irrationals, and attaches no meaning to symbols like $(a/b)+(c/d)$. A number of mathematicians, beginning with al-Māhānī (*c.* 880) sought to remove these defects by comparing, essentially, the continued fractions of a/b and c/d. His work has pre-Eudoxan origins, cites Thābit ibn Qurra (836–901), and was adopted by his Iraqi contemporary al-Nayrīzī (*c.* 897–922).[14] However, it was not until Ibn al-Haytham (965–*c.* 1040)[15] and al-Khayyāmī ('Umar Khayyam, 1048 (?)–1131 (?)) that a proof of the equivalence of the al-Māhānī and Euclidean theories was carried out.[16] Naṣir al-Dīn al-Ṭūsī (1210–1274), whose work on computational mathematics saw fruition in the hands of al-Kāshī, extended this work.[17] In the Muslim West, al-Jayyānī of Cordoba (*c.* 889/990–1079+) tried a different approach: taking ratio and proportion as primitive, undefined terms. This has elements in common with Isaac Barrow's work on Euclid V.[18] However, Aḥmad Ibn Yūsuf (Ametus) (fl. *c.* 900) was the only one of these scholars to have direct influence on Western mathematics. His treatise on Euclid V is mentioned in Fibonacci's *Liber abacci*, and either adapted or cited by Jordanus de Nemore (*Arithmetica in decem libris demonstrata*) and Thomas Bradwardine.[19] Unfortunately, his efforts are all directed toward elucidating proportion as a problem solving technique (the 18 cases of the divided figure).

At any rate, Middle Eastern calculators, or "numerical analysts" as we shall refer to them, came closer to a real number system than their Greek predecessors, but still did not have a completely satisfactory theory. However, from a practical point of view, they did possess a most effective tool: the base ten decimal system. This may have been the discovery of al-Uqlīdīsī (fl. 952–953).[19a] As Saidan has noted,[19b] his arithmetic (*Kitāb al-fuṣūl fī᾿l-ḥisāb al-hindī*) is the first to treat decimal fractions with a clear perception of the decimal idea. Others before him had made the mechanical transition from sexagesimal to decimal scale, and Kūshyār Ibn Labbān (*c.* 1000) in his arithmetic (*Uṣūl Ḥisāb al-Hind*) does a number of computations which suggest that he was very close to discovery.[20] The discovery of al-Uqlīdīsī is elucidated again in al-Kāshī's *Arithmetic* (*Miftāḥ al-ḥisāb* (1427)), a veritable compendium of computational techniques. Rosenfeld and Youschkevitch attempt to show how this work may have had direct European influence by tracing it through Turkey by means of al-Kāshī's colleague ᶜAli Qūshjī and thence to Vienna in 1562 in a Byzantine problem collection.[21] However this may be, Simon Stevin usually is credited with the independent European discovery in his *Decimal Arithmetic* ("La Disme", 1585),[22] even though Rudolff (*c.* 1500–*c.* 1545) and Vièta (1540–1603) had preceded him.

Although al-Khwārizmī's *Algebra* (*c.* 840) provides "what is easiest in arithmetic",[23] and his *Treatise on Calculation with Hindu Numerals* (Kitāb ḥisāb al-ᶜadad al-hindī) discusses the place value system and rules for performing the four elementary operations, the computational schemes which appear in medieval Latin works, especially multiplication schemes, seem to originate in the later *Arithmetic* (*Kitāb al-fuṣūl fī᾿l-ḥisāb al-hindī*) of Al-Uqlīdīsī.[24] This work borrows freely from Hindu arithmetic, but goes to great length to adapt the Hindu dust board techniques to methods suitable for pen and paper. Its four part arrangement: introduction of Hindu numerals, explanation of place value concept, arithmetical operations, and extraction of square root in part one; casting out 9 and further operational explanations in part two; justifications in three; dust abacus to pen and paper transition in four; provides an excellent overview of the computational art in the tenth century Middle East.[25] This was preceded by al-Dīnawarī's Indian arithmetic for businessment (before 895) and followed by Kūshyār ibn Labbān late tenth century treatise for astronomers.[26] A later work by the Spanish Arab al-Umawī (fl. fourteenth century), *Marāsim al-intisāb fī ᶜilm al-ḥisāb* is of more interest to the history of number theory than to numerical computation.[27] The capstone of all computational texts is the *Miftāḥ* of al-Kāshī, to which we return in our discussion of numerical solutions of equations.

We close this section with a few comments on round-off errors. No Middle Eastern numerical analyst seems to have paid theoretical attention to this subject. Al-Bīrunī, for example, who is noted for his highly accurate approximation of π, generally rounds off correctly, but, as Kennedy notes, is careless when combining numbers accurate to different precision.[28] In this respect, he can claim Ptolemy as a model.[29] Yet, even the great table maker Napier, who had an explicit rule for rounding, still was not totally aware of the problem of rounding errors. The general problem, and the phenomenon now called numerical stability was discussed for the first time centuries later by Gauss in *Theoria Motus*.[30]

III. Solution of Equations

Let us now consider the first of Henrici's divisions of numerical analysis: solution of equations, iteration, and differencing schemes. The discussion will be devoted to the story of the development of the so-called Ruffini–Horner method by Middle Eastern numerical analysts, the iterative schemes for solving equations implicit in the works of al-Bīrūnī and others, and the use of first and second order differences in astronomical computations.

A common method used to solve simple linear equations in the Middle East, the method of false position—with "one false position"—appears in the Ahmes Papyrus (problem 24), which dates to 1650 B.C.,[31] while the Bakhshālī Ms. (India, third or fourth century A.D.) contains the rule with two "false positions".[32] In one form or another, it was also known in China as early as 250 B.C. (*Chiu-chang suan-shu*).[33] It was explained by al-Khwārizmī, and Qusṭā ibn Lūqā even devoted a long treatise *Kitāb fīᵓ l-burhān ᶜalā 'amal ḥisāb al-khaṭa'ayn*) to it.[34–5] The method is essentially nothing more than inverse linear interpolation, and can form the basis for an iterative procedure for solving polynomial equations, but this does not seem to have been exploited in numerical texts before al-Samawᵓal (1172).[35a]

Many Middle Eastern arithmetics also treat the numerical solution of the quadratic $x^2 = A$, and the cubic $x^3 = B$, by discussing approximate methods of extracting square and cube roots. The Banū Mūsā in *Verba Filorum*, for example, give a simple method for extracting cube roots which has many parallels in other arithmetics: one expresses the number whose cube root is required as an improper sexagesimal, and finds the perfect cube nearest to the numerator[36]: $\sqrt[3]{r} = \sqrt[3]{r\cdot(60)^{3k}}/60^k$. Their explanation suggests an awareness that the method could be used to approximate cube roots to any degree of accuracy. Of course, the defect of the method is that it requires large tables of cubes. Saidan [21] surveys various approximate formulae current in these arithmetics from the time of al-Khwārizmī to al-Kāshī.[37] In modern terms, for example, Kūshyār gives

$$\sqrt{N} = a + \frac{r}{2a+1} \quad \text{if} \quad N = a^2 + r,$$

where a is the largest square in N.[38] This is exactly the same as the rule given by al-Uqlīdisī, and al-Nasawī, who "justifies" it by observing that $(a+1)^2 - a^2 = 2a+1$, $N - a^2 = r$, and using linear interpolation.[39] Similar formulae are given for $\sqrt[3]{a^3 + r}$, e.g.,[40]

$$a + \frac{r}{3a^2+1} \quad \text{and} \quad a + \frac{r}{3a^2+3a+1};$$

and Ibn Ṭāhir (d. 1037) gives an analogous justification of the latter of these (in *Al-Takmila fī al-Ḥisāb*).[41] By the time of al-Kāshī (*Miftāḥ*), a general approximation

$$\sqrt[n]{a^n + r} \approx a + \frac{r}{(a+1)^n - a^n},$$

which includes the above as special cases, was being used. Curiously enough the approximations

$$\sqrt{a^2 + r} = a + \frac{r}{2a},$$

$$\sqrt[3]{a^3 + r} = a + \frac{r}{3a^2},$$

which are special cases of the first iteration of the Newton formula itself were also current, though not so widely used.[42]

The crowning achievement of this work is, however, the discovery of the so-called Horner method (1819) by Middle Eastern numerical analysts, and its application to the finding of approximate solutions of $X^n = A > 0$, with arbitrary precision. This is essentially a tableau, similar to synthetic division, for systematically transforming an equation into another with a smaller root. The processes may be initiated by application of the rule of double false position. Luckey [11] established conclusively that the method appears in detail in the first book of al-Kāshī's textbook *Miftāḥ al-ḥisāb* (*Key of Arithmetic*) (1427). Recent studies by Rashed [23] prove that the method appears two and a half centuries earlier in al-Samawʾal's treatise on algebra (1172), which also contains an account of decimal fractions.[43] Moreover, there are indications that the method was not even new at that time. Rashed also speculates on the possible origin of this work in the school of al-Karkhī (*c.* 1030), and its possible transmission to al-Kāshī through the works of al-Ṭusī.[44–5] Evidently al-Samawʾal was even aware of the fact that the method produces a chain of rational approximates which "converge" to the required root.[46] In an earlier work Rashed [24] has explored the relation between the work of al-Ṭūsī on the theory of equations, and that of Vièta.

Since the al-Samawʾal method is essentially a specialization of the Newton–Raphson method to polynomials, which in turn is an iterative method, it should not be surprising to find that interation, as a computational technique, was well known to Middle Eastern analysts. As Kennedy [19] points out, iteration methods go back at least to Ptolemy (150 A.D.).[47] Specifically, one wishes to solve an equation of the form $x = f(x)$. The process starts by choosing an arbitrary x_0 and generating recursively a sequence $x_n = f(x_{n-1})$, $n = 1, 2, 3, \ldots$. Under certain condition, $\lim_{n\to\infty} x_n$ exists and $x = f(x)$ has a unique solution, independently of the initial choice of x_0.[48] For example, the Kepler equation, which plays an important role in dynamical astronomy: $x = m - e \sin x$, where m and e are real and $|e| < 1$ has a unique solution in $[m - \pi, m + \pi]$. Kepler solved it (1600) by this technique, but so did Ḥabash al-Ḥāsib al-Marwazī in the ninth century. Moreover, the method was not original with him.[49–51] Al-Kāshī applied it, with an $f(x)$ equivalent to $[4x^3 + \sin(3°)]/4$ in *Risāla al-watar waʾl-jaib* (Treatise on the Chord and Sine) to approximate sin (1°) with 12 decimal place accuracy.[52–3] Variants of his method of solution were given by Qāḍī Zāda al-Rūmi (*c.* 1364; d., *c.* 1436),[54–5] Ulugh Beg (1394–1449),[56] and others. However, al-Kāshī's computational *tour de force* reached Europe for the first time in the nineteenth century as a part of mathematical history. As a last application of the iterative technique, in the intuitive sense of the term, we mention al-Bīrūnī's approximation of the chord of 40°, using chord 12°, chord 30°, and the known rules for finding chord $\frac{1}{2}A$ and chord $\frac{1}{8}A$ when chord A is given, as well the rule for chord $(A + B)$. (Qānūn al-Mas'ūdī, *c.* 1000).[57]

We conclude this section with a few remarks on the use of first and second order difference schemes in astronomical computations. Recall that if we have a table of values (x_i, y_i) obtained from a function $y = f(x)$ where the $x_i < x_{i+1}$ are equally spaced the first differences are formed by subtracting $f(x_i)$ from $f(x_{i+1})$ for each i. The second differences arise from subtracting successive first differences, etc.. Such difference schemes have several uses in numerical work. For example, if the tabulation interval is small, rounding errors of order ε are propagated through successive differences with binomial coefficient factors, e.g. ε, -2ε, ε (2nd order); ε, -3ε, 3ε, $-\varepsilon$ (3rd order), etc.. Hence larger fluctuations than this can be used to detect non-systematic errors—tabular errors which may have occurred for reasons unrelated to rounding. Secondly, since the nth order differences of a degree n polynomial are all constant, and those of higher order zero, the scheme may be used to detect polynomial dependence between x and y, and to recover a representation for such a polynomial. There appears to be no evidence for the first usage in Middle Eastern work; however, the second is certainly implicit in the "zig-zag functions" of Babylonian astronomy.[58] (First order differences.) Kennedy has found explicit evidence of this also in al-Kāshī's *Zīj-i Khāqānī*, where a second order scheme is applied. Al-Kāshī refers to it as the method of "difference arcs" (qaws al-ikhtilāf), and uses it to prepare tables of true longitude of planets over short periods from three observations.[59]

IV. Interpolation and Approximation

Middle Eastern numerical analysts contributed to "interpolation and approximation", the second of Henrici's divisions, with their work on integration. By the time of the Banū Mūsā (9th century) the *Elements* and part of the work of Archimedes, including his application of the method of Eudoxus (exhaustion) to finding areas and volumes were known. In their treatise *Book on the Measurements of Plane and Spherical Figures*, which was later edited by Naṣīr al-Dīn al-Ṭūsī, and translated into Latin by Gerard of Cremona (*Liber trium fratrum de geometria*, 12th century), the Banū Mūsā reproduced results from Archimedes works *Measurement of the Circle* and *On the Sphere and Cylinder*.[60] Their proofs of formulae for the area of a circle and the surface area of a sphere also use the method of exhaustion but omit the usual Archimedean arguments by appealing to *Elements*, Book 12, proposition 16 and Book 12, proposition 11 respectively.[61–61a] In other words, although they use the traditional inscribed polygons, there is no "passage" to the limit. In a sense they were attempting to formalize Archimedean "infinitesimals". In further contrast to Archimedes, who expressed areas and volumes in the traditional way: e.g. volume of sphere equals four times the volume of a cone whose base is a great circle of the sphere and height is equal to its radius, the Banū Mūsā wrote: volume of a sphere is the product of its radius and one third its surface area.[62] This shows a break from the Greek, geometrical concept of number.

Even more remarkable is the work of the Banū Mūsā's contemporary Thābit ibn Qurra (836–901). As Youschkevitch [20] has pointed out, his quadrature of the parabola is the first to use the method of upper and lower sums. Archimedes was certainly familiar with arguments for estimating $\int_0^a x^2\,dx$ by dividing the interval $[0, a]$ into n equal segments and using the method of exhaustion for

$$\text{“passing to the }\lim_{n\to\infty} a^3 \sum_{k=1}^{n} \left(\frac{k^2}{n^3}\right)\text{.”}$$

However, he does not make use of such arguments in his treatise *Quadrature of the Parabola*. Youschkevitch speculates on why, but more importantly, points out that Thābit does treat the problem via integral sums, essentially involving the integral

$$\int_0^{a^2} (a - \sqrt{y})\,dy.$$

To handle the resulting sums, he uses the ingenious device of partitioning at squares 1, 4, 9, 16, ..., an idea generalized by Fermat in the seventeenth century for the purpose of handling integrals of the type $\int_0^a x^r \, dx$, r a positive rational number.[63] The integration then requires the formula,

$$\sum_{k=1}^{n} (2k-1) = n^2,$$

which Thābit proves as proposition 15, in a set of 20. Finally Thābit uses the method of contradiction to deduce the required result. In the course of this argument, he proves a proposition equivalent to[64]

$$\lim_{n \to \infty} \frac{1}{n^2} = 0.$$

In a second treatise *Book on the Measurement of Parabolic Bodies* (Maqāla fī Misāḥat al-mujassamāt al-mukāfiya) he attempts to find the volume of various solids generated by rotation of parabolic segments about various axes. He is not so successful, but in the course of his work, proves the important result: if $0 < \alpha < 1$, $\lim_{n \to \infty} \alpha^n = 0$.[65]

One of Thābit's contemporaries, al-Māhānī later gave a shorter proof of his quadrature of the parabola, and rather amusingly, Thābit's grandson Ibrāhīm ibn Sinān (908(?)–946) took the new proof as a direct challenge to the family's scientific honour:[66] He then sought a more economical proof which would eliminate his grandfather's use of *reductio ad absurdum*, and succeeded in giving an elegant one based on his proven proposition that affine transformations preserve proportionality of areas.[67–8] Later al-Qūhī (fl. 970–1000), who was familiar with Thābīt's treatise on paraboloids, wrote a three proposition proof (*Risāla fī istikhrāj misāḥat al-mujassam al-mukāfī*) for computing the volume generated by rotating $y = x^2$ about the y-axis (the cupola-paraboloid of Thābit). This is equivalent to computing the

$$\int_0^{a^2} y \, dy.$$

His solution is even shorter and more elegant than Archimedes'.[69–70] Finally Ibn al-Haytham (965–1039) completed Thābit's and al-Qūhī's abortive attempts to find the volume of the "spherical paraboloid", obtained by rotating $y = x^2$ around the line $y = a$. This is equivalent to computing[71–2]

$$\int_0^{\sqrt{a}} (a - x^2)^2 \, dx.$$

Further progress was to await the seventeenth century precalculus savants: Fermat, Wallis, and Cavalieri. It is interesting to speculate, as Youschkevitch does,[73] how much more rapidly calculus might have grown, if Thābit and Ibn al-Haytham had possessed the notion of derivative. Finally, we should note that this Middle Eastern work is more a chapter in the theory of integration than one in numerical integration, since Thābit and others sought to approximate the area directly, rather than to approximate the bounding curve. Such a refinement was made much later by Cavalieri (1639), who also deserves credit for discovery of Simpson's rule (the Newton–Cotes formula for $n = 2$).[74]

V. Comparisons and failures

So far we have a fair picture of the "state of the numerical art" in the Middle East, up to the fifteenth century. What was the situation in contemporary Europe? That era Boyer calls "Prelude to Modern Mathematics", which saw the flourishing of Galileo Galilei (1564–1642), Cavalieri (1598–1647), Briggs (1561–1639), Napier (1550–1617), Bürgi (1552–1632), Kepler (1571–1630), and Vièta (1540–1603) was a century away.[75] Even Pacioli's three treatises: *Summa de Arithmetica, Proportione et Proportionalita*, and *Geometria* did not appear until the end of the fifteenth century (1494).[76] Methods of plane trigonometry were just starting to replace the geometric methods of the Romans by mid-century,[77] and Peuerbach (1423–1461) and his pupil Johannes Müller (Regiomontanus, 1436–1476) were engaged, respectively in a Latin translation of the *Almagest*, and a five place table of sines and tangents, with decimal divisions of angles (*Tabulae Directionum*).[78] The first printed Euclid appeared six years after the death of Müller,[79] and the decade from 1484 to 1494 saw the writing of Chuquet's *Triparty*, the use of plus and minus signs (Widmann, 1489), and the use of the decimal point (Pellos, 1492).[80] In short, for fifteenth century Europe, it was a time of preparation for the long "prelude".

What effect did Middle Eastern numerical analysis have on all this? We have already mentioned the influence of Ibn Yūsuf on such thirteenth century figures as Fibonacci, de Nemore, and Bradwardine, and Rosenfeld and Youschkevitch's case for the possible transmission of part of Al-Kāshī's *Miftāḥ* to sixteenth century Europe. The story of the translations from Arabic of Gerard of Cremona, Robert of Chester, and others is well known. Its hard to say more than this. The work of the Banū Mūsā was translated into Latin, but the numerical integration papers of Thābit,[81–2] Al-Qūhī,[83–4] Ibrāhīm ibn Sinān,[85–6] and ibn al-Haytham[87–8] do not seem to have been. In fact, by the time Europe could have used such works, the translation impulse of the pre-Renaissance was dead, at least as far as Arabic was concerned, and scholars were either turning to Greek texts, or looking inward for new knowledge. The two numerical tools which helped foster much of the early European scientific Renaissance are of mixed origin. The first, tables of trigonometric functions, owes much to Middle Eastern numerical analysts. The second, logarithms, seems to have no counterpart in the Middle East or the Orient.[89–92] Finally we note that the method of prosthaphaeresis, which played some role in European computational schemes before the advent of logarithms, was actually unknown to the Middle East, despite the claims of Braunmühl.[93] As David King points out, Braunmühl's source Delambre[94] only infers the formula $2\cos x \cos y = \cos(x + y) + \cos(x - y)$ from a work of Ibn Yunūs (d. 1009). In reality, the formula itself does not appear in the text Delambre cites, nor does it occur in any of the known corpus of Ibn Yunūs.[95]

Notes

1. Henrici (25), 3.
2. Neugebauer (1), 155ff.
3. Chandrasekhar and Swerdlow (33), 428–44.
4. Goldstine (7), 1–2, 64(n) (al-Kāshī); 23 (al-Bīrūnī); 50n. (al-Khwarizmi).
5. Henrici (25), 3–4.
6. Henrici (25), *passim*.
7. Goldstine (7), 1–50.
8. Goldstine (7), 51–260. Even such a meticulous table maker as Napier had no clear conception of rounding procedures. See Goldstine (7), 9; 16n.
9. Goldstine (7), 261–315.
10. Kline (27), 5–7.
11. Kline (27), 6, 183ff.
12. Saidan (34), 532.
13. Kline (27), 56, 68–73.
14. Dold-Samplonius (29), 21; see also Sabra (35), 6.
15. Dold-Samplonius (29), 22; see also Becker, O., "Eudoxos Studien I", *Quellen und Studien zur Geschichte der Mathematik, Astronomie und Physik*, Abt. B 2 (1933), 311–33.
16. Youschkevitch and Rosenfeld (28), 326–7.
17. Nasr (36), 510.
18. Dold-Samplonius and Hermelink (37), 82.
19. Schrader (38), 82–83; see also Murdoch, J. E., "The Medieval language of proportions", in A. C. Crombie, ed., *Scientific Change* (Oxford, 1961), 237–72.

19a. Saidan (22), 475–90; see also Saidan (21).

19b. Rashed (23) argues in favour of the school of al-Karajī.

20. Saidan (34), 532–3; see also Saidan (31), 614.
21. Rosenfeld and Youschkevitch (30), 257.
22. Kline (27), 254.
23. Toomer (39), 358–9.
24. Saidan (40), 544.
25. Saidan (40), 544–6.
26. Saidan (41), 614. This was used by his student al-Nasawī as a model for his *al-Muqniᶜ fīʾl-Ḥisāb al-Hind*. See also Pingree (42), 556.
27. Saidan (43), 539–40.
28. Kennedy (13), 228–9.
29. Kennedy, *loc. cit.*. See also Neugebauer (6), Index.
30. Goldstine (7), 258–60.
31. Boyer (44), 16.
32. Bose, *et al.* (4), 164, 190.
33. Boyer (44), 218.
34. Suter (9), 111–12.
35. Harvey (45), 246.

35a. Rashed (23), 218.

36. Clagett (46), 349–50, 367.
37. Saidan (21), 437–67.
38. Kūshyār (26), 22.
39. Saidan (21), 447.
40. Saidan (21), 454; the latter is also used by Fibonacci. See Kūshyār (26), 32.
41. Saidan (21), 460; $a + (r/3a^2)$ is used by Cardano. See Kūshyār (26), 32.
42. Saidan (21), 445, 454. The Newton–Raphson method applied to the polynomial

$F(x) = x^n - a^n - r$, starting with $x_0 = a$ yields

$$x_1 = a + \frac{r}{na^{n-1}},$$

just as does the bionomial theorem for

$$(a^n + r)^{1/n} = a\left(1 + \frac{r}{a^n}\right)^{1/n}.$$

43. Rashed (23), 196ff.
44. Rashed (23), 208; see also Anbouba (47), 91–5.
45. Anbouba (48), 515–17.
46. Rashed (23), 216ff.
47. Kennedy (19), 355–6.
48. Henrici (25), 65. On the finite interval $[a, b]$, $f(x)$ is continuous, $f([a, b]) \subset [a, b]$, and $|f(x_1) - f(x_2)| < k|x_1 - x_2|$ for all x_1, x_2 distinct points of $[a, b]$, $k < 1$.
49. Kennedy (17), 248.
50. Kennedy and Transue (16), 80–3.
51. Kennedy (18), 51–2.
52. Kennedy (17), 250.
53. Rosenfeld and Youschkevitch (30), 259.
54. Dilgan (49), 228.
55. Rosenfeld and Youschkevitch (50), *passim.*
56. Kari-Niazov (51), 536; see also Rosenfeld and Youschkevitch (30), 259.
57. Schoy (52), 95–6.
58. Neugebauer (1), 97ff.
59. Kennedy (15), 117–20.
60. Al-Dabbagh (53), 444.
61. Al-Dabbagh (53), 444–5.

61a. Heath (54), volume 3, 423–5. Proposition 16: Given two circles about the same centre, to inscribe in the greater circle an equilateral polygon with an even number of sides which does not touch the lesser circle. Proposition 17: Given two spheres about the same centre, to inscribe in the greater sphere a polyhederal solid which does not touch the lesser sphere at its surface.

62. Al-Dabbagh (53), 444–5.
63. Toeplitz (55), chapter on "Integration".
64. Rosenfeld and Grigorian (56), 289.
65. Rosenfeld and Grigorian (56), 290; see also Youschkevitch (20), 44–45.
66. Rashed (57), 3.
67. Rashed (57), *loc. cit.*. See also "Measurement of the parabola" ("Fī misāḥat qat ʿal-Maḥrūt al-mukāfī") in Rāsaʾ ilu Ibn-i-Sinān (Hyderabad, 1948).
68. Suter (8), *passim.*
69. Dold-Samplonius (58), 240.
70. Suter (10), *passim.*
71. Dold-Samplonius (58), 240.
72. Suter (59), *passim.*
73. Youschkevitch (20), 45.
74. Goldstine (7), 77ff.
75. Boyer (44), 333.
76. Kline (27), 236.
77. Kline (27), 237.

78. Kline (27), 238.
79. Boyer (44), 688.
80. Boyer (44), 305–7, 688.
81. Rosenfeld and Grigorian (56), *passim.*
82. Sezgin (60), 264–72.
83. Suter (10), *passim.*
84. Dold-Samplonius (58), 239–241.
85. Rashed (57), 2–3.
86. Suter (8), 214–28.
87. Suter (59), *passim.*
88. Sabra (61), 206.
89. Needham (2), 52–3.
90. Libbrecht (3), *passim.*
91. Bose, *et al.* (4), 103.
92. Mikami (62), *passim.*
93. Braunmühl (63), 105–6.
94. Delambre (65), 76–156; especially 128.
95. King, oral communication. See our essay on "The Myth of Logarithms".

7 The Geometric Theory of Equations in the Near East in the Middle Ages

I. Introduction

This study is devoted to the development of the geometric theory of equations in one variable as it appears in the works of Near Eastern mathematicians in the Middle Ages. Roughly speaking, it focuses attention on those geometric methods which lead to ᶜUmar al-Khayyamī's representation of positive roots of cubic equations in one variable as intersection points of conics. At the same time, it offers a parallel, though less detailed treatment of the rival method of numerical approximation so artfully articulated in al-Kāshī's *Miftāh al-Ḥisab*. This latter method is the older, and historically, at least, finds its origins in the Babylonian mathematical tradition, while the former owes its impetus to Greek geometry, particularly to the works of Euclid, Archimedes, and Apollonius.

After sketching these historical traditions for perspective, we give an account of the known works of Near Eastern mathematicians, examining each for influence and originality. In particular, we are interested in the validity of Heath's contention that Archimedes had essentially solved the problem of cubics completely in his treatise on the sphere. Moreover, since nomographic methods were inherently available to the Near Eastern mathematician, for example, in the plates of the astrolabe, we examine their writings for any such connection between numerical methods and geometric ones via nomography. We also seek to delimit the achievements of Middle Eastern geometric theory within the framework of the theory of geometric construction, as expounded in various contemporary references, e.g. Bieberbach's *Theorie der Geometrischen Konstruktionen*. This is done, we hope, without introducing anachronisms into the works of the mathematicians we discuss. For example, although we adhere to modern algebraic notation, we continually stress the fact that our sources are, in many cases, rhetorical, and that for their authors, quadratic equations are primarily problems in relating areas, while cubic ones are problems in volume. This is highlighted, for example, by Umar's expressions for cubic equations as homogeneous, when viewed as polynomials in both coefficient *and* variable. Thus, in modern notation, he writes: $x^3 + a^2x + b^3 = 0$, not $x^3 + cx + d = 0$. It also accounts, to some extent, for the slowness with which the geometric theorists approached polynomial equations of higher degree. Some, in fact, dismiss them, as did ʿUmar, as being physically irrelevant. We also re-examine the evidence for the work of Medieval Near Eastern mathematicians on equations of degree four.

We, of course, leave unanswered some tantalizing questions. How detailed was ʿUmar's knowledge of the *Conics* of Apollonius? What did the versions of this work available to him contain? Such questions must be carefully answered if one is to gain an accurate picture of ʿUmar's work. Using the Apollonius of Heiberg or Heath just won't do. This is to commit a grave error, much as the use of Heiberg's Euclid to decide finer points on the contents of the "Arab–Euclid". For example, *Elements* I, 45, as is well known, is missing from the Arabic tradition. Hopefully, the announced edition of the Arabic Apollonius by Toomer will help to settle this and other questions, while the proposed edition of the Arabic Euclid by students of Sabra may give us a complete overview of the Arabian *Elements*. Unfortunately, there is currently no available substitute for the work of Toomer, while, for Euclid, we do have, at least, the edition of Codex Leidensis 399, I of Besthorn, Heiberg, and Junge upon which to make some judgements.

Similar questions arise in the later section of this work, in which we consider the works of al-Haytham and Sharaf al-Dīn al-Ṭūsī. For the former, we've had to rely primarily on Risner's edition of the *Optics* to draw conclusions on al-Haytham's work on fourth degree equations, and for the latter, upon rather sketchy accounts by Rashed and Anbouba. A more thorough treatment of these two Near Eastern mathematicians may be possible soon, however. For al-Haytham, Sabra's Arabic edition is promised; but Rashed's edition of the unique manuscript of al-Ṭūsī's *Algebra*, promised in 1974 is still unavailable.

Our study proceeds, for the most part, chronologically. We first present an account of Babylonian work on the theory of equations, and look for evidence of any such work in Egyptian mathematics, with an eye for "borrowings" by Egypt from Babylonia, as established in the general mathematical context by Parker. In the following section, we outline the Greek contribution, in particular, that of Euclid, Archimedes, Diophantus, Apollonius, and Eutocius. At the beginning of the next section we sketch the relation between al-Khwarizmī's *Algebra* and other similar contemporary compendia, and the work of Babylonian algebraists as argued by Gandz. This brings us to the heart of our treatise: the writings of Thābit, ibn al-Haytham, and others on the geometric theory. Incidentally, it is worth pointing out, as a preliminary, that although later writers on the subject of geometric solutions of equations, e.g. ʿUmar and Sharaf al-Dīn al-Ṭūsī, could be reasonably expected to adopt a Euclidean treatment and to incorporate some account of Thābit's argument justifying the geometric method, at least in their theory of quadratics, we find repeated, wholesale, in their works, al-Khwarizmī's definitely un-Euclidean work. At appropriate places, we also mention Chinese and Indian work on the theory of equations. Final sections set the geometric theory within the context of the general theory of geometric constructions, examine it, as viewed by primarily numerically oriented writers such as Sharaf al-Dīn, and compare it with the numerical theory. The very last section examines known evidence for transmission of either theory to the Latin West.

II. Babylonian and Egyptian Theories of Equations

In 1938 Gandz[1] examined the published editions of Babylonian cuneiform texts by Neugebauer[2] and Thureau-Dangin,[3] and classified the work of Babylonian mathematicians on quadratic equations under eight types. We reproduce this as Table 1, preserving Gandz's designations, BI–BVIII.

As Gandz observed, the x and y are "length" and "width" respectively, with the usual convention being that x exceeds y. Thus the larger of the two bracketed quantities is x, the smaller y. The texts from which this classification scheme is extracted are rhetorical, and for the most part are example problems chosen either for drill in computation, or to indicate how variants of the basic types can be transformed into the standard ones. Moreover the texts cover a time span of roughly 2500 B.C. to 100 A.D.. Even a cursory examination of types BI through BVI would lead one to conjecture that they arise from an attempt to correlate perimeters and areas of rectangles (I–II), perimeters and diagonals (III–VI). In fact, Gandz, in a later paper does indeed make a strong case for the origin of quadratic equations among the Babylonians as an approach to the isoperimetric problem for rectangles, primarily rectangular sections of land, and perhaps of cloth and hides.[4] The modern notation also points up the fact, which is totally alien to Babylonian mathematics, that all solutions are intersections of lines and conics; I–II, V–VI being lines and hyperbolas, III–IV, lines and circles. More

Table 1. Babylonian types of quadratic equations (© Uitgeverij "De Tempel". Reproduced with permission from S. Grandz, "The origin and development of the Quadratic Equation in Babylonian, Greek and early Arabic Algebra", *Osiris* **3** (1938), 405, table A.)

BI	$x+y=a;\ xy=b$	$\left.\begin{matrix}x\\y\end{matrix}\right\}=\frac{a}{2}\pm\sqrt{\left(\frac{a}{2}\right)^2-b}$
BII	$x-y=a;\ xy=b$	$\left.\begin{matrix}x\\y\end{matrix}\right\}=\sqrt{\left(\frac{a}{2}\right)^2+b}\pm\frac{a}{2}$
BIII	$x+y=a,\ x^2+y^2=b$	$\left.\begin{matrix}x\\y\end{matrix}\right\}=\frac{a}{2}\pm\sqrt{\frac{b}{2}-\left(\frac{a}{2}\right)^2}$
BIV	$x-y=a;\ x^2+y^2=b$	$\left.\begin{matrix}x\\y\end{matrix}\right\}=\sqrt{\frac{b}{2}-\left(\frac{a}{2}\right)^2}\pm\frac{a}{2}$
BV	$x+y=a;\ x^2-y^2=b$	$\left.\begin{matrix}x\\y\end{matrix}\right\}=\frac{a}{2}\pm\frac{b}{2a}=\frac{a^2\pm b}{2a}$
BVI	$x-y=a;\ x^2-y^2=b$	$\left.\begin{matrix}x\\y\end{matrix}\right\}=\frac{b}{2a}\pm\frac{a}{2}=\frac{b\pm a^2}{2a}$
BVII	$x^2+ax=b\quad(=\text{AI})$	
BVIII	$x^2-ax=b\quad(=\text{AIII})$	
[BIX	$x^2+b=ax\quad(=\text{AII})]$	

relevant is the fact that since I–VI are expressed as systems of equations in two variables, length and width, the question of how many "roots" a quadratic in one variable must have does not arise. Indeed, the method of solution, for type BI, for example, consists not in transforming BI to a quadratic in one variable, but in setting $(x-y)/2=z$, so that since $x+y=a$, $x=(a/2)+z$, $y=(a/2)-z$. Hence $xy=(a/2)^2-z^2=b$ or $z=\sqrt{(a/2)^2-b}$. Whence

$$x=\frac{a}{2}+\sqrt{\left(\frac{a}{2}\right)^2-b} \quad \text{and} \quad y=\frac{a}{2}-\sqrt{\left(\frac{a}{2}\right)^2-b}.$$

Gandz even goes so far as to postulate a primitive tabular origin of the method, which developed with exceeding elegance and ingenuity over the long span of Babylonian activity in mathematics.[5] Similar reconstructions can be offered for each of types BII–BVI.[6]

As is easily seen, the types BI–BVI are not independent in a loose sense; for example, type BII readily transforms into BIV, etc., and BI changes into BIX, etc. Gandz also observed cases in which the Babylonians do solve quadratics in one variable, which in his table are dually labeled as BVII (AI), BVIII (AIII). The A is used to indicate that this type is common to the algebra of Al-Khwarizmī. He found no evidence for AII, however; and a recent survey of Babylonian algebra of Goetsch,[7] which incorporates additional material from the Susa texts[8] and the *Mathematical Cuneiform Texts* of Neugebauer and Sachs, does nothing to change this state of affairs. Indeed, Gandz presents strong evidence that the quadratic equations in Euclid (*Elements* II, 5–6, 9–10) are Babylonian types in geometric disguise, while Heron's works contain BI and BVII, and that Diophantus' *Arithmetic* contains all of BI–BVIII, as well as BIX (= AII). We consider this in its proper section.

Given the apparent isoperimetric origin of the quadratic among the Babylonians, one might also expect to find problem texts concerned with similar problems for volume, e.g. $x+y+z=a$, $xyz=b$, $xy=c$ say. Indeed there are some volume problems, e.g. (BM 85200, VAT 6599, No. 27; MKT, p. 218): $v=xyz=b$, $x=b$, $z=c(x-y)$.[9] But none illustrates a simple isoperimetric interest.

The few instances of cubic and higher degree equations cited by Neugebauer[10] in cuneiform texts were, for the most part, challanged successfully by Gandz, who showed them to be solvable by application of methods common to the standard Babylonian types. Bruins,[11] however, claims evidence for cubic equations in at least three texts published by Thureau-Dangin.[12] These are of the form $(ax)^3+(ax)^2=b$ or $ax^3+bx^2=c$, the latter being reduced to $(ax/b)^3+(ax/b)^2=ca^2/b^3$ by the simple device of multiplying through by a^2/b^3, giving both the same canonical form $x^3+x^2=d$, for which, as Bruins points out, tables for solution were available, e.g. VAT 8492.[13] Goetsch[14] provides a few further examples of cubics, as well as biquadratics, solvable by quadratic methods. Both he and Bruins also discuss the Babylonian solution of exponential equations, e.g. $a=b^x$, with Bruins giving the more convincing explanation of the solution as an application of linear interpolation.[15] This type of example is our clearest evidence that the Babylonians used approximation techniques in solving equations.

One looks in vain for evidence of such mathematical sophistication in the theory of equations in existing Egyptian documents. The Rhind Mathematical Papyrus, for example, contains solutions of equations of degree one (nos. 24–7) by the methods of false position, while the Moscow Mathematical Papyrus, studied by Struve, contains such equations as $x+2x=9$.[16] Gillings[17] offers evidence for three quadratic equations:

$$\begin{cases} x^2+y^2=100 \\ 4x-3y=0 \end{cases} \qquad \begin{cases} x^2+y^2=400 \\ 4x-3y=0 \end{cases}$$

$$\begin{cases} xy=a \\ x=by \end{cases}$$

All three, however, come from papyri that have been restored, and for which much of the reconstruction is conjectural. Interestingly enough, all three reduce to pure quadratics, $x^2=d$. The first two arise from an interpretation by Schack-Schackenburg[18] of Berlin Papyrus 6619, while the third arises from Gillings' own reinterpretation of Griffith's translation of problem LV, 4 of the Kahun papyrus.[19] Both reconstructions essentially reduce the problems to $x^2=d$ and take the square root.

Given the demonstrated influence of Babylonian mathematics on the Egyptian, for example, in Parker's study of certain third century demotic texts,[20] one might expect to find a higher degree of development in the theory of equations; however, there seems to be no evidence to support such expectations.

III. The Greek theory of equations

As we mentioned in the previous section, Gandz has shown that the Babylonian types of quadratics reappear in Diophantus, in particular, as problem 27–30 of Book I of the *Arithmetic*.[21] These are, following Heath,[22]

> "27. To find two numbers such that their sum and product are given numbers.
> 28. To find two numbers such that their sum and the sum of their squares are given numbers.
> 29. To find two numbers such that their sum and the difference of their squares are given numbers.
> 30. To find two numbers such that their difference and product are given numbers."

These are, of course, just BI, III, V, and II respectively. Moreover, as Gandz also notes, the mode of solution is also Babylonian, e.g. after giving a necessary condition for solution in 27 ("The square of half the sum must exceed the product by a square number"),[23] Diophantus proceeds to give as an example of his method:

> "Given sum 20, given product 96. $2x$ the difference of the required numbers. Therefore the numbers are $10+x$, $10-x$. Hence $100-x^2=96$. Therefore $x=2$, and the required numbers are 12, 8."[24]

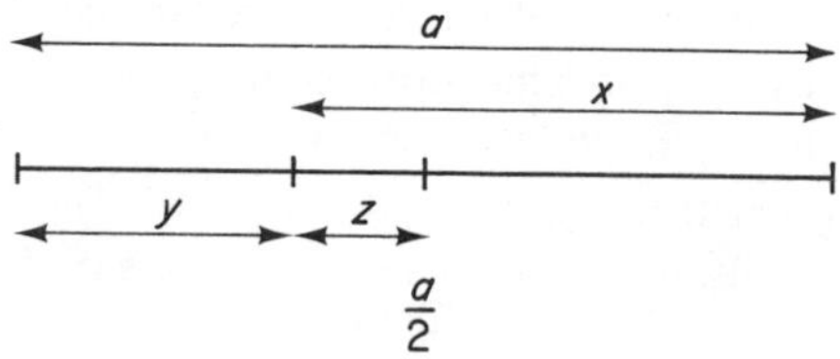

Figure 1

This same study by Gandz goes on to establish that the geometric theory of quadratics in *Elements* II, 5–6, 9–10 is just the Babylonian method for solving BI, II, III, and IV in geometric language, in exactly that order. Proposition II, 9 (BIII) will suffice as an example:

> "If a straight line be cut into equal and unequal segments, the squares on the unequal segments of the whole are double of the square on the half and of the square on the straight line between the points of section."[25]

Algebraically, this says that if $x + y = a$ and $x^2 + y^2 = b$, then $x^2 + y^2 = 2[(a/2)^2 + z^2]$, i.e., $z = \sqrt{(b/2) - (a/2)^2}$ and

$$\left.\begin{matrix} x \\ y \end{matrix}\right\} = \frac{a}{2} \pm \sqrt{\frac{b}{2} - \left(\frac{a}{2}\right)^2}$$

(see Figure 1).

Gandz gives further evidence for such Babylonian problems and methods in other Greek authors, e.g. Heron (the commentary on Euclid as quoted in Curtze's edition of al-Nayrīzī's commentary on the *Elements*, and in the Schone edition of his *Metrica*).[26] Moreover, there is, in Heron's *Opera* (IV, 380–381), a quadratic equation of the form $ax^2 + bx = c$ (BVII), solved in a distinctly Babylonian–Diophantine fashion.[27]

So far, however, all of this is quite far, mathematically, from techniques for representing solutions of polynomial equations in one variable as intersections of conics. To this end, we turn to the next development in the Greek geometric theory of equations.[28] Before the time of Archimedes (c. 287 B.C.–212 B.C.),[29] a number of special problems we now associate with cubic equations, e.g., the duplication of the cube and the trisection of an angle, were solved by means of special transcendental curves such as the quadratrix, invented by Hippias of Elis (*c.* 420 B.C.).[30] With the discovery of conics by Menaechmus, a pupil of Eudoxus, around 350 B.C., the picture changes. As Heath argues,[31] it is quite plausible that Greek geometers of Menaechmus' time, concerned as they were with such problems, may have discovered the means of reducing by means of the *νεῦσις* figure for trisection of an angle, via conic sections (in this case, a hyperbola and a circle), the cubic equation for this problem.[32] In any case, existing evidence strongly suggests that these two problems, duplication and trisection, are the only ones studied by the Greeks which led to cubic equations up to the time of the invention of conics. As Heath points out,[33] it was Menaechmus himself who showed that conics could be used to solve the problem of two mean proportionals, i.e. of constructing segments x, y, such that for two given segments a and b, $a:x = x:y = y:b$ consequently a device for solving the pure cubic $x^3 = a^2 b$ by means of the parabola $ay = x^2$ and the hyperbola $ab = xy$.[34] Moreover, he argues that it seems unlikely that the trisection problem would have been reduced by conics before 320 B.C. (the time of Aristaeus' *Solid Loci*), owing to the fact that such a reduction would have required the use of conics $xy = a^2$, $bx = 3a^2 + 3by - y^2$, a construction "decidedly more difficult than that used by Archimedes in connection with his cubic"[35] less than 100 years later. Heath goes on to present further evidence that the use of conics to reduce and solve cubic problems in Archimedes' time must still have been novel. The particular application of this technique by Archimedes appears in Book II of his treatise *On the Sphere and the Cylinder*, as proposition 4: "To cut a given sphere by a plane so that the volumes of the segments are to one another in a given ratio."[36] Archimedes, in the text, reduces this problem, by means that we sketch below, to a cubic equation, [37] and promises a solution of this subsidiary problem at the end of his treatise. By the time of Eutocius of Ascalon (b. Palestine, *c.* 480 A.D.),[38] however, this promised solution had, seemingly been lost, and Eutocius tells us, in his commentary on this Archimedian treatise, how he finally located a much corrupted manuscript, in the old Doric dialect used by Archimedes, which allowed him to reconstruct the solution.[39] This commentary of Eutocius has further value, not only because it preserves two additional solutions to the problem by Dionysodorus and Diocles respectively, but also because it enters directly into the Middle Eastern tradition through translation into Arabic.

In our paraphrase (which follows the translation of Thomas),[40] Archimedes proceeds as follows. Suppose the problem is solved. Let ΑΒΓΔ be the given sphere with centre K (Figure 2), cut by the plane ΑΓ perpendicular to the diameter ΔΒ. Let Λ and Ρ be on ΔΒ produced such that (ΚΔ + ΔΧ): ΔΧ = ΡΧ:ΧΒ and (ΚΒ + ΒΧ): ΒΧ = ΛΧ:ΧΔ. Join ΑΛ, ΛΓ, ΑΡ, and ΡΓ; (by proposition 2 of the treatise),[41] the cone ΑΛΓ equals the segment ΑΔΓ and the cone ΑΡΓ equals the segment ΑΒΓ of the sphere. Since the two cones have the same base, the ratio of

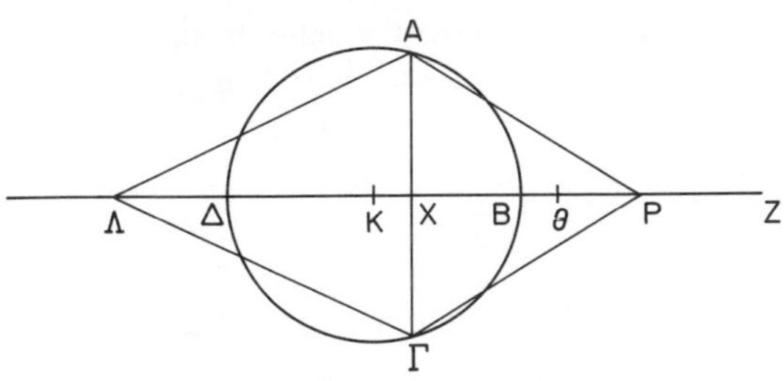

Figure 2. Reproduced with permission from I. Thomas, ed., *Greek Mathematics*, Harvard University Press: Heinemann, London, 1968, vol. II, figure, p. 129

their altitudes $\Lambda X:XP$ is equal to the ratio of the volumes of the segments. Using the standard operations with proportions from the *Elements* (Books V–VI), and the above data, the goal is to link the point X with a known point P′, i.e. to express the segment XP′ as a length.

First $\Lambda\Delta:K\Delta = KB:BP = \Delta X:XB$;[42] moreover, since $PB/BK = K\Delta/\Lambda\Delta$, $(PB + BK)/BK = (K\Delta + \Lambda\Delta)/\Lambda\Delta$, i.e. $PK/BK = K\Lambda/\Lambda\Delta$, i.e. $PK/K\Delta = K\Lambda/\Lambda\Delta$ since $BK = K\Delta$. Thus $P\Lambda/K\Lambda = K\Lambda/\Lambda\Delta$.[43] Thus $(P\Lambda)(\Lambda\Delta) = (\Lambda K)^2$, and $P\Lambda/\Lambda\Delta = (\Lambda K)^2/(\Lambda\Delta)^2$. Also, since $\Lambda\Delta/\Delta K = \Delta X/XB$ (above), $\Delta K/\Lambda\Delta = XB/\Delta X$, or $(\Delta K + \Delta\Lambda)/\Delta\Lambda = (XB + \Delta X)/\Delta X$, i.e. $K\Lambda/\Lambda\Delta = B\Delta/\Delta X$. Now extend KB to Z making $BZ = KB$. Then Z lies beyond P since $X\Delta/XB = KB/BP$ (above) and $\Delta X > XB$ implies that $KB > BP$. Next, since the ratio of $\Delta\Lambda/\Lambda X$ is given, so is that of $P\Lambda$ to ΛX,[44] and consequently that of $P\Lambda$ to $\Lambda\Delta$. Now recall that $\Delta X/XB = \Lambda\Delta/\Delta K = KB/BP$, and note that since $P\Lambda/\Lambda X = (P\Lambda/\Lambda\Delta)(\Delta\Lambda/\Lambda X)$, while $P\Lambda/\Lambda\Delta = (K\Lambda)^2/(\Lambda\Delta)^2 = (B\Delta)^2/(\Delta X)^2$ (above), and $\Delta\Lambda/\Lambda X = BZ/ZX$,[45] $P\Lambda/\Lambda X$ is a composite of $(B\Delta)^2/(\Delta X)^2$, BZ/ZX. Now take θ so that $P\Lambda/\Lambda X = BZ/Z\theta$. Note, however, that the ratio $P\Lambda/\Lambda X$ is given; hence so is $BZ/Z\theta$. But BZ, the radius is given; thus $Z\theta$ is too. "Therefore the ratio $BZ/Z\theta$ is composed of the ratios $(B\Delta)^2/(\Delta X)^2$ and BZ/ZX. But the ratio $BZ/Z\theta$ is composed of the ratios BZ/ZX and $ZX/Z\theta$. Therefore $(B\Delta)^2/(\Delta X)^2 = ZX/Z\theta$ in which $(B\Delta)^2$ and $Z\theta$ are given. And the straight line $Z\Delta$ is given; therefore it is required to cut the given straight line ΔZ at X so that XZ bears to the straight line $Z\theta$ as the area $(B\Delta)^2$ bears to the square on (ΔX)."[46]

As Thomas notes at this point, the corresponding algebraic condition, for $x = \Delta X$, $a = \Delta Z$, b the given straight line, and $B\Delta = c$ is[47]

$$\frac{c^2}{x^2} = \frac{a - x}{b} \quad \text{or} \quad x^2(a - x) = bc^2.$$

Archimedes concludes by remarking that this latter condition must, in the general case, be investigated for possibility (*διορισμός*), but that in the present situation there is no need since (presumably) the problem arises in a "natural way".

Essentially this elegant example of geometric analysis (which we have distorted by trying to make the presentation as self-contained as possible) uses proposition 2 of Archimedes' *On the Sphere and the Cylinder*, Book II, to express the fact that the two spherical segments $A\Delta\Gamma$ and $AB\Gamma$ with heights $\Delta X = h$ and $BX = h'$ have volume ratio

$$\frac{p}{q} = \frac{h(3r - h)}{(2r - h)} \Big/ \frac{h'(3r - h')}{(2r - h')}.$$

Then argument eliminates $h' = 2r - h$, to reduce the problem to a cubic

$$h^3 - 3rh^2 + \frac{4pr^3}{p + q} = 0,$$

i.e.

$$h^2(3r - h) = \frac{4pr^3}{p + q},$$

or more generally[48]

$$x^2(a-x)=bc^2,$$

and discusses the διορισμός (conditions for a solution). Before outlining Archimedes' solution, as reconstructed by Eutocius, we should also note that such geometric analysis did not spring full-bloom with Archimedes. Euclid's *Elements*, particularly Book II, 5, 6 and Book VI, 28, 29 contain earlier examples, though only involving constructions for equations of degree two. Moreover, as Mahoney[49] has pointed out, such analysis is very structured, in the sense that numerous works of geometry were designed to guide the mathematician in finding a way of "connecting" a required point to some ratios of segments or areas in a construction that determines the point. Indeed, to prove his point, Mahoney examines Book VII of Pappus' *Mathematical Collection*, which contains parts of Euclid's *Data, Porisms, Surface Loci,* Aristaeus' *Solid Loci,* Apollonius' *Conics, Section of Area, Section of Ratio, Contacts, Determinate Sections, Inclinations* and *Plane Loci,* and the *Means* of Eratosthenes,[50] and shows how the parts included act as outlines to show one how to proceed from the given knowns to the unknown by some minimal sequence. The subject of geometrical analysis itself is worth considering further, but we must simply refer the reader to Mahoney and Gulley.[51] Finally, we should observe that διορισμός also has a long tradition in Greek mathematics. A first instance in the *Elements*, for example, occurs in Book I, 22: "Out of three straight lines, which are equal to three given straight lines, to construct a triangle: thus it is *necessary* that two of the straight lines taken together in any manner should be greater than the remaining one."[52] Moreover, it appears again, implicitly, in Euclid's treatment of the quadratic equation $x^2+b^2=ax$ (*Elements* II, 5),[53] where the geometry requires that $b^2 \leqslant (a/2)^2$.

Now, let us return to Eutocius' account, again following Thomas. Recall that we are given segments AB, AΓ and an area Δ, and want to divide AB into subsegments at a point E, say, so that $\mathrm{AE}:\mathrm{A\Gamma}=\Delta:(\mathrm{EB})^2$. Suppose there is such a point. Construct AΓ perpendicular to AB, pass a line through Γ and E, and a line perpendicular to AΓ at Γ, and another perpendicular to AB at B. Call the intersection of these perpendiculars H, and the intersection of the latter with ΓE extended Z. Then ΓH, HZ from adjacent sides of a rectangle; complete it, calling the corner opposite H, θ (Figure 3). Also construct KΛ through E and parallel to AΓ. On ΓH extended, take M so that

$$(\mathrm{\Gamma H})(\mathrm{HM})=\Delta.$$[54]

Now, by hypothesis $\mathrm{EA}/\mathrm{A\Gamma}=\Delta/(\mathrm{EB})^2$, while $\mathrm{EA}/\mathrm{A\Gamma}=\mathrm{\Gamma H}/\mathrm{HZ}$ (similar rectangles). Moreover, $\mathrm{\Gamma H}/\mathrm{HZ}=(\mathrm{\Gamma H})^2/(\mathrm{\Gamma H})(\mathrm{HZ})$. Hence $(\mathrm{\Gamma H})^2/(\mathrm{\Gamma H})(\mathrm{HZ}) =\Delta/(\mathrm{EB})^2=\Delta/(\mathrm{KZ})^2$, since $\mathrm{KZ}=\mathrm{EB}$. Thus $(\mathrm{\Gamma H})^2/\Delta=(\mathrm{\Gamma H})(\mathrm{HZ})/(\mathrm{KZ})^2$, or since $\Delta=(\mathrm{\Gamma H})(\mathrm{HM})$, $(\mathrm{\Gamma H})^2/(\mathrm{\Gamma H})(\mathrm{HM})=(\mathrm{\Gamma H})(\mathrm{HZ})/(\mathrm{ZK})^2$. However, $(\mathrm{\Gamma H})^2/(\mathrm{\Gamma H})(\mathrm{HM})=(\mathrm{\Gamma H})/(\mathrm{HM})=(\mathrm{\Gamma H})(\mathrm{HZ})/(\mathrm{HM})(\mathrm{HZ})$, so $(\mathrm{MH})(\mathrm{HZ})=(\mathrm{ZK})^2$.

Now draw a parabola through H with parameter HM and axis HZ. [In modern language, the vertex of the parabola is at $\mathrm{H}=(a,0)$, taking ΓH as the x-axis, $\theta\Gamma$ as the y-axis and Γ the origin, with $\mathrm{\Gamma H}=a$, and its equation is $4py=(x-a)^2$ where $4p=\mathrm{HM}$. Also since $\mathrm{A\Gamma}=b$ say, and $\Delta=c^2=(\mathrm{\Gamma H})(\mathrm{HM})$, $(\mathrm{HM})=c^2/a$.

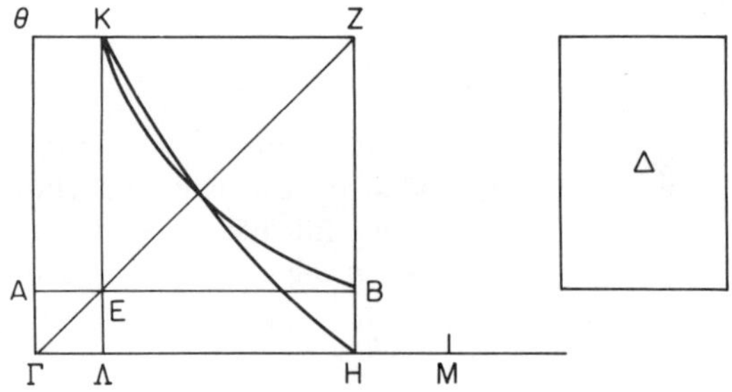

Figure 3. Reproduced with permission from I. Thomas, ed., *Greek Mathematics*, Harvard University Press: Heinemann, London, 1968, vol. II, figure, p. 139

Archimedes would probably have justified this, as Thomas does, with Apollonius' *Conics* I, 11 and *Data*, 57.[55]] Hence the parabola must pass through K; (for $(KZ)^2 = (MH)(HZ)$ is just $(x-a)^2 = (c^2/a)y$).

Note that the areas of rectangles $\theta\Lambda$ and ΓB are equal since triangles on opposite sides of the diagonal ΓZ are equal.[56] Consequently $(\theta K)(K\Lambda) = (AB)(BH)$. Thus one may pass a hyperbola through B with asymptotes $\Gamma\theta$ and ΓH. [By definition, this would be $xy = ab$].[57] Since K lies on both the parabola and hyperbola, it is given. Thus E is determined; for, KΛ is perpendicular to AB.

Also, since $(EA)/(A\Gamma) = \Delta/(EB)^2$, the rectangular solids with bases $(EB)^2$, Δ and heights EA, AF respectively are equal in volume. Recall, too, that both Δ and AΓ are given, and we have

$$(EA)(EB)^2 = (A\Gamma)(\Delta).$$

At this point Archimedes (Eutocius) claims that of all figures so taken on AB, $(BE)^2(EA)$[58] is of greatest volume when $BE = 2(EA)$, and consequently that it is necessary that $(A\Gamma)(\Delta)$ not exceed $4(EA)^3$, i.e. $(BE)^2(EA)$ when $BE = 2(EA)$.[59] As we remark in the notes (58), this is the limiting condition for the solutions of $x(x-a)^2 = bc^2$. The reader is invited to examine our summary for details. It will be noted that our particular choice of coordinate system leads to three positive solutions, rather than the two positive and one negative when $bc^2 = 4a^3/27$, as Thomas' choice of coordinate system gives.[60] We return to this point momentarily.

Let us now continue with Eutocius' account of Archimedes' synthesis. He takes AB, AΓ as given line segments, Δ as a given area, and proposes to cut AB at E, say, so that $(EA)/(A\Gamma) = \Delta/(EB)^2$. Suppose EA is chosen so that it is one third of AB. $\Delta(A\Gamma)$ may, of course, be less than, equal to, or greater than $(BE)^2(EA)$. If it were greater, by our analysis, the problem has no solution. (See note 59.) If it is equal, E satisfies the conditions, i.e. $(EA)/(A\Gamma) = \Delta/(BE)^2$. So suppose that it is less. Construct a segment equal to AΓ perpendicular to AB; pass a line through Γ parallel to AB, and one through B parallel to AΓ. Extend ΓE and denote its intersection with AΓ by H. Then complete the parallelogram, and construct KΛ

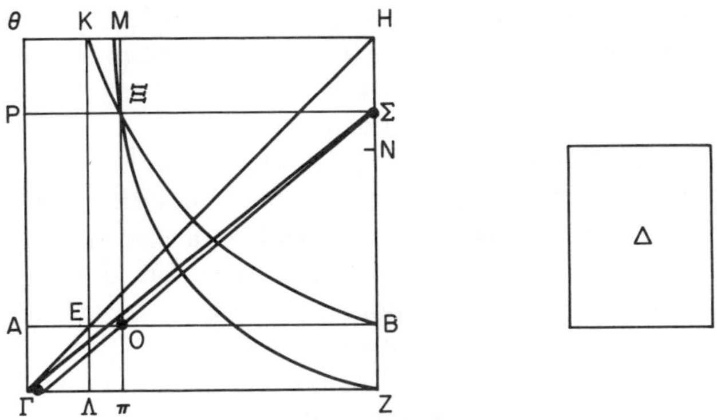

Figure 4. Reproduced with permission from I. Thomas, ed., *Greek Mathematics*, Harvard University Press: Heinemann, London, 1968, vol. II, figure, p. 146

parallel to ΛΓ and passing through E. (See Figure 4.) By hypothesis $\Delta(\mathrm{A\Gamma}) < (\mathrm{BE})^2(\mathrm{EA})$. Hence $(\mathrm{EA})/(\mathrm{A\Gamma}) = \Delta/(T)^2$, where $T < \overline{\mathrm{BE}} = \overline{\mathrm{HK}}$, call it HM. Then $(\mathrm{EA})/(\mathrm{A\Gamma}) = \Delta/(\mathrm{HM})^2$; set $\Delta = (\mathrm{\Gamma Z})(\mathrm{ZN})$, getting $(\mathrm{EA})/(\mathrm{A\Gamma}) = (\mathrm{\Gamma Z})(\mathrm{ZN})/(\mathrm{HM})^2 = (\mathrm{\Gamma Z})/(\mathrm{ZH})$, since parallelograms AΛ and θZ are similar. But $(\mathrm{\Gamma Z})/(\mathrm{ZH}) = (\mathrm{\Gamma Z})^2/(\mathrm{\Gamma Z})(\mathrm{ZH})$, so $(\mathrm{\Gamma Z})^2/(\mathrm{\Gamma Z})(\mathrm{ZH}) = (\mathrm{\Gamma Z})(\mathrm{ZN})/(\mathrm{HM})^2$, i.e., $(\mathrm{\Gamma Z})^2/(\mathrm{\Gamma Z})(\mathrm{ZN}) = (\mathrm{\Gamma Z})(\mathrm{ZH})/(\mathrm{HM})^2$ and $(\mathrm{\Gamma Z})/(\mathrm{ZN}) = (\mathrm{\Gamma Z})(\mathrm{ZH})/(\mathrm{NZ})(\mathrm{ZH})$. Finally $(\mathrm{\Gamma Z})(\mathrm{ZH})/(\mathrm{NZ})(\mathrm{ZH}) = (\mathrm{\Gamma Z})(\mathrm{ZH})/(\mathrm{HM})^2$. Hence $(\mathrm{HM})^2 = (\mathrm{NZ})(\mathrm{ZH})$.

As in the previous argument, pass a parabola through Z as vertex with $4p = \mathrm{ZN}$ and axis ZH, and call it MΞZ. Again $\theta\Lambda = \mathrm{AZ}$ and hence $(\theta\mathrm{K})(\mathrm{K\Lambda}) = (\mathrm{AB})(\mathrm{BZ})$. Now pass a hyperbola through B with asymptotes θΓ and ΓZ. It cuts our parabola at Ξ say, and (by definition) passes through K. Construct PΣ through Ξ and parallel to AB, and ΞOπ parallel to AΓ. Again (by definition) rectangle PO = OZ; whence ΓΣ passes through O.[61] Then $(\mathrm{OA})/(\mathrm{A\Gamma}) = (\mathrm{OB})/(\mathrm{B\Sigma})$ (similar parallelograms)[62] $= (\mathrm{\Gamma Z})/(\mathrm{\Sigma Z}) = (\mathrm{\Gamma Z})(\mathrm{ZN})/(\mathrm{\Sigma Z})(\mathrm{ZN})$. But $(\mathrm{\Gamma Z})(\mathrm{ZN}) = \Delta$, $(\mathrm{\Sigma Z})(\mathrm{ZN}) = (\mathrm{OB})^2$, since MΞZ is a parabola. Thus $(\mathrm{OA})/(\mathrm{A\Gamma}) = \Delta/(\mathrm{OB})^2$, i.e. O satisfies the required condition.

At this point Eutocius' commentary goes on to establish the maximum condition on $(\mathrm{BE})^2(\mathrm{EA})$. He then observes that when $(\mathrm{A\Gamma})(\Delta) < (\mathrm{BE})^2(\mathrm{EA})$ with $\mathrm{BE} = 2(\mathrm{EA})$ that there are two points satisfying the required conditions for E. Thomas[63] then makes an irrelevant comment to the effect that intersection equation for x will in any case have three real solutions, but the Greeks didn't recognize the *negative* one. As we show in note 58, this third root is negative only because Thomas has chosen his coordinate system as he did. Negativity has nothing to do with the omission of the third root. It is omitted because it lies outside AB. In fact, in our choice of coordinate system, all three roots are positive, and one is always bigger than $a = \mathrm{AB}$, and has no bearing on the problem Archimedes is trying to solve, viz., to cut a sphere into segments whose volumes are in a given ratio. Two things are to be learned from this: firstly, we must never lose sight of the fact that the Greeks, Archimedes included, did not set out to *solve* cubic equations. They merely used those intersections of conics which represent the solutions that are relevant to a particular geometric problem. Secondly, in converting the language of geometric analysis to coordinate geometry, we must be wary of assertions that depend on choice of origin.

At this point in the text Eutocius offers solutions to Archimedes' proposition 4 by Dionysodorus and Diocles. As he rightly observes, neither seems to have had any access to Archimedes' own solution, and both provide solutions to the proposition directly, rather than Archimedes' reduction. In fact, as Thomas[64] observes, the former uses a parabola and rectangular hyperbola for his intersections, albeit different from Archimedes' choice, and solves only the particular case, not the more general one, while the latter solves the original problem using a rectangular hyperbola and an ellipse.

Two final remarks on the Greek contribution to the geometric theory of equations before we close this section, and start the proper subject of this study. As Thomas remarks,[65] there are no other known Greek solutions of cubics, apart from Menaechmus' solution of the pure cubic $x^3 = a^2 b$, via conics;[66] of course this loses sight of the fact that Pappus' direct solution of the trisection problem is *equivalent* to solving a cubic.[67] Also, as Thomas does observe, Archimedes makes further references to problems that lead to cubics in his treatise *On Conoids and Spheroids.*[68] In particular, in the opening pages of this treatise, Archimedes poses some additional exercises that follow from the main text-methods, e.g. "Such also is the problem, from a given spheroidal figure or conoids to cut off a segment by a plane drawn parallel to a given plane so that the segment cut off is equal to a given cone or cylinder or to a given sphere."[69] Since a spheroidal figure and a conoid (right angled, obtuse-angled) are respectively an ellipsoid of revolution, paraboloid, respectively hyperboloid of revolution, one sees, via the calculus, for example that these problems reduce to

$$\pi \int_{-a}^{t \leqslant a} \left(1 - \frac{x^2}{a^2}\right) dx = k_1,$$

where the ellipse

$$\frac{x^2}{a^2} + \frac{y^2}{b^2} = 1$$

is rotated about the x-axis;

$$\pi \int_{0}^{t} (\sqrt{y})^2 dy = k_2,$$

where the parabola $y = x^2$ is rotated about the y-axis;

$$\pi \int_{a}^{t} b^2 \left(\frac{x^2}{a^2} - 1\right) dx = k_3,$$

where the hyperbola

$$\frac{x^2}{a^2} - \frac{y^2}{b^2} = 1$$

is rotated about the x-axis. Thus, one has the equations

$$t-\frac{t^3}{3a^2}+\frac{2a}{3}=\frac{k_1}{\pi}$$

$$t^2=\frac{2k_2}{\pi}$$

$$\frac{t^3}{3a^2}-t=\frac{k_3}{\pi b^2},$$

with both the first and third being cubics.[70]

Our second remark concerns the following statement of Heath:[71]

> "There can be no doubt that Archimedes solved this equation as well as the similar one with a negative sign, i.e., he solved the two equations
>
> $$x^3 \pm ax^2 \mp b^2c=0,$$
>
> obtaining all their positive real roots. In other words, he solved completely, so far as the real roots are concerned, a cubic equation in which the term in x is absent, although the determination of the positive and negative roots of one and the same equation meant for him two separate problems. And it is clear that all cubic equations can be easily reduced to the type which Archimedes solved."

It is, as Heath remarks, true that all cubics can be reduced to those with x-term missing.[72] But this has no bearing whatsoever on the general problems of Greek geometric analysis, simply because there was *no* attempt to solve cubics specifically. Moreover, whether the "cubic", translated from geometric to analytic language has or doesn't have an x-term is simply a matter of choice of origin, as we pointed out above. As for the contention that Archimedes solved such general equations "as far as the real roots are concerned", it simply doesn't hold water, even in the instance of proposition 4 of Book II (*On the Sphere and the Cylinder*). Finally, there is no evidence that Archimedes actually solved the "cubic" problems suggested in the introduction to *On Conoids and Spheroids*. In fact, if he had, we contend that he probably would have dismissed the case of sectioning the right-angled conoid since it is just a "plane" problem, as Heath admits.[73] This in no way detracts from Archimedes' real achievements, which are fundamental and lasting.

In our next section we shall begin to examine just how profound an impact these achievements had on Near Eastern mathematicians of the Middle Ages, in the guise of Arabic translations of Eutocius.

IV. Near Eastern geometric theory of Equations in the Middle Ages

The Arabic work in the geometric theory of equations, geometric in the sense that the quadratic equations which appear are solved by relating the terms to areas of rectangles, and geometrically "completing the square", is usually supposed to begin with Muḥammad ibn Musa al-Khwarizmī's *al-Kitāb al-mukhtaṣar fī ḥisāb al-djabr wa'l-mukābala*, written about 850. Current work suggests that other compendia similar to al-Khwarizmī's may have been written contemporaneously.[74] For example, Sayili has studied a similar, though apparently independent manuscript of ʿAbdalḥamid ibn Wāsi ʿ ibn Turk,[75] written, possibly, around the mid-ninth century. This suggests that the study of quadratic equations in the manner of al-Khwarizmī may have been a fairly well established tradition by his time, and that he may simply have put some final touches on a developed theory. In any case, it is al-Khwarizmī whose version of this theory was transmitted to the Latin West, first in the translation of Robert of Chester (1145) as *Liber algebrae et almucabola*, and later by Gerard of Cremona as *De jebra et almucabola*—both being restricted to the first part which deals with quadratics, and omitting the second which applies linear equations to practical problems of inheritance.[76] This latter part seems to have made its scholarly debut in the poor, indeed, error ridden edition of Rosen (1831), which did more harm than good to al-Khwarizmī's reputation. Fortunately, the corrective studies of Ruska,[77] and later of Gandz,[78] have helped bring about al-Khwarizmī's rehabilitation.

In fact Gandz's study[79] of the *Mishnat ha Middot*, a Hebrew geometry of around 150 A.D., gave scholars the first inkling of the complex origins of al-Khwarizmī's algebraic work. The geometry of the *Kitāb al-mukhtasar fi ḥisāb* turns out to be entirely based on the *Mishnat*, in fact, a literal translation, with only minor rearrangements, rather than upon any of the Arabic editions of the *Elements* that al-Khwarizmī's contemporaries were preparing. This accounts for the primitive look of al-Khwarizmī's geometry, and his definitely un-Euclidean treatment of quadratics. Gandz's second study[80] on the source of al-Khwarizmī's algebra itself was already mentioned in our earlier section on Babylonian algebra. The evidence Gandz's finds for Babylonian influence is indeed very strong. Coupled with the fact that al-Khwarizmī's geometry is also filtered through an Eastern tradition, this provides support for Gandz's contention that al-Khwarizmī was definitely not a pupil of Greek methods, and, in fact, that he may actually have been antagonistic to their introduction. It is not our intention to pursue this matter further, nor are we particularly interested in reviewing al-Khwarizmī's work on quadratic equations. This has been admirably done by Gandz himself.[81] Moreover, a renewal of the discussion would contribute nothing to our principal task—the representation of solutions of polynomial equations by intersections of conics. However, we would be remiss in not stating al-Khwarizmī's real achievement. It is simply this:[82]

> "...the reverse of the Babylonian attitude. [In his work] we find that the three Arabic types are used, regularly and exclusively.[83] The old Babylonian types and methods are entirely rejected and repudiated.... In al-Khwarizmī's algebra we, very frequently, find the same problems as in the Babylonian texts. But the Babylonian methods of solution, though near at hand and though very convenient, are systematically avoided. Herein lies the great merit of al-Khwarizmī, his great contribution to the progress of algebra. He does away with all those brilliant ideas, ingenious devices, and clever tricks adopted by the Babylonians for the solution of their individual problems.... He introduces and originates what we may call the 'classic period in algebra'. The methods are, so to say, 'standardized'. There are only three types and all the quadratic equations and all the

problems may be reduced to these standard types and solved according to their rules. "And we found", so he says at the outset of his book, ..., after the six types[84] have been described and explained by him, "that all the problems handled by algebra will necessarily be reduced to one of these six types just described and commented upon. So bear them in mind." "[85]

Following al-Khwarizmī is Abu Kamil Shujā' ibn Aslam ibn Muḥammad ibn Shujā' (*c.* 850–*c.* 930), considered one of the greatest algebraists of his time.[86] In his own algebra, he sometimes follows al-Khwarizmī, but owing to his interest in Diophantine problems, he often reverts to "classical" methods that are common to Diophantus and the Babylonians.[87] Since his interest is heavily in quadratic equations, and indeterminate problems in several variables that reduce to quadratics, we merely refer the reader to the copious literature dealing with him.[88] We shall, however, have occasion to return his treatise *Kitāb ... al-mukhammas wa' al-mu ͨ ashshar* when we consider equations of higher degree. The algebra of the ninth and tenth century has, of course, considerable interest in its own right. For this we refer the reader to a survey paper of Anbouba.[89]

Not long after al-Khwarizmī's time attempts were made to shore up the weak geometric underpinnings of his method with arguments from the *Elements.* Abū Kamil, for example, cites Euclid in his algebra, while Thābit ibn Qurra (836–901), much earlier, wrote a short Euclidean based justification of the methods of *Kitāb al-mukh̬tasar*, the *Qaul fī taṣḥīh masā'il al-jabr bi-l-barāhīn al-handasīya* (Concerning the correct solution of algebraic problems by geometric means).[90] For example, he uses *Elements* II, 6 to justify the algebraic treatment of $x^2 = px + q$.[91]

The thurst of Near Eastern algebra about this time also began to change quite rapidly with the translation of Eurocius of Ascalon's commentary on Archimedes' *Treatise on the Sphere and the Cylinder* by Isḥāq ibn Hunain (*Kitāb al-Kura wa-l-usṭuwana*), and improvements by Thabit.[92] At about the same time, the *Conics* of Apollonius (*Kitāb al-mah̬rūṭāt*) also became available in Arabic. Hilāl ibn Abi Hilāl al-Himsī,[93] who worked as a translator for Ahmad ibn Musa ibn Shakir, prepared the first four parts, while Thābit translated books V–VII. All parts were apparently revised by the Banū Mūsā.[94] Both texts were studied carefully, and various mathematicians began to apply these methods for "solving" various cubic and quartic equations as intersections of conics. The theory of cubics culminates in the exhaustive account given by ͨUmar al-Khayyamī (1038–48?/1123–4).[95] We treat it immediately below, since a chronological approach would probably serve only to detract from ͨUmar's actual achievement. Where known to us, we shall insert, either in the text, or notes, works of his predecessors. We intend to offer only an explanation of ͨUmar's constructions, with a statement of their limitations on his solutions from an algebraic point of view. Comments, in the context of the theory of geometric construction, are reserved for section five. This section closes with a discussion of ibn al-Haytham's use of conics to solve equations of degree four (Alhazen's problem). Of his geometric solution of a pure quintic by the device of inserting four proportionals, little of a non-conjectural nature can be said, since it survives only as a passing reference in ʿUmar's *Algebra* and Usaybiʿa's ʿ*Uyun al-abā fī ṭabaqāt al-aṭibbā*ʾ.[96]

ʿUmar Al-Khayyamī, known for his work in astronomy, philosophy, and mathematics,[97] begins his treatise on algebra[98] with a few brief remarks related to the history of the problem he sets himself: the study of cubic equations by the methods of al-jabr, and credits al-Māhānī (*c.* 860–880)[99] with the first step in this direction, an attempt to use al-jabr to solve the cubic ($x^3 + a^3 = cx^2$) suggested by proposition 4 of Eutocius' commentary on Archimedes' *Sphere and the Cylinder*.[100] Al-Māhānī's failure was followed, according to ʿUmar, by a successful solution using conics by al-Khāzin (d. 961/971).[101] This is a curious statement, for as Hartner[102] points out, the complete Arabic version of Eutocius' commentary, which, as we have seen, contains his reconstruction of Archimedes' own solution, had been available since the time of Iṣhāq ibn Hunain (d. 910).[103] Since ʿUmar later says that his own treatise will be understandable only to those who have mastered the *Elements*, the *Data* and Apollonius' *Conics*, the explanation may be that ʿUmar simply did not know Eutocius' commentary at first hand, or perhaps had seen a version containing only the propositions of Archimedes. In any case, he emphasizes the fact that his own treatise is based wholly on these three books.[104]

ʿUmar then presents this list of the admissible unknowns: line, plane, solid, and time; and remarks that square of a square represents areas in a metaphorical sense, but has no units in the sense that line (length), plane (area), solid (cube) do. This is followed by his enumerative classification of all equations of degrees from one to three. The actual classification scheme, however, is based not on degree, but on number of terms, with the added constraint that all coefficients be positive. It will emerge from ʿUmar's solutions that all equations are homogeneous when considered as polynomials in variable and coefficient. His list of equations is rhetorical algebra. We represent it in modern notation in Table 2.

He notes also that 4 and 5 are the same as 1, and 6 the same as 2 in the simple groups, while 4–6 are the same as 1–3 in the compound group. Then he remarks that 7–12 require conics for solution, while 13–19 require "geometric" solutions and that simple equation three may be solved either by calculation or by conics. For our purposes, only simple equation 3, and equations 7–19 are of interest. Suffice it to say that the quadratics (compound equations 1–3) are solved in the manner of al-Khwarizmī, with Euclidean justifications for the steps. The type $x^2 + ax = b$, for example, is first illustrated by al-Khwarizmī's equation $x^2 + 10x = 39$, and then the geometric justification is offered. The type $x^2 + b = ax$ is given, again with al-Khwarizmī's geometric method, shored up by a citation of *Elements* II, 5 and the example $a = 10$, b unspecified. (Compare this with al-Khwarizmī's $a = 10$, $b = 21$.) Finally, the type $ax + b = x^2$ has example $5x + 6 = x^2$ (Al-Khwarizmī has $x^2 = 3x + 4$), and ʿUmar's accompanying figure is merely a dissection of al-Khwarizmī's, with the implied justification, *Elements* II, 6 as in Thābit (*supra*). Two things are worth remarking upon: ʿUmar does not go back and recast al-Khwarizmī's treatment in the manner of the *Elements*; the methods of al-jabr, seemingly, were not to be sacrificed to pure geometric

Table 2. ʿUmar's 25 cases

	Simple equations	
1.	$c = x$	2 terms, 1–1[105]
2.	$c = x^2$	
3.	$c = x^3$	
4.	$cx = x^2$	
5.	$cx^2 = x^3$	
6.	$cx = x^3$	
	Compound equations[106]	
1.	$x^2 + ax = b$	
2.	$x^2 + b = ax$	
3.	$ax + b = x^2$	
4.	$x^3 + ax^2 = bx$	
5.	$x^3 + bx = ax^2$	
6.	$x^3 = bx + ax^2$	
7.	$x^3 + ax = b$	3 terms, 2–1
8.	$x^3 + b = ax$	
9.	$b + ax = x^3$	
10.	$x^3 + ax^2 = b$	
11.	$x^3 + b = ax^2$	
12.	$b + ax^2 = x^3$	
13.	$x^3 + ax^2 + bx = c$	4 terms, 3–1
14.	$x^3 + ax^2 + c = bx$	
15.	$x^3 + bx + c = ax$	
16.	$x^3 = bx + ax^2 + c$	
17.	$x^3 + ax^2 = bx + c$	4 terms, 2–2
18.	$x^3 + bx = ax^2 + c$	
19.	$x^3 + c = bx + cx^2$	

consideration. Secondly, ʿUmar does not attempt to uniformize his treatment of equations by solving quadratics as intersections of conics, as he later does for cubics.[107]

The actual solution of cubics begins with what we choose to call lemma 1.[108]

Lemma 1. Given two line segments AB and BG to construct segments x, y so that

$$\frac{\mathrm{AB}}{x} = \frac{x}{y} = \frac{y}{\mathrm{BG}}.$$

The construction of ʿUmar uses *Conics* I, 32, 56, and the *Data* (without specific citation). One draws AB perpendicular to BG. Then one constructs a parabola with vertex B, axis BG and "parameter" BG.[109] Call it BDH. AB is tangent to BDH at B.[110] Similarly, a second parabola is constructed with vertex at B, axis AB and parameter AB, say BDZ. It is tangent to BG at B (see Figure 5). Both constructions are justified by *Conics* I, 56 [= Heiberg's edition I, 52]. These parabolas necessarily intersect, say at D. ʿUmar offers no justification, evidently

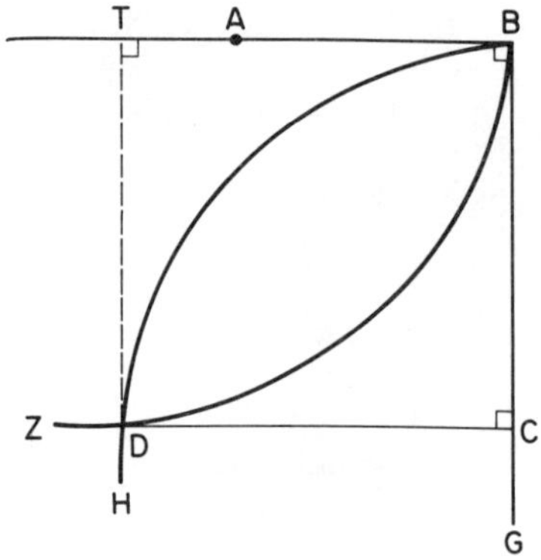

Figure 5

relying on the picture for a proof. The construction continues by dropping perpendiculars from D to AB (extended) and from D to BG (extended). ʿUmar cites the *Data*, but could just as well have quoted *Elements* I, 12. The perpendiculars determine DT and DC respectively. Then AB/BC = BC/BT = BT/BG, i.e. $x = $ BC, $y = $ BT. For the proof, $\overline{CD}^2 = (BC)(BG)$.[111] Thus BG/CD = CD/BC = BT/BC, since CD = BT. Similarly $\overline{DT}^2 = (AB)(BT)$, i.e. DT/AB = BT/DT = BT/BC. Hence

$$AB/BC = AB/BT = BC/BT = BT/BG.$$

ʿUmar then establishes:

Lemma 2. Square ABGD is given as the base of a rectangular solid ABGDH. Square MC is given (see Figure 6). To construct on MC a rectangular solid with volume equal to that of ABGDH.

His solution is to construct a line segment K such that: (a) AB/MC = MZ/K; and then a segment ZT with (b) AB/K = ZT/HD, letting ZT be perpendicular to the face MC at Z. Complete the solid MZTC. Then

$$\text{volume of ABGD} = \overline{AB}^2(HD)$$
$$\text{volume of MZTC} = \overline{MZ}^2(ZT) = \overline{AB}\,\overline{(K)(ZT)}$$
$$= (AB)(AB)(HD) = \overline{AB}^2(HD).$$

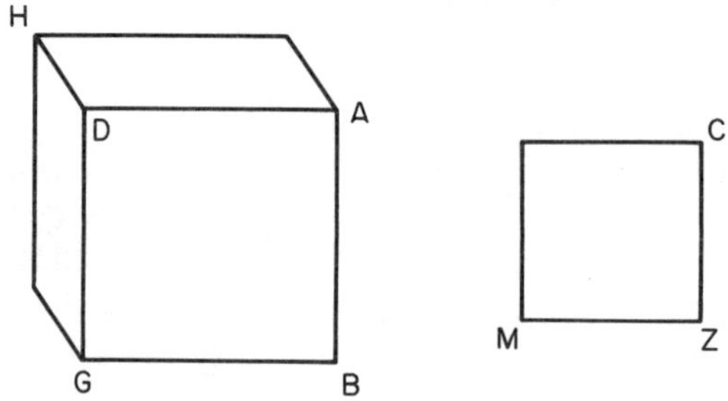

Figure 6

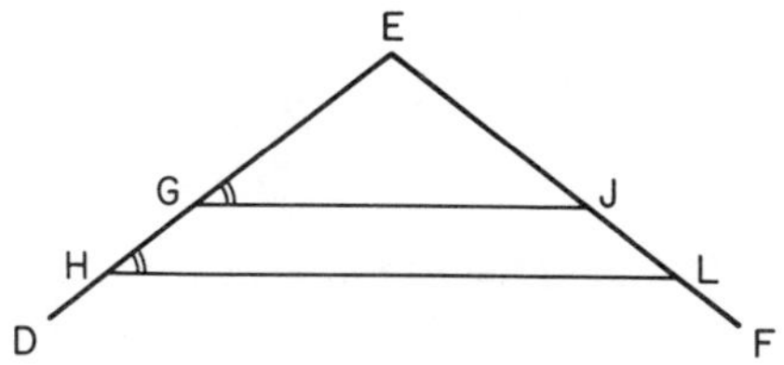

Figure 7

ʿUmar handles the volume computations using Euclid XI, 34.[112] The Euclidean constructions for (b) and (a) are based on similar triangles, e.g. (b) is equivalent to $K/\text{AB} = \text{HD}/\text{ZT}$, i.e. ZT is the 4th proportional to K, AB and HD (given). The standard construction is to draw any angle, $\langle$DEF (see Figure 7) with $\overline{\text{EG}} = K$, $\text{GH} = \overline{\text{AB}}$, $\text{EJ} = \overline{\text{HD}}$, and HL$\|$GJ, then JL is ZT; (a) proceeds in the same way.

Lemma 3, the next preliminary result is: A rectangular solid ABGD is given with square base AG. To construct a rectangular solid of given height HT and square base with the same volume as ABGD.

The construction ʿUmar gives requires the construction of a fourth proportional K:$\text{HT}/\text{BD} = \text{AB}/K$ as in Lemma 2. This is followed by a mean proportional,[113] say HZ to AB and K, i.e. $\text{AB}/\text{HZ} = \text{HZ}/K$. Then HZ is the required square base, for

$$\begin{aligned} \text{volume ABGD} &= (\text{AB})^2\text{DB} \\ \text{volume ZHTC} &= (\text{HZ})^2\,\text{HT} = \text{AB}(K)(\text{HT}) \\ &= (\text{AB})(\text{AB})(\text{BD}) = (\text{AB})^2(\text{DB}) \end{aligned}$$

ʿUmar's fourth lemma is the geometrical solution of $x^3 = c$ via two parabolas. We note that although he mentioned earlier that this can be solved numerically as well, he offers no numerical example.

Lemma 4. Let the rectangular solid ABGD with unit square base and height c represent the "number". To construct a solid cube of equal volume.

His construction is shown in Figure 8. By Lemma 1, construct segments H, Z so that $\text{AB}/H = H/Z = Z/\text{BD}$.[114] Then set $\text{CT} = H$ and construct a cube with this as its side. The volume $\text{ABGD} = (\text{AB})(\text{BG})(\text{BD}) = \text{BD}$, while volume $\text{TCKL} = H^3$.

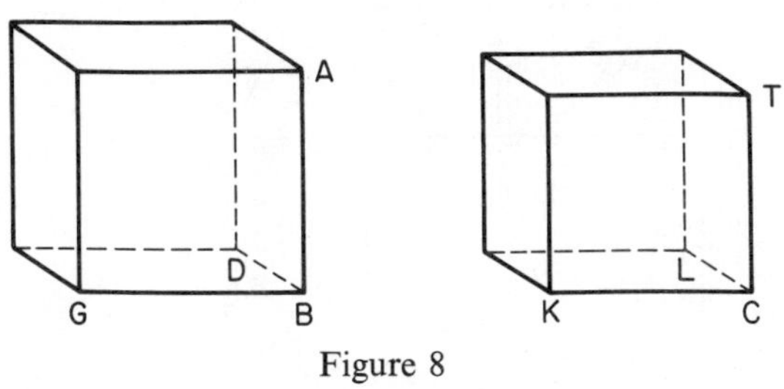

Figure 8

From the construction we have

$$\frac{1}{H}=\frac{H}{Z}=\frac{Z}{\mathrm{BD}}.$$

Thus $H^2=Z$, $Z=\mathrm{BD}/H$ and $H^3=\mathrm{BD}$ as required.

Now ʿUmar's text begins his discussion of the solutions of 3 term equations (degree 3), starting with compound equation number 7.

Lemma 5. Geometrical solution $x^3+ax=b$.[115]

Construction (for 5): Let AB equal the side of a square equal to the number of roots, i.e. $\mathrm{AB}=\sqrt{a}$.[116] Now construct a rectangular solid with base a square of side AB and height BG such that its volume is b.[117] Take BG perpendicular to AB; extend BA to Z and draw a parabola with vertex B, axis BZ and parameter AB. Call it CBD. (See Figure 9). It's tangent to BG.[118] Draw semicircle on BG. Then ʿUmar claims, again without argument, that this semicircle intersects his parabola, say at D, and draws perpendiculars from D to GB and DZ respectively. By the defining property of the parabola,[119] $\overline{\mathrm{DZ}}^2=(\mathrm{BZ})(\mathrm{AB})$, so $\mathrm{AB}/\mathrm{DZ}=\mathrm{DZ}/\mathrm{BZ}$. But $\mathrm{DZ}=\mathrm{BH}$, $\mathrm{HD}=\mathrm{BZ}$, so $\mathrm{AB}/\mathrm{BH}=\mathrm{BH}/\mathrm{HD}$. However $\mathrm{BH}/\mathrm{HD}=\mathrm{HD}/\mathrm{HG}$ (since HD is the mean proportional to BH and HG; see note 113). Thus $\mathrm{AB}/\mathrm{BH}=\mathrm{BH}/\mathrm{HD}=\mathrm{HD}/\mathrm{HG}$, and $(\mathrm{AB})^2/(\mathrm{BH})^2=(\mathrm{BH}/\mathrm{HD})(\mathrm{HD}/\mathrm{HG})=\mathrm{BH}/\mathrm{HG}$. Whence, $(\mathrm{AB})^2(\mathrm{HG})=(\mathrm{BH})^3$. We construct a rectangular solid with square base AB and height BH. Then

$$(\mathrm{BH})^3+(\mathrm{AB})^2\mathrm{BH}=(\mathrm{AB})^2\mathrm{BG}=b,$$

for

$$(\mathrm{AB})^2(\mathrm{HG}+\mathrm{HB})=(\mathrm{AB})^2\mathrm{BG}.$$

Let $\mathrm{BH}=x$. Then

$$x^3+(\sqrt{a})^2x=x^3+ax=b.$$

ʿUmar then remarks that all problems of this type may be so solved, as

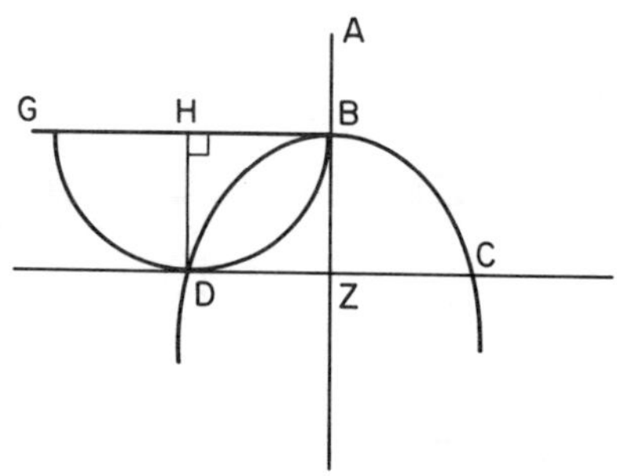

Figure 9

intersections of a circle and a parabola. As a matter of fact, since $x^3 + ax - b = 0$ has only one sign change, while $-x^3 - ax - b = 0$ has none, there is only one real root, which is positive, and hence ʿUmar's solution is complete.[120]

In (our) Lemma 6, he examines the equation $x^3 + c = ax, a, c > 0$ for a solution; his construction contains the following preliminary steps.

Let AB be a side of a square equal to the number of roots, i.e. as above $AB = \sqrt{a}$. Construct a solid with base a square of edge AB, volume c.[121]

Lemma 6. Geometrical solution $x^3 + c = ax$.

As in the preceding lemmas, ʿUmar lets AB be a side of a square equal to the number of roots, i.e. $AB = \sqrt{a}$, and constructs a solid with base a square with edge AB and volume c. Let the height of the solid be BG and take $BG \perp AB$. Then construct a parabola with vertex B, axis AB and parameter AB, and a hyperbola with vertex G, axis along BG, with parameter and diameter BG.[122] (See Figure 10.) (That is, in an appropriate coordinate system x, y, the hyperbola has defining equation

$$\frac{(x + \frac{1}{2}BG)^2}{(\frac{1}{2}BG)^2} - \frac{y^2}{(\frac{1}{2}BG)^2} = 1.$$ [123]

Then ʿUmar observes that the problem is solvable or not, just as the hyperbola ZG and parabola BD intersect or not. If there is an intersection, he claims that it may be tangential or in two points.[124] If either of these conditions is obtained, let H be one of the intersection points. Drop perpendiculars from H to GB extended and AB extended, say HT and HC respectively.

The next line then states that: "the proportion of the square of HT to the product of BT and TG would equal the proportion of the parameter to the diameter."[125] Symbolically, $\overline{HT}^2 = (BT)(TG)$, which is just the property of the hyperbola, i.e.

$$y^2 = x(x + BG)$$

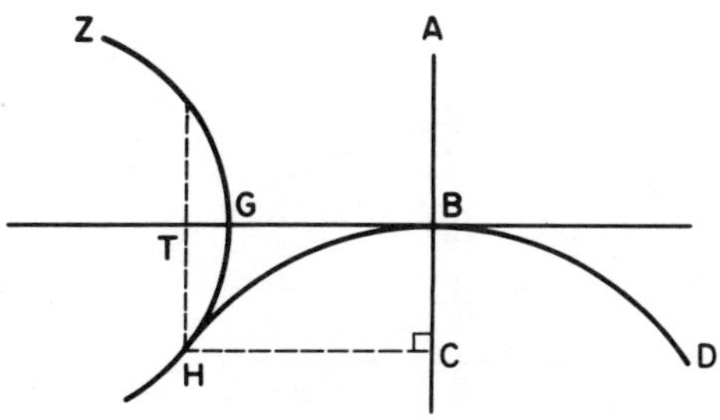

Figure 10

or[126]

$$\frac{(x+\frac{1}{2}\mathrm{BG})^2}{(\frac{1}{2}\mathrm{BG})^2}-\frac{y^2}{(\frac{1}{2}\mathrm{BG})^2}=1,$$

the hyperbola with vertex (− BG, 0) and conjugate axis (BG), in the more modern sense.[127] Hence (BT)/(HT) = (HT)/(TG). But $\overline{\mathrm{HC}}^2 = (\mathrm{AB})(\mathrm{BC})$ by the defining property of the parabola, and HC = BT, while BC = TH, so (AB)/(BT) = (BT)/(TH), or (AB)/(BT) = (BT)/(HT) = (HT)/(TG). Consequently $(\mathrm{BT})^3 = (\mathrm{AB})^2(\mathrm{GT})$; for, $(\mathrm{BT})^2 = (\mathrm{AB})(\mathrm{HT})$, $(\mathrm{BT})(\mathrm{BT})^2 = (\mathrm{AB})(\mathrm{BT})(\mathrm{HT}) = (\mathrm{AB})(\mathrm{AB})(\mathrm{TG})$. Also $(\mathrm{BT})^3 + (\mathrm{AB})^2(\mathrm{BG}) = (\mathrm{AB})^2(\mathrm{GT}) + (\mathrm{AB})^2(\mathrm{BG}) = (\mathrm{AB})^2\mathrm{BT}$, i.e.

$$x^3 + c = ax$$

ʿUmar then continues in (our) Lemma 7 to consider the solution of $x^3 = ax + c$. His solution is again via an intersecting parabola and branch of a hyperbola. Noteworthy, however, is the fact that he takes the branch he *did not use* in Lemma 6. His conclusion that nothing is impossible in this case is justified since $x^3 - ax - c = 0$ has indeed just one positive root. Moreover his *διορισμός* for Lemma 7 would provide a geometric condition for the existence of negative roots.[128]

His next construction (Lemma 8) deals with $x^3 + ax^2 = c$. Since this and the remaining constructions add little to the already complete picture of ʿUmar's accomplishments, (in the sense that all cubics reduce to one of the three previous forms $x^3 + ax = c$, $x^3 + c = ax$, $x^3 = ax + c$ via a linear transformation of the variable),[129] we present them only in so far as they offer noteworthy features of detail. ʿUmar was, of course, unaware that his work was done with Lemma 7.

Lemma 8. $x^3 + ax^2 = c$.

Departing from the usual course, he takes AB equal to the number of squares (*a*), and constructs a solid cube in volume equal to *c* (Lemma 4). Then he instructs us to extend AB, making BT equal to the edge of this solid cube. Complete the square BTDG, and draw through D a hyperbola with asymptotes BG and BT (Figure 11). The hyperbola is determined (well-defined) following Apollonius.[130] (Rather than trace ʿUmar's references to Apollonius we refer the reader to

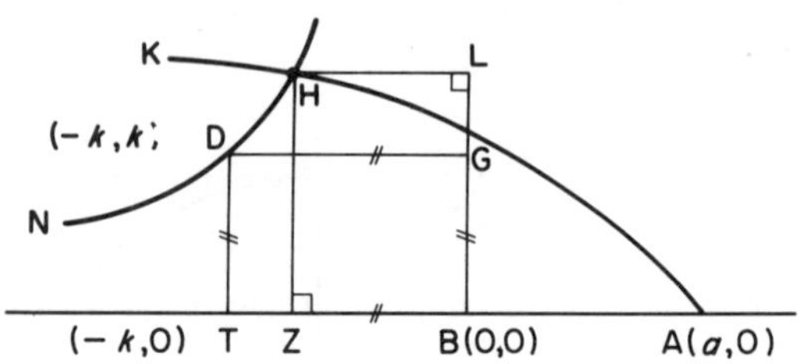

Figure 11

Woepcke (189), note (***), p. 38, and remark that with the coordinate system added in Figure 11 BT and BG are x, y axes (segments of) respectively. The hyperbola is then *determined*, since it is $xy = -k^2, k = \mathrm{BT} = \sqrt[3]{c}$). Next draw a parabola with vertex A, axis AT and parameter BG. This is also determined; for example, as $y^2 = -k(x-a) = k(a-x)$. Then ʿUmar claims that the curves necessarily intersect, say at H.[131] Drop perpendiculars from H to AB extended and to BG extended. Call these HZ and HL respectively.

Now Z cannot coincide with T or lie to the left of it. For, in ʿUmar's words, "Let it coincide with T. Then $(\mathrm{HZ})^2 = (\mathrm{AT})/(\mathrm{TB})$ [i.e. $y^2 = k(a+k)$] $= (\mathrm{AT})(\mathrm{BG})$. But HZ would then coincide with DT. Therefore $(\mathrm{DT})^2 = (\mathrm{AT})(\mathrm{TB}) = (\mathrm{TB})^2$, impossible, [i.e. $k(a+k) = k^2 \Rightarrow a = 0$]."[132] Now if Z is to the left of T, $\mathrm{HZ} < \mathrm{DT}$ which is impossible.[133] Hence Z falls between A and T. Also $(\mathrm{HZ})^2 = (\mathrm{AZ})(\mathrm{BG})$ by the defining condition for our parabola. Thus $(\mathrm{AZ})/(\mathrm{HZ}) = (\mathrm{HZ})/(\mathrm{BG})$ and the area of HB, i.e. HLBZ, equals the area of DB, i.e. (DGBT) by *Conics* II, 8.[134]

Hence $(\mathrm{HZ})/(\mathrm{BG}) = (\mathrm{BG})/(\mathrm{BZ})$, i.e. $(\mathrm{AZ})/(\mathrm{HZ}) = (\mathrm{HZ})/(\mathrm{BG}) = (\mathrm{BG})/(\mathrm{BZ})$. Consequently $c = (\mathrm{BG})^3 = (\mathrm{BZ})^2(\mathrm{AZ})$,[135] and $(\mathrm{BZ})^2\mathrm{AZ} = (\mathrm{BZ})^3 + (\mathrm{BZ})^2\mathrm{AB}$;[136] thus for $\mathrm{BZ} = x$.

$$x^3 + ax^2 = c$$

ʿUmar concludes this example by remarking that nothing is impossible, which is indeed the case, since his branch (reflected to make it positive in x, or as is, in the Arabic style) does give the guaranteed single positive root.

In Lemma 9, he considers $x^3 + c = ax^2$, which, of course has a guaranteed negative root, but either two or no positive roots. ʿUmar struggles to find geometric conditions for solvability, but his actual contribution to this problem is pale compared with Archimedes' exact διορισμός noted in Section III above.[137] His first condition is: $c^{1/3}$ is necessarily less than a, but this does not guarantee a positive root. He then provides subcases; for example, $c^{1/3} = a/2$. Since this corresponds to $\Delta = (a^6/2) - (27a^6/64) > 0$, he is indeed correct in claiming a solution. In fact, the first of his accompanying three figures for this case suggest that he may have been aware of both positive solutions. However, he makes no mention of it. His other two subcases are $c^{1/3} > a/2$, $c^{1/3} < a/2$. In the first of these he remarks that a solution is still possible, since the auxiliary conics may still intersect in two points, or share a common tangent, and notes that Abuʾl Jud (*c.* 950) missed this possibility in his own analysis. But his bounds are insufficient to separate the possible from impossible. In the second case, he fares better, since, as he notes (geometrically), his condition guarantees two intersections.[138]

His next case (Lemma 10), $x^3 = ax^2 + c$ contains nothing remarkable. There is only one positive root, and his construction (intersection of hyperbola and parabola) determines it.

The next seven lemmas detail his work on the four term cubic. He runs into no difficulty in handling Lemma 11, $x^3 + ax^2 + b^2x = c$. There is one real, positive root, and he finds it using a hyperbola and circle (actually a branch of the hyperbola, and a semicircle). Lemma 12, $x^3 + ax^2 + c = bx$, as expected, gives him difficulties, since only a negative root is guaranteed. His solution, via two hyperbolas is curious, since he sets the hyperbola up in such a way that an intersection is guranteed (the negative root), but remarks, "If there is another point of intersection, the problem is solvable, otherwise not. The meeting (intersection) is either by touching, or by intersection at two points."[139] This would correspond to no positive roots ($\Delta < 0$), two equal positive roots ($\Delta = 0$), or to two distinct positive roots ($\Delta > 0$).

Lemmas 14–16 ($x^3 = bx^2 + ax + c; x^3 + ax^2 = bx + c; x^3 + ax = bx^2 + c$) offer him no intrinsic difficulties. Each has a guaranteed positive root, and ᶜUmar constructs it using two hyperbolas, in the first two examples, and a hyperbola and circle in the last, which is of some interest since the arrangement of the figure forces ᶜUmar, for the only time in his discussion of cubics, to subtract volume terms.[140] The two remaining cases (Lemma 13 and 17), treat respectively $x^3 + bx + c = ax^2$ and $x^3 + c = ax + bx^2$, and cause difficulty since both have only guaranteed negative roots. The latter is handled in his typically through manner, but the former seems muddled, so we examine it in detail, highlighting the features that seem curious to us.

Lemma 13. The equation $x^3 + bx + c = ax^2$.

The construction is as follows (see Figure 12). Let BH $= a$, BG $= \sqrt{b}$, and place BG perpendicular to BH. Construct a rectangular solid of square base, edge BG, and volume c (Lemma 2). Let its height be AB, marked on BH produced $[(\sqrt{b})^2 \mathrm{AB} = c$, when $|\mathrm{AB}| = c/b = k$, say$]$.

Draw a semicircle AZH

$$\left[\left(x - \frac{k-a}{2}\right)^2 + y^2 = \left(\frac{a+k}{2}\right)^2\right].$$

Now G can be inside, on, or outside the semicircle. First suppose G is inside. Produce BG until it intersects the semicircle at Z. Complete the rectangle ABGD. Construct on ZG an area GC having area equal to that of rectangle ABGD.[141] Then C is known, and lies inside, on, or outside the semicircle. Suppose first that it

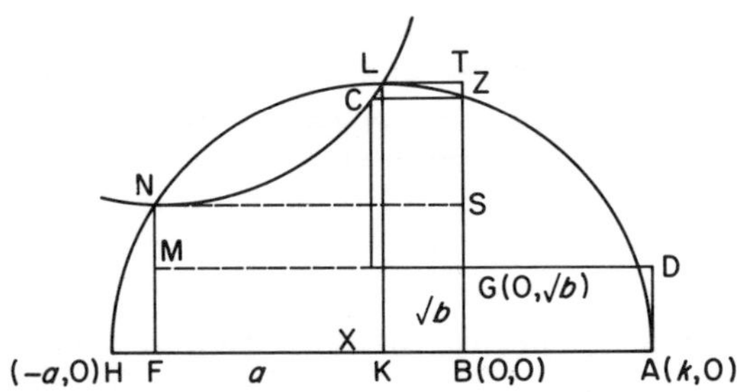

Figure 12

is inside. Draw through C a hyperbola with asymptotes ZG, GD [e.g., $x(y - \sqrt{b}) = -k\sqrt{b}$}. Then the hyperbola intersects the circle in two points, say L, N. Drop perpendiculars LK, NF onto AH, and LT onto BZ extended. Then the area LG equals the area GC (property of hyperbola, both being $k\sqrt{b}$), and area GC = area AG (by construction). Therefore, area LG + GK = area GA + GK or area TK = area DK, i.e. (LT)(LK) = (KA)(BG) whence (LK)/(KA) = (BG)/(LT), $(LK)^2/(KA)^2 = (BG)^2/(LT)^2 = (BG)^2/(BK)^2$. But $(LK)^2/(KA)^2 = (HK)/(KA)$.[142] Thus $(BG)^2(KA) = (BK)^2(HK)$. But $(BG)^2(KA) = (BG)^2(KB + AB) = b(KB) + c$. Then $(BG)^2(KA) + (BK)^3 = (BK)^3 + b(KB) + c = (BK)^3 + (BK)^2(HK) = (BK)^2 \times (BK + HK) = (BK)^2(HB) = a(BK)^2$. Let $x = (BK)$, then

$$x^3 + bx + c = ax^2.$$

Then comes a curious comment: "The cube of BF will also satisfy the same equation and can be proved by the same demonstration when the two points C and H fall inside the circle."[143] At first glance, this seems redundant, since *we have been* considering this case. But then one realizes that ʿUmar is giving a rare instance of the construction of a second positive root of a cubic.[144] (See note 144 for reconstruction of proof.)

With G still inside, ʿUmar considers the possibility of C being outside, and notes that there is a solution provided the hyperbola meets the semicircle either tangentially or at two points, and credits Abūʾl Jūd with this approach. Then he gives a useless suggestion: to construct the area on a subsegment or extension of ZG in case there's no intersection. In fact, the next few lines seem to indicate his frustration at being unable to isolate the possible cases. For example, he alludes to a construction for G on or outside the circle, the sense of which seems to be given in the two parts of Figure 13. Here area GC equals area BD and the hyperbola through C has asymptotes MG, MZ. In each case BF is the candidate for the positive root. But the whole business is irrelevant,[145] for consider $a = b = c = 1, k = c/b = 1$. Then $x^3 + x + 1 = x^2$ has one negative root and since $\Delta < 0$ no other real roots, while the circle and hyperbola intersect. This is the case Z = G, on the circle. Similar examples may be constructed for the first figure.[146]

ʿUmar at first only hints at this, leaving it for the reader's discovery, saying that all of this comes from an easy rule, and that any one who cannot discover it for himself is certainly wasting his time with the *Algebra*.[147] He finally relents and gives the rule as described above.

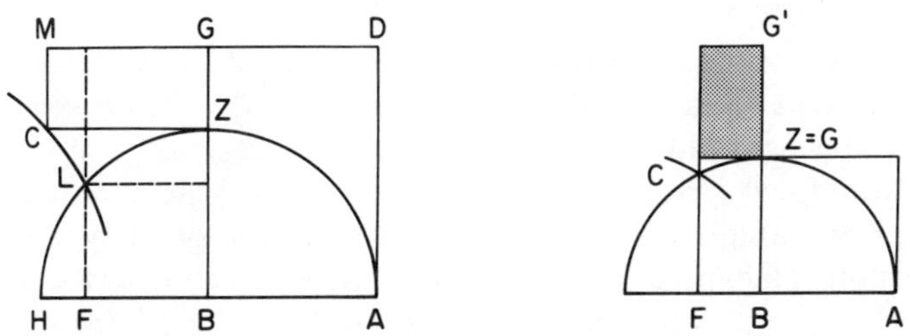

Figure 13

Table 3. Summarizes ʿUmar's achievement

		Number of roots $x > 0$	ʿUmar
L.4	$x^3 - c = 0$	1	1
L.5	$x^3 + ax - b = 0$	1	1
L.6	$x^3 - ax + c = 0$	2 or zero, 1 negative	indicates possibility of none
L.7	$x^3 - ax - c = 0$	1	1
L.8	$x^3 + ax^2 - c = 0$	1	1
L.9	$x^3 - ax^2 + c = 0$	2 or zero, 1 negative	tries to distinguish 2 or zero by inequalities
L.10	$x^3 - ax^2 - c = 0$	1	1
L.11	$x^3 + ax^2 + bx - c = 0$	1	1
L.12	$x^3 + ax^2 - bx + c = 0$	2 or zero, 1 negative	tries to distinguish cases
L.13	$x^3 - ax^2 + bx + c = 0$	2 or zero, 1 negative	tries to distinguish cases, but falters. Does offer a clear determination of 2 positive roots in one case.
L.14	$x^3 - bx^2 - ax - c = 0$	1	1
L.15	$x^3 + ax^2 - bx - c = 0$	1	1
L.16	$x^3 - bx^2 + ax - c = 0$	3 or 1	1
L.17	$x^3 - bx^2 - ax + c = 0$	2 or zero, 1 negative	tries to isolate impossible cases, with some throughness

ʿUmar's real accomplishment is his systematic attempt to solve all cubics by the intersection of conics. In retrospect, his work on L.5–L.7 provides the best measure of his success. Although his *διορισμός* for the possible case of say L.9 is weak, it is none the less an indication that his work may well have been done independently of Eutocius' commentary on the *Sphere and the Cylinder*.[148] As for the meaning of his "impossible cases", one thing should be clear to the reader at this point: every line segment is for ʿUmar a positive measure. Intersections are rejected as roots, not because they are negative, but rather, because they are positive magnitudes which do not satisfy the rules of al-jabr and al-maqabala in balancing the positive volumes (monomials) to which they give rise. Similarly, there is no rejection of "imaginary" roots; they just don't occur. Indeed, ʿUmar does not even list $x^3 + bx^2 + cx + d = 0$ in his classification scheme. (See Medovoy (110) for analysis of a treatise of Abū-l-Wefa purportedly containing negative numbers.)

ʿUmar himself tells us that his work was influenced by al-Māhānī (d. *c.* 880), Abū Ja ʿfar al-Khāzin (d. 961/971), ibn al-Haytham (d. *c.* 1040), Abūʾl Jūd (fl. *c.* 1000), and Abū Sahl al-Qūhī (*c.* 988). To trace the actual debts and lines of influence, would however take a study of nearly twice the length than we have so far offered. For example, Ibn al-Haytham solves a cubic of the type $x^3 + x = a$ as the intersection of a hyperbola and parabola, while ʿUmar does the more general equation $x^3 + bx = a$ as the intersection of parabolas.[149] Moreover, there is an extensive literature, of which Woepcke gave a survey, giving some of the solutions (always less general in scope than ᶜUmar's) on solutions of the cubic.[150] On the other hand, there is evidence that other Near Eastern mathematicians arrived at slightly better conditions for positive roots for L.9.[151] Finally, there is an extensive literature by Near Eastern mathematicians dealing with problems in geometry which lead to cubic equations, in particular the problems of angle trisection, and construction of a regular nonagon by conics.[152] This literature has recently been elegantly surveyed by Hogendijk,[153] who corrects some earlier commentaries of Schoy, and provides us with some insight into the competitive nature of certain of the mathematicians prominent in developing the theory. He cites, for example, a text of al-Shannī *Book of the Revealing of the Falsification by Abūʾl-Jūd in the Matter of Two Lemmata for an Alleged Construction of the Heptagon of His*, who contends that Abūʾl-Jūd was a trickster and plagiarist who even attempted to pass on work of Menaechmus as his own.[154] Unfortunately, Abūʾl-Jūd's own work *Book on Geometrical Subjects*, is lost. Hogendijk makes a case for its being the collection of tracts mentioned by ᶜUmar in his *Algebra*, and rightly laments its loss, since it could throw considerable light, not only on Abūʾl-Jūd, but also on the whole problem of influence and originality in the Near Eastern theory of geometric solutions of equations.[155] In any case, there are no such charges against ᶜUmar, and his accomplishments are universally agreed upon.

We close this section with an account of Medieval Near Eastern work on quartics. Examples of equations of this type appear in Babylonian work, as problems that are solvable by double application of one of their quadratic procedures,[156] but as we noted above, even ᶜUmar thought them strange and without "real significance". Nonetheless, isolated examples do appear in Arabic treatises; for example, Abū Kamil (fl. 850–930) solves a pure equation of fourth degree: $x^4 = 8000 - \sqrt{51200000}$ in his treatise on the pentagon and decagon. As a matter of fact, he considers this equation in propositions 16 and 20, and the equation $x^4 = 1600 - \sqrt{2048000}$ (propositions 17–18).[157] Abū Kamil's first equation, itself, is reminiscent of the Babylonain type mentioned above, namely $x^4 + 3125 = 125x^2$.

Ibn al-Haytham presented no general theory of equations of degree four. In Book V of his *Kitāb al-Manāẓir*, he sets himself the problem of solving a problem on reflection from mirror surfaces that appears, in part, in Ptolemy's *Optics*, namely, given a spherical, cylindrical, conical surface, viewed either from the inside, or outside, and two points not on the surface, to characterize the points on the surface for which light from point A (say) is reflected to point B.[158] His solution, by conics, was effectively the best result until the time of Huygens.[159] Given the comments of Barrow on al-Haytham's proofs ("horribly prolix"),[160] it is really a moot question as to whether al-Haytham's fame as a mathematician rests on his solution, or the general mathematical appeal of the problem. Among writers in European languages, the remark of Bode (1891) is probably still true,"...in der neuren zeit scheint niemand die Lösung Alhazens wirklich durchgearbeitet zu haben: die Notizen, die sich daruber finden, sind ungenau und falsch."[161] In Arabic, however, there is a clear, careful analysis by Mustafa Nazīf.[162] Sabra[163]

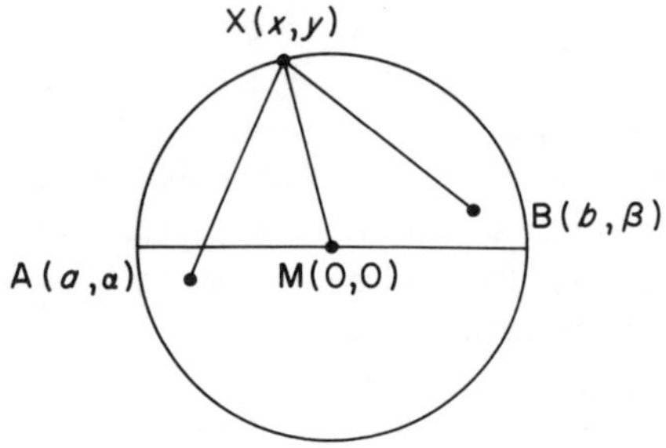

Figure 14

also presents a short account in English, based on Nazīf. This indicates that al-Haytham resolved certain cases of the problem using a circle and hyperbola as auxiliaries, in particular, that of a spherical mirror (inside), and considered the possibility of four reflection points. Although Barrow is correct in his assessment of the case-wise sequence of solutions, none the less, a modern, annotated edition of the *Kitāb al-Manāẓir* would be of considerable interest to the historian of mathematics. Unfortunately we cannot present ibn al-Haytham's arguments in a short space, but we do offer an account of his solution of part of the problem, based on translations from Risner's Latin edition (1572). First, to give the reader at least an understanding of how conics arise in the solution, we reproduce a short "proof" of Bruins for the spherical mirror (points A and B inside).[164] (See Figure 14.) This actually corresponds to Huygen's second proof, and was done by Bruins in 1927, as a final form gymnasium student.

Evidently we want X(x, y) so that ⟨AxM = ⟨MxB. This is obtained just in case[165]

$$\frac{\dfrac{y-\alpha}{x-a}-\dfrac{y}{x}}{1+\dfrac{y(y-\alpha)}{x(x-a)}}=\frac{\dfrac{y}{x}-\dfrac{y-\beta}{x-b}}{1+\dfrac{y(y-\beta)}{x(x-b)}}$$

Clearing fractions gives

$$(x^2+y^2)[(a+b)y-(\alpha+\beta)x]+(\alpha b+a\beta)(x^2-y^2)$$
$$+2(\alpha\beta-ab)xy=0,$$

which reduces to a hyperbola on taking $x^2+y^2=r^2$.

However, it requires a considerable stretch of the imagination to consider this analytic transformation of the problem as a "proof" in the sense of al-Haytham, especially, when one actually examines his own solution. Granted that one may actually solve the resulting system of equations, it still must be done, and the mechanics of the work simply transform a perfectly natural problem from geometry into a messy piece of algebra.

To illustrate the ideas and techniques inherent is al-Haytham's work, we now offer his proof of proposition 87: "If a straight line connecting two points on different diameters of a circle (which is the section common to the surface of reflection and of a concave spherical mirror), at equal distances from the centre, extended, cuts the circle, reflection between the said points is possible from one, two or four points of the mirror, but never from three, only."[166]

The proof dependency seems to be the one indicated below:[167]

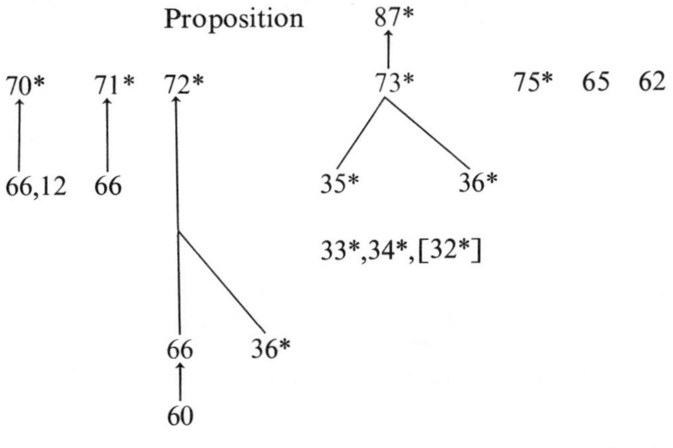

The propositions marked with an asterisk (*) are translated in detail, the others, of lesser mathematical interest, we merely state. The bracket around 32 indicates that it is used, as a case of 33, but that it is not explicitly mentioned later in the text. The reader whose patience is sorely tried by geometric constructions may wish to skip to the part of this section, where we discuss the possible sources of al-Haytham's work.[168]

Proposition 12, which al-Haytham, or at least Risner, uses in 70, is simply: "The eye and the visible object given, to find the point of reflection in a plane mirror."[169]

We go, consequently to al-Haytham's proposition 32. Additions to the text by Risner, usually in the form of theoretical justifications are inclosed by square brackets [ ], while additions of our own are introduced in parenthesis.[170] Other comments are reserved for the notes. Statements of the propositions appear in the appendix to this paper; these are usually rather imprecise. In fact, the best indication of the content of a particular proposition comes from reading its proof.

Proof of Proposition 32: Given the diameter of the circle, and the point on the circumference: to draw from it to the diameter extended, a line, which from the point on which it cuts the circle, upto its intersection with the diameter, makes a line segment equal to the given line (Figure 15). In other words, let qe be the given line, gb the diameter of the circle abg, a the given point. I say that a line drawn from the point a, from the point where it cuts the circle to (that at which) it cuts the diameter gb equal to line qe, is to be constructed as follows. Let there be drawn two lines ab, ag: they will either be equal or not. Let them be equal; and let there be adjoined to line qe such a line so that the product of the adjoined part to the whole be equal to the square on ag. [It may indeed be done: if line qe makes the

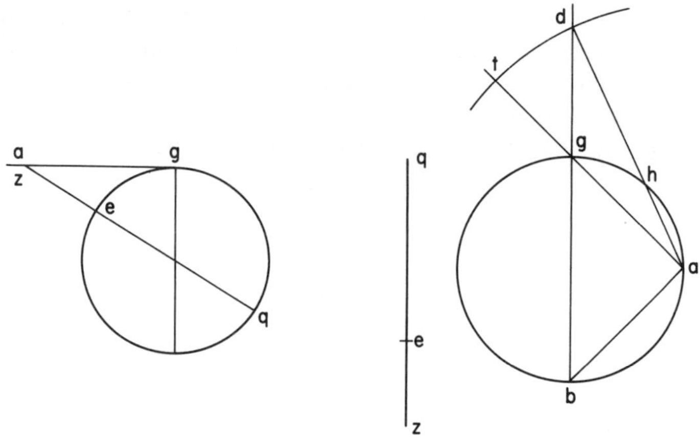

Figure 15. Figure for Proposition 32. (Reproduced with permission from D. C. Lindberg, ed., *Opticae Thesaurus Alhazeni Arabis*..., Johnson Reprint, N.Y., 1972, p. 142 (bottom figures).). © by Johnson Reprint Corporation, New York, N.Y.

diameter of a circle, at the end of which a line segment equal to ag is tangent, intersecting with the extension of diameter qe, then, indeed, the rectangle composed of the extended diameter and its exterior segment will be equal to the square on line ag by 36 p3.][171] And let the adjoined segment be ez. Therefore since the rectangle on qz into ez is equal to that on ag into itself: it must be that qz is larger than ag, and ez smaller. For if ez should be equal to, or larger than ag: it would be impossible that the said qz into ez be equal to the square of ag [for, indeed, the rectangle contained by qz and ez would always exceed the square on ag: because the line qz is larger than ez, as the whole exceeds its part.] If, however, it is smaller; clearly qz is greater than ag. Let it (ag) therefore be extended to equality: say, as agt: and with the foot of the compass set at a, make a circle with respect to the quantity agt (i.e., as radius) which circle will cut the diameter bg [continued passed t], say in point d, and let line ad be drawn, which will necessarily cut the circle. For, if it is tangent with it at point a, it would be parallel to bg and never intersect with it. [For, by hypothesis, ga, ba are equal. And also a radius drawn from the centre to a makes by 8 p10, d1, right angles with bg. Similarly, the angle of the tangent line da and the semidiameter is a right angle by 18 p3; therefore by 28 pl bd, ad are parallel (segments) which, however, intersect at point d, by construction.] Therefore, it cuts in point h: and let line gh be drawn. Clearly [by 22 p3] since abgh is a quadrilateral in a circle: abg, ahg as opposite angles sum to two right angles: but [by 5 p1] agb is equal to angle abg, since they cut off equal arcs (lit. sides) by hypothesis. It will follow therefore that angle ahg is equal to angle dga, [by 13 p1]; and angle hag is common[172] to triangle adg, and[173] to triangle ahg: it follows therefore [by 32 p1] that angle hdg is equal to angle hga: and one triangle is similar to the other [by 4 p1, d6]. Consequently the ratio of da to ag, is as ag to ah; therefore [by 17 p6] the product[174] of da and ah is equal to the square on ag. But [by 15 d1] da is equal to ta: therefore [by 1 ax] is equal to qz; and ah will be equal to ez [since from the initial construction the rectangle contained by qz and ez equals the square on ag, and evidently equals to the rectangle contained by da & ah; and da is itself equal to qz] and [by 3 ax] dh equals qe, which is the given line. And thus is the proposition established.

Proof of Proposition 33 (Figure 16):[175] If, in fact, ab and ag be unequal, let there be drawn a line, say gn from point g parallel to ab [by 31 p1], and let there be given a line, any one whatsoever: zt; and [by 23 p1] at point z make an angle

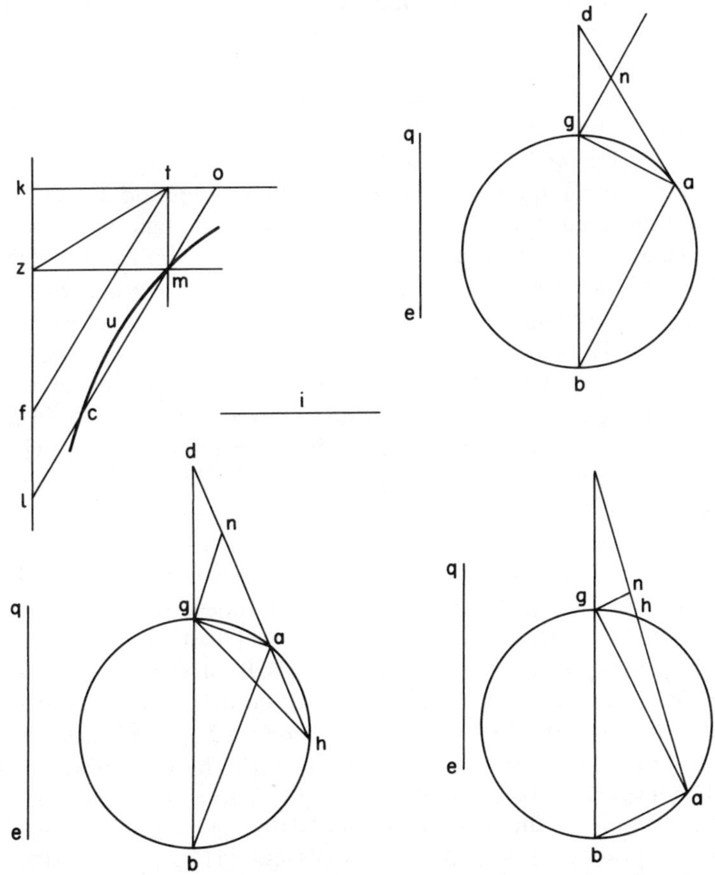

Figure 16. Figure for Proposition 33. Note: This figure in Risner, showing the hyperbola, has it curving away from the asymptotes! This also seems to be the case in a number of Latin mss. See Clagett, *Archimedes*, vol. 4, pt. 2, 473 for additional sketches. Also, it is clear from the Apollonian tradition (oblique coordinates), that kt and kz are not usually, and in fact need not be, perpendicular. (Reproduced with permission from D. C. Lindberg, ed., *Opticae Thesaurus Alhazeni Arabis*..., Johnson Reprint, N.Y., 1972, figures, pp. 143–4.).
© by Johnson Reprint Corporation, New York, N.Y.

equal to agd with terminal side zf; and [by 31 p1] let there be drawn a line from point t parallel to zf, say tm; and [by 23 p1] let there be cut from angle tzf an angle equal to dgn by line zm. This line, therefore, of necessity intersects with tm: [by Proclus' Lemma 29 p1] since it lies between parallels. Let the intersection point be m. It follows therefore [by 3 ax] that angle mzf equals angle agn. And through point t let there be drawn a line parallel to zm [by 31 p1] say to, which necessarily intersects with fz: [by Proclus' Lemma 29 p1] and let k be the point of intersection: and [by 12 p6] let there be given a line, whose ratio to the line zt stands as bg to qe, say line i. Then make, through point m a conic section[176] in the manner that Apollonius prescribes in book 2, Proposition 4 of his *Conics*;[177] and let it be denoted by ucm, which section, indeed, does not cut the lines ko, kf, and on this section let a line be drawn equal to i, say mc,[178] and let it be extended upto the lines kt, kf, with o and l the points of intersection. Therefore, just as it is demonstrated in that same place [8 th2 *Conics*],[179] om will be equal to cl; and from point t let a line be drawn parallel to cm: [by 31 p1], say tf; and [by 23 p1] at point a make an angle equal to angle zft with terminal side and. It is clear that this line intersects with gd; since angle agn is equal to angle fzm (by deduction) and angle gan is equal to angle zft [by construction, and the whole angle fzt is equal to the whole angle dga, and by 32 p1 the angles at z and f are less than two right angles. Therefore, the angles at g and a, themselves equal to these, will be less than two right angles. Thus by 2 ax, gd, ad are concurrent.] Therefore line ad is either tangent to the circle, or will cut it. Whereas, if it were not tangent, and arc ab greater than arc ag, it would cut arc ab; and if ab were smaller it would cut arc ag. Therefore (first) let it be tangent at a. Since, therefore, [by construction] angle gan is equal to angle zft, and angle agn is equal to angle fzy, [by 32 p1] the third angles are equal, and the triangle agn will be similar to triangle zfy. Similarly since [by construction] angle agd is equal to angle fzt, [by 32 p1, 4 p1, d6] triangle agd will be similar to triangle fzt. Therefore an is to ag as fy is to fz, and ag is to gd as fz is to zt. Consequently [by 22 p5] an is to gd as fy to zt. Indeed since [by construction] tm is parallel to fl, and ft is parallel to lm, [by 34 p1] ft equals lm. Hence [by 2 ax] it will be equal to co, since [by Apollonius' *Conics*, Book II, theorem 8], mo equals lc; but [by 34 p1] mo equals yt: since [by construction] it is itself equidistant from it, and ym is equidistant from to. It follows, therefore, [by 3 ax] that fy equals cm; but [by construction] cm is equal to i. Consequently [by 1 ax] fy is equal to i; but [by construction] the ratio of i [that is, by 7 p5] fy to zt is as bg to eq. Therefore [by 11 p5] the ratio of an to gd is as bg to eq.[180] In fact angle gan is equal to angle gba: as Euclid proves in (book) three [proposition 32]. But [by 29 p1] angle ngd is equal to angle abg, since [by construction] ng is parallel to ab. Therefore [by 1 ax] angle ngd is equal to angle nag, and angle ndg is common. Consequently [by 32 p1] the third angles are equal. Thus [by 4 p1 d6] triangle ngd is similar to triangle adg. Therefore the ratio of ad to dg is as gd to dn. Hence [by 17 p6] it follows that ad times dn is equal to dg square. In fact the square on ad is equal to that rectangle drawn with sides bd and dg as Euclid proves in proposition 36 [Book III], and the square on ad is equal to sums of rectangles with sides ad, dn and ad, na [by 2 p2], and the rectangle with sides bd, dg is equal to the square on dg added to the rectangle with sides bg, gd: as Euclid proves [3 p2]. Therefore on removing equals from equals [to be sure dg square and rectangle adn][181] it follows [by 3 ax] that ad times an equals bg times gd. Therefore [by 16 p6] the ratio of the first line to the second is as the third to the fourth [certainly as ad is gd, so is bg to an, and alternately [by 16 p5] as ad is to bg, so is gd to an]. Hence [by consequence of 4 p5] an is to gd, as bg to ad. But now it was stated that an to gd is as bg to eq. Therefore [by 9 p5] eq is equal to ad. That is what was proposed. What if ad were not tangent to the circle (at a), but cut it, and ag happened to be larger than ab; it would indeed cut the arc ag. Therefore, let it cut at point h, and let line hg be drawn. It is clear [by 22 p3] that the two angles ahg, abg (together) equal two right angles; but angle ngd is equal to abg [by 29 p1 since ng is itself parallel to ab.] Therefore angles ahg and ngd form two right angles. Whence [by 13 p1 3 ax] angle ngd is equal to angle nhg, and angle ndg is common. Thus [by 32 p1] the third angle is equal to the third: and triangle ngd is similar to triangle ndg [by 4 p1 d6.] Therefore the ratio of hd to dg is as the ratio of dg is to dn. It follows [by 17 p6] that the rectangle on hd, dn is equal to the square on dg; but the rectangle on ad, hd is equal to that on bd, dg, as Euclid proves [consequence of 36 p3]; and [by 1 p2] the rectangle on ad, dh is equal to the sum of rectangles on dh, dn and dh, an; and [by 3 p2], the rectangle on bd, dg is equal to the sum of the rectangle on bg, gd and the square on dg. Therefore with equals subtracted, that is the square on dg and the rectangle on dh, dn, it follows [by 3 ax] that the rectangle on hd, an is equal to that on bg, dg. Thus the ratio of the second line to the fourth, that is an to gd, is as the third to the first, that is, bg to dh [for by 16 p6 it follows that dh to dg is as bg to an, and by 16 p5, that dh to bg is as dg to an, and as a consequence of 4 p5, that an to dg is as bg to dh.] But, now, it was proven[182] that an to dg is, as bg to eq. Therefore [by 9 p5] eq is equal to dh. And thus the proposition is established. If, indeed, ag is less than ab, and ad cuts arc ab, let the point of intersection be h, and let line hg be drawn. It is clear [by the first construction and 29 p1] that angle ngd is equal to angle abg; but [by 27 p3] angles abg, ahg are equal since they fall on the same arc. Therefore [by 1 ax] angle ngd is equal to angle ahg, and angle ndg is common. Whence [by 32 p1] the third angle is equal to the third, and the triangles are similar [by 4 p1 d6.] Therefore the ratio of hd to dg is, as dg to dn. Thus [by 17 p6] the rectangle on hd, dn is equal to the square on dg; but the rectangle on hd, da is equal to that on bd, dg [by Campanus' completion of 36 p3] and [by 1 p2] the rectangle on hd, da is equal to the sum of that on dn, hd and an, hd; and [by 3 p2] the rectangle bd, dg equals the square on dg added to the rectangle on bg, dg. Therefore with equals removed: [undoubtedly the rectangle hdn, and square dg], it will follow that the rectangle on hd, an is equal to that on bg, dg. Therefore [by 16 p6, 16 p13 d5] the ratio of an to dg is as bg is to hd. But, now, it has been stated, that an is to dg, as bg to eq. Thus [by 9 p5] eq is equal to hd. That is the proposition. Whence from a given point a, we draw a line, cutting the circle, and from the point of intersection (with the circle) to the diameter, it is equal to a given line.

Proof of Proposition 34. From a given point on a circle, and not on a (certain) diameter, a line is to be drawn through the diameter (and) to the circle, so that its part from the diameter to the (intersection point with the) circle is equal to a given line. For example, let abg be the given circle, bg the diameter, a the given point, hz the given line (Figure 17). I say, that from point a there is to be drawn a straight line, cutting through the diameter bg, whose segment from the diameter to the circumference makes a line equal to hz. Let lines ab, ag be drawn, and [by 23 p1] (draw) at point h an angle equal to agb with terminal side mh, and through the same point let there be an angle equal to abg with terminal side hl. And [by 31 p1] from point z let a line, say zn be drawn parallel to hm; this same line will cut hl [by lemma of Proculus' Commentary on 29 p1].[183] Also from point z let zt be drawn parallel to hl, and cutting hm in point (say) t. Then through point t let the conic section tp be drawn, which Apollonius marked out in his book of *Conics* [4th. 2], taking zn, hl as asymptotes.[184] Similarly let a conic be constructed opposite (this branch) between these lines (as asymptotes), say cu.[185] Therefore when the minimal line segment from among the lines drawn from point t to the section cu is

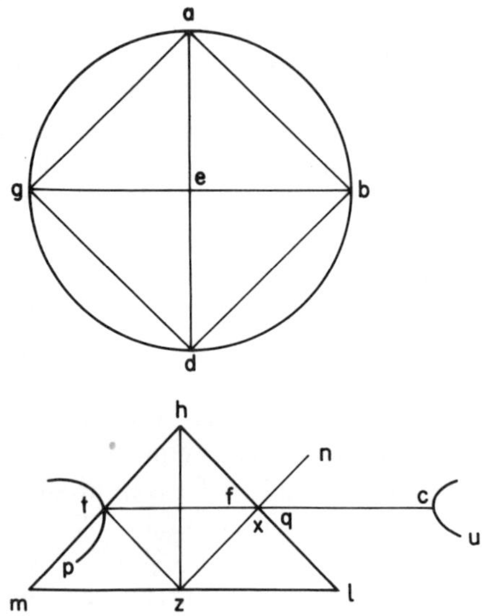

Figure 17. Figure for Proposition 34. Note: The figures for 34–36 reproduce the essential features of those in Risner's edition to give the reader some idea of constructions as represented in early printed books. Clagett (204), vol. 4, pt. 2, offers variants. (Reproduced with permission from D. C. Lindberg, ed., *Opticae Thesaurus Alhazeni Arabis*..., Johnson Reprint, N.Y., 1972, figures, p. 145.). © by Johnson Reprint Corporation, New York, N.Y.

equal to the diameter bg, the circle made (when) I place this minimal line as the opening of a compass with foot on point t, will touch the section cu. However, if the minimum from among the lines drawn from point t to section cu be less than bg, the circle made in the prescribed manner with bg as radius will cut the section in two points. Therefore, let tc be the minimum, and be equal to diameter bg: it will cut zn and zl, since it is drawn to a conic section, which is inclosed between them. And [by 31 p1] let there be drawn through z this line which will cut hm, hl [by lemma of Proclus' Commentary on 29 p1] and be parallel to tc. Therefore, let it intersect in points m, l, and call it mzl; and let the point of the conic in which tc cuts zn be q. Then [by 21 p1] on diameter gb let an angle say dgb be constructed equal to angle hlm, and let lines ad, db be drawn. Evidently, therefore, since [by 31 p3] gab is a right angle, the sum of the other two angles of triangle agb makes a right angle [by 32 p1]. Consequently angle lhm is right [indeed it follows from constructing two angles equal to the complementary angles[186] agb, abg], and it equals angle gdb, and [by construction] angle hlm is equal to angle dgb. Therefore [by 32 p1] the third angles are equal, and the triangles are similar [by 4 p1 d6]. Consequently the ratio of bg to bd is as lm to mh. But since [by 27 p3] angle adb is equal to angle bga, because they cut the same arc, and angle bga equals angle mhz [by construction] it follows therefore [by 1 ax] that angle adb equals angle mhz. And now we have that angle dbg is equal to angle hmz. Therefore [by 32 p1] the third angles are equal, and triangle deb is similar to triangle mhz [by 4 p1 d6]. Moreover, let e be the point at which line ad cuts the diameter bg. Then the ratio of bd to de is as mh to hz. As a matter of fact, Apollonius proves [16 th2][187] that when there be two opposite branches, and a line be drawn from one to the other, the part of it which lies between one of the branches and an asymptote (lit. line) is equal to the other part which lies between the other branch and the other asymptote. Consequently qt is equal to cf. But [by 34 p1] qt is equal to mz, since it be parallel to it, and between parallels. Therefore [by 1 ax] mz equals fc, and [by 34 p1] zl equals tf. Therefore [by 2 ax.] ml equals tc. Thus the ratio of tc to hz is as ml to hz [by 7 p5]. Hence gb to ed is as tc to hz. [For it has been demonstrated that gb is to bd as lm to mh, (and) in like manner that bd is to de as mh to hz; therefore, by 22 p5, that gb is to de as lm to hz. But as lm is to hz, so is tc to hz; consequently by 2 p5 as gb is to de, so is tc to hz]. And, since tc is equal to bg [by hypothesis] ed will be equal [by 14 p5] to hz. That is the proposition. Moreover if line tc be drawn to branch cu, and the minimum, but less than diameter bg, let it be produced beyond the branch, until it be equal (to bg), and let there be a circle made with this second quantity (as a radius), which will cut the branch in two points, from which lines drawn to t will be equal to bg [by 15 d1] and let parallels be drawn to each line through z. And then it will (only) remain to draw from the point a, in the prescribed manner two lines equal to the given line, with the same method of proof.[188]

Al-Haytham now proceeds to use Propositions 32–34 to prove 35, and 35, in turn, to establish 36. We offer these, in translation, with some notes, before making comments on his possible sources.

To bring the reader back into a more modern mathematical world for a

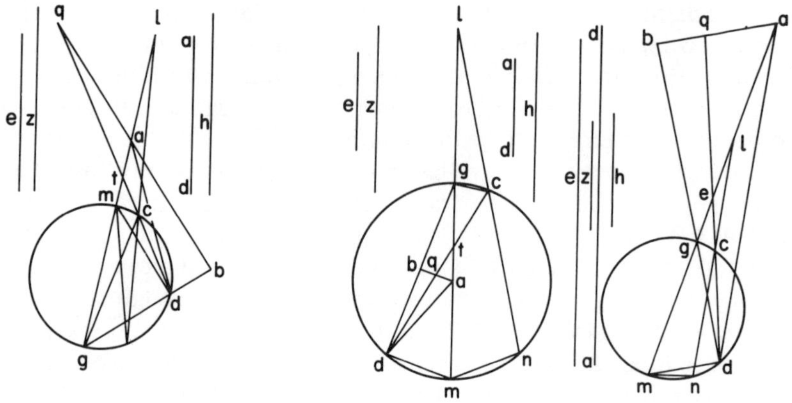

Figure 18. Figure for Proposition 35. Reproduced with permission D. C. Lindberg, ed., *Opticae Thesaurus Alhazeni Arabis*..., Johnson Reprint, N.Y., 1972, figures, p. 146.). © by Johnson Reprint Corporation, New York, N.Y.

moment, we note that Proposition 35 is essentially the following problem (Figure 19). $a, b, c > 0$ are given, and we want (α, β), or what amounts to the same thing, $(d, 0)$, $d > 0$, so that

$$\frac{\sqrt{\alpha^2 + (b - \beta)^2}}{\sqrt{(d - \alpha)^2 + \beta^2}} = \frac{A}{B} = r,$$

say, a given ratio. This is certainly easy to say with a combination of drawings and symbolic notation, but more difficult to express in words, particularly when those words have been filtered through the peculiarities of two intermediate languages. Moreover, one must keep in mind the proven prolixity of many Medieval Latin translations from Arabic, an issue discussed in the first paper of these essays.

Proof of Proposition 35. Amplification (see Figure 18): With a right triangle given, and from a given point on one of the sides containing the right angle, one is to draw from that point a line to the other side containing the right angle, cutting (first) the third side opposite to the right angle, so that the part of the line lying between the point of intersection and the side on which no point is given, has to

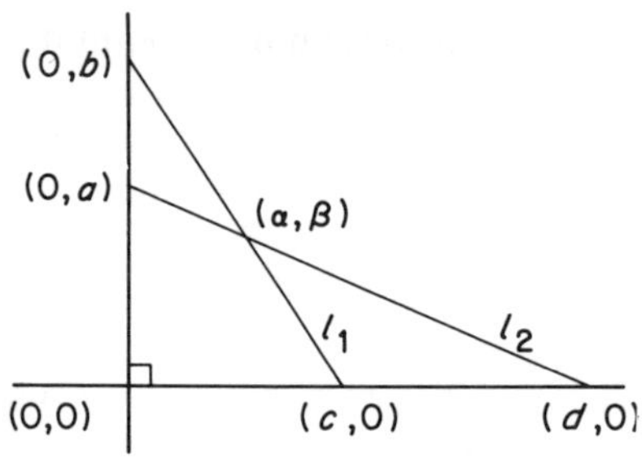

Figure 19

the part of the side opposite to the right angle, which is from the section to the side on which the given point lies, the same ratio as a given line to a given line. In other words, triangle abg is given, whose angle abg is right, and on side gb a point d is given, outside, or inside the triangle. I say that from point d, one is to draw a line, cutting side ag, and intersecting with side ab (extended) so that the part of it lying between sides ab, ag makes it ratio to the part of the side ag, which is from that line upto the point g, as does e to z, which are the given lines. Let d be the point on the triangle abg, and [by 31 p1] let there be drawn from it a line equidistant from ab, say dm, and [by 5 p4] through the three points g, m, d, let there be a circle: and let line ad be drawn. Since it is clear [by 29 p1] that angle gmd is equal to angle gab, gmd will be greater than angle gad. Let there be cut from it by line mn (an angle) equal (to angle gad), say dmn[189] and [by 12 p6] let h be a line to which ad stands as e to z. And from point n, which is a point of the circle, let a line be drawn toward the diameter gm and equal to line h, in accordance with what was said above [32 or 33n] and let it be denoted by nl, [so that segment lc between (its intersection with) the extended diameter and the periphery (of the circle) equals line h]. And let the point in which it cuts the circle be c, and let line gc be drawn, and from point d let a line be drawn to point c. Now, since it falls between two parallel lines, making an acute angle with one of them, if extended, it necessarily meets the other [by Lemma 29 p1 of Proclus]. Therefore they (cd and ab) intersect: and let q be the point of intersection. It is clear [by 27 p3] that angle gmd is equal to angle gcd, because they cut the same arc, and [by 29 p1] angle gmd is equal to angle gab; therefore, it follows that [by 1 ax 13 p1 3 ax] that angle gcq equals angle gaq. Let t be the point at which dq cuts ag. [By 15 p1] angle gtc is equal to angle atq; therefore, [by 32 p1] the third angle is equal to the third (i.e. $\langle$tqa = $\langle$tgc). Consequently triangle atq is similar to triangle tcq [by 4 p1 d6]. Therefore, qt and tg stand in the same ratio as at to tc. Indeed [by construction] angle nmd is equal to angle tad, and [by 27 p3] to ncd [that is by 15 p1 angle lct.] On this account [by 1 ax] angle lct equals angle tad: and angle ctl is common to two triangles (i.e. $\triangle$ctl, $\triangle$tad); whence [by 32 p1], the third angles are equal, and the triangles are similar. So, evidently, is triangle tlc to triangle tad [by 4 p1 d6]. Therefore, the ratio of ta to tc is as ad to lc. Whence [by 2 p5], the ratio of ad to lc is as qt to tg [it follows for, as qt is to tg, so is at to tc]. But [by construction] lc is equal to line h, and [by construction] the ratio of ad to h is that of e to z. Therefore [by 7, 11 p5] qt is to tg as e to z. That is our intention. If, in fact, d is taken "on that side" (but) outside the triangle,[190] let a line be drawn from point d [by 31 p1], equidistant from ab, say dm, and let ag be extended, until it intersects with dm in point m. [It does indeed intersect by Proclus 29 p1.] And let a circle pass through the three points g, d, m, and let line ad be drawn; it will indeed follow that [by 16 p1] angle gad is greater than angle gmd. Let it be made equal to this (gad), say as angle nmd; and from point n, which is a point of the circle, let a line be drawn equal to line h, [it will indeed be done by 33 or 34n: so that not all of the line drawn from point n, but rather the part of it coterminus with the extended diameter is itself equal to h], to which line ad stands as e to z, and let it be ncl [the whole line, undoubtedly, whose part cl equals h] on (intersecting with) diameter mg, with intersection l.

Since, therefore [by 22 p3] angle nmd and angle ncd form two right angles, and [by construction] angle nmd is equal to angle tad, the two triangles tcl, tad will be similar. [For, indeed, angle ncd and angle lcd form two right angles by 13 p1, from which one concludes then that lcd and tad do also, and therefore with the common part ncd removed, the remaining angles lcd and tad are equal: and the angles at the vertex t are equal by 15 p1, and by 32 p1, the third angles are equal. Consequently triangles tcl, tad are similar by 4 p1 d6.] And since [by 27 p3] the two angles gcd, gmd are equal, the triangles gtc, taq will be similar; [for, since angle gmd equals angle taq by 29 p1 (indeed, dm and ba are parallel by construction, angle taq is equal to angle tcg by 1 ax, and at vertex t the angles are equal by 15 p1, and on that account, by 32 p1 triangles gtc, taq are equiangular, and by 4 p1 d6 similar] and the ratio of ad to cl (which is equal to h)[191] will be as qt to tg: and e is to z, as qt to tg. [For indeed, triangles tad, tcl are equiangular; it will follow (then) by 4 p6 that ad is to cl as at to tc. On the otherhand triangles atq, tcg are equiangular; hence by 4 p6 at to tc will be as qt to tg. Consequently by 11 p5 it follows that ad is to cl (that is e to z) as qt to tg]. That is the proposition.

Proof of Proposition 36 (Figure 20). Clarification: From two given points, say e, d, and a given circle, one is to find a point on it, so that the angle contained by lines drawn from the prescribed points to that point, is bisected by the tangent line to

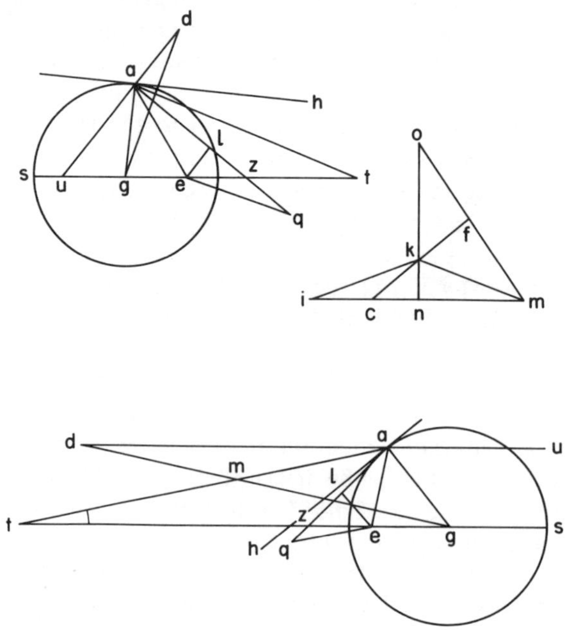

Figure 20. Figure for Proposition 36. (Reproduced with permission from D. C. Lindberg, ed., *Opticae Thesaurus Alhazeni Arabis*..., Johnson Reprint, N.Y., 1972, figures, pp. 147–8 (top).) © by Johnson Reprint Corporation, New York, N.Y.

the circle at that point. In other words, let there be drawn from point e to the center of the given circle, line eg: and let it be extended to the circumference, and call it es. Then let line gd be drawn. And [by 10 p6] let a line mi be divided at point c, so that ic is to cm,[192] as eg is to gd, and [by 10 p1] let mi be bisected at n, and [by 11 p1] let the perpendicular no be drawn, and on point m make an angle equal to half of angle dgs [by 9 and 23 p1] with terminal side mo. It is clear that it will be less than a right angle. [For the angles at g are taken one after the other (i.e. additively) equal to two right angles by 13 p1. Thus, dgs is itself less; consequently dgs halved (lit. diminished) is less than a right angle] and onm is right. Therefore, [by 2 ax] it intersects with no. Let it intersect in point o, and let a line be drawn from point c [by the preceding number (evidently 35)] to the triangle, say ckf, so that the ratio of kf to mf is as that of eg to gs. And [by 23 p1] on point g [end of side eg] make an angle equal to angle mfk by a line drawn up to the circle, say ag; and let the angle be age, and let lines ag, ad be drawn. I say that a is the point which we require. Let line ea be drawn. Since, therefore, angle mfk [by construction] be equal to angle age, and [by construction] the ratio of kf to mf, be as ge to ga, and since [by 15 d1] ga be equal to gs, triangle age will be similar to triangle mfk [by 6.4.1 d6]. Therefore angle fmk is equal to angle eag, and angle aeg is equal to angle mkf. Now [by 23 p1], let a line be drawn from point a, making with line ae an angle equal to angle nmk, say line az, which, necessarily, intersects with line eg. Note: (lit. *Quonian*) whatever the ratio of kf to mk, this is that of ea to eg, and angle gaz is equal to angle fmc. [Equal; for, angle fmk is contained in (i.e. equal) to angle eag.] Therefore just as line mo intersects with fk, so az intersects with eg. Let the point of intersection be at z, and extend az upto a point q; so that line az is to zq as mc is to ci; [by 12 p6] and let line eq be drawn. Then [by 31 p1] let a line be drawn through point a parallel to eq: say at; indeed [by 29 p1] angle aqe will be equal to angle qat. And since the two angles zea, eat are less than two right angles [because by 29 p1 angles qea, eat are equal to two right angles],[193] at necessarily intersects with ez [by 2 ax]. Let t be the point of intersection. It is clear [from earlier demonstrations] that angle aeg is equal to angle mkf. On the other hand, let el be a line drawn from point e perpendicular to az; [by 32 p1] angle ael will be equal to angle mkn, since [by construction] angle eal be equal to angle kmn, and angle alc is equal to angle mnk: as both are right. Therefore, it follows [by 13 p1. 3 ax] that lez equals angle nkc, and angle elz is right, and equal to angle knc. It follows [by 32 p1] that angle ezl be equal to angle kcn; therefore, [by 13 p1. 3 ax] angle ezq is equal to angle kci. It is clear, therefore, [by 4 p.1 d6] that triangle eag is similar to triangle fmk, and triangle eal is similar to triangle kmn; and triangle elz is similar to triangle knc and triangle eaz is similar to triangle kmc. [For by construction angle eal is equal to angle kmn, and angle ezl is evidently equal to angle kcn; therefore by 32 p1, the remaining angles are equal, and on that account, by 4 p.1 d6 triangles eaz, kmc are similar.] Therefore the ratio of az to ze is as mc to kc, and [by construction] the ratio of az to zq, is as mc to ci. Therefore [by 22 p5] qz is to ez, as ic to ck. Whence [by 6.4 p.1 d6] triangle qze is similar to triangle ick, and triangle qle is similar to triangle ink, [since triangle elz was shown to be similar to triangle knc, and thus, since the parts are similar, the entire triangles qle and ikn will be similar. Whence by 1 d6, as ql is to le, so is in to nk: and similarly from the similitude of triangles ael, kmn, as el is to al, so is kn to mn]. Therefore [by 22 p and the consequences of 4 p5] the ratio of mn to ni is as al to lq, and thus al equals lq [since mn is equal to ni], and [by 4 p1] eq will equal ea; and angle eqz equals angle lae, and [by construction and 29 p1] angle eqz is equal to angle zat. Therefore [by 15.32 p1] the third angle is equal to the third (i.e. $\langle$ezq = $\langle$azt, so $\langle$zeq = $\langle$atz). Whence [by 4 p6] the ratio of qz to za is as ez to zt, and as eq is to at, and [by 7 p5] as ae is to at. But qz to za, is as eg to dg [for by construction eg to gd is as ic to cm; likewise, as cm is to ic, so is za to zq, and by consequence of 4 p5, as ic is to cm, so is zq to az. Therefore by 11 p5, as eg is to gd, so is zq to za.] Therefore [by 11 p5] ae is to at, as eg to gd. Moreover [by 23 p1], through point a let an angle, say uag be constructed equal to angle gae. It is clear that angle gal is equal to half of angle uat. [For, indeed, from above, angles zat, zae are both equal to angle zqe; hence, they are equal to each other. Also if equals be added to equals, angle gal is equal to (the sum of) the two angles uag, zat. Whence uat will be the double of angle gal]. But (gal) is half angle dgu, [since angle gal by the above argument is equal to angle fmc, which, by construction is one half angle dgu.] Whence, angle uat is equal to angle dgu: [by 6 ax], but angles uat and tua are (together) less than two right angles [by 17 p1], since at and tu intersect. Hence the two angles tua and dgu are less than two right angles; therefore, [by 2 ax] au intersects with dg. I say that it does at point d, since it (au) effects, with lines ug, gd a triangle similar to triangle aut; for, they have aug in common, and angle tau is equal to angle dgu [by deduction]. Therefore [by 4 p6] the ratio of au to at is, as ug to the line which au (extended) cuts from dg: and [by 3 p6] the ratio of ea to au is, as eg to gu, since angle uag be equal to angle gae [by construction]. Since, therefore, the ratio of ea to at be the same, as eg to gd [by deduction] and the ratio of ea to at, be the composition of ea to au and au to at [for the ratio of the extremes is composed from all the intermediate ratios, as Theon demonstrated in 5 d6], the ratio of eg to ad will be composed in the same (way).[194] Whence it will be composed from the ratios eg to gu, and gu to the line which au cuts from dg. But [the ratio of eg to gd] it is composed of the ratios, eg to gu and gu to gd. Therefore, the line which au cuts from gd is the line gd; therefore, au cuts dg in point d. Therefore [by 17 p3] let a tangent be drawn from point a, say ha; [by 18 p3] gah will be a right angle, but gal is half angle dgu. Therefore angle lah is half angle dge:[195] since these two [dgu, dge] sum to two right angles [by 13 p1.] But since angle tau be equal to angle dgu, angle tad will be equal to angle dge[196] [by 13 p1. 3 ax]. Therefore angle lah is one half angle tad, and angle eal is half angle eat [for, as indicated, angle eal is itself equal to angle lat]. Thus angle eah is half angle ead. Whence ah bisects angle ead. That is the proposition. If, in fact, au (since the angle at a be equal to angle gae) does not fall on es outside the circle, or inside: it is therefore parallel. Hence [by 29 p1] angle uag is equal to angle age: but the same angle (i.e. uag) is equal to angle gae [by thesis, (actually by construction)]. Consequently [by 1 ax.] angle gae is equal to angle age. Therefore, [by 6 p1] eg is equal to ae. Similarly angle tad will be equal to angle atg [by 29 p1]. But it was already stated [in the first part of this number] that angle tad is equal to angle dgt. Therefore angle atg is equal to angle dgt, and similarly [by 29 p1] the two angles adg, dgt are equal. Therefore, the two angles adg, atg are equal. It follows, therefore, from these things, that the line, which au cuts from dg be equal to the line at [for, since angles atg, dgt, and also adg, tad are equal, by 6 p1 tm will be equal to mg and also md to ma. Also if equals be added to equals, (equals result): dg will equal at.] And it was already noted, that eg be equal to ae. Thus [by 7 p5] the ratio of eg to the line, which au cuts from dg, is as ae to at. But as already noted ae is to at as eg to gd; therefore, the line which au cuts from dg, is dg. And since tad be equal to dgt, lah is half angle tad, as was said above, and angle eal half angle eat. Therefore eah will be half angle ead. That is the proposition.

Logically we should continue with a statement of Propositions 60, 62, 65, 66 and proofs of 70–75, culminating in the Proof of Proposition 87. But it is time that we digressed, and made a serious search for possible sources of al-Haytham's constructions, and the sources of the problem he discusses. Mathematically, al-Haytham has offered the following results:

1 (Proposition 32): Given a circle with diameter gb and point a not on the diameter, to construct a line from a to the diameter extended so that hd = qz, a given line segment (Figure 21). Here al-Haytham specifically assumes that ag = ab.

2 (Proposition 33): This is the same as 32 with the equality restriction on ag, ab removed, i.e. ag > ab or ag < ab in the construction.

3 (Proposition 34): Given a point a on a circle agb with diameter gb, and a not on gb, and a segment hz, to construct a line through a, making ed = hz (Figure 22). Al-Haytham places the natural restriction hz < gb.

4 (Proposition 35): Given a right triangle ABC with a point d on a side containing the right angle, to construct a line through d, cutting the hypotenuse and the other side (extended) so that ab/Ab = r, a given ratio (Figure 23). Actually

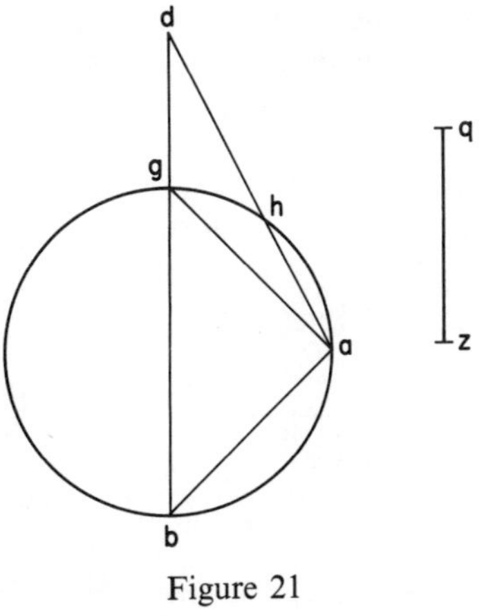

Figure 21

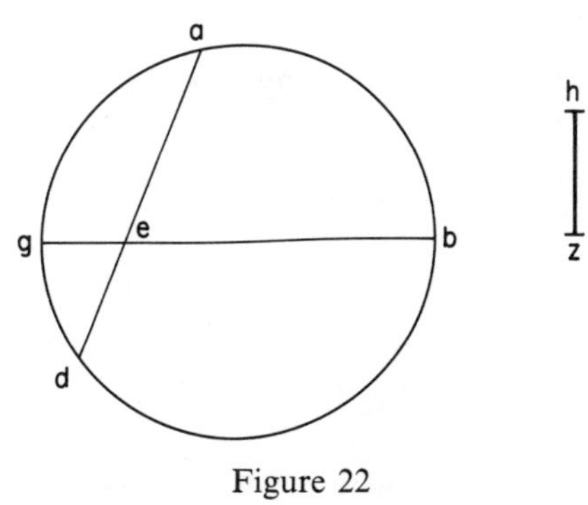

Figure 22

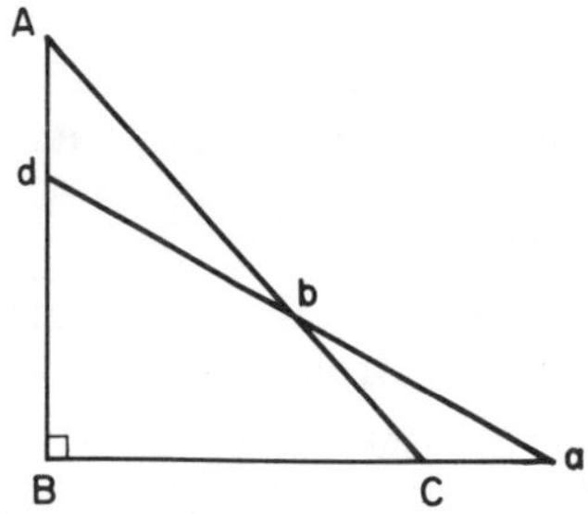

Figure 23

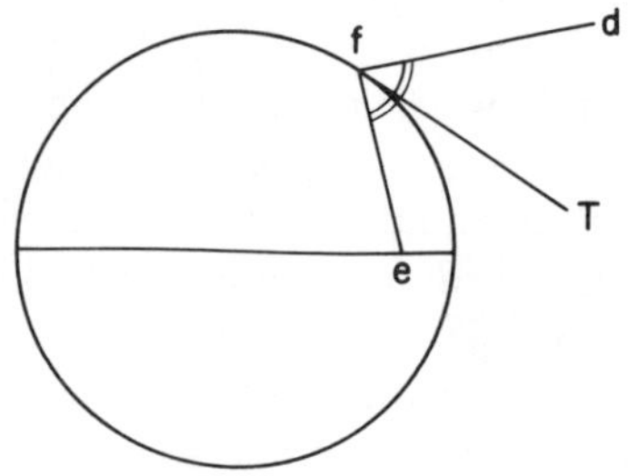

Figure 24

the construction is more general, and allows d to be "inside" or "outside" the triangle, in a sense prescribed by al-Haytham.

5 (Proposition 36): Given points d, e, and a circle with e inside and d outside, to find a point f on the circle so that the tangent T at f bisects the angle dfe (Figure 24).

6 (Proposition 37): Given a circle with centre g, radius gb and external point e, to draw a line from e cutting the circle so that dz equals gz (Figure 25).[197]

7 (Proposition 38): Given a right triangle abg with right angle at b, and point d on bg (or ab) to draw a line through d making a segment qt so that qt/gq = e/z, where e, z are given segments (Figure 26).

Recall that number one requires only the intersection of a circle and a line for its solution, while 2 uses a circle and a hyperbola; 3 does the same; 4 uses 2 and 3; 5 uses 4. Six and seven require three.

Of course constructions of this general type are not original to al-Haytham. We have already mentioned the Greek school of geometrical analysis discussed by Gulley and Mahoney.[198] In fact, within this body of analysis, these problems are known as νεῦσις, or problems of "inclination" or "verging". Both Heath and Zeuthen[199] present discussions and refer to a passage in Pappus' *Collection* in which the author tells of a (now lost) work of Apollonius on the subject.[200] Pappus' account seems to imply that the method of neusis was highly structured. He begins with a definition and follows this with subcases.[201]

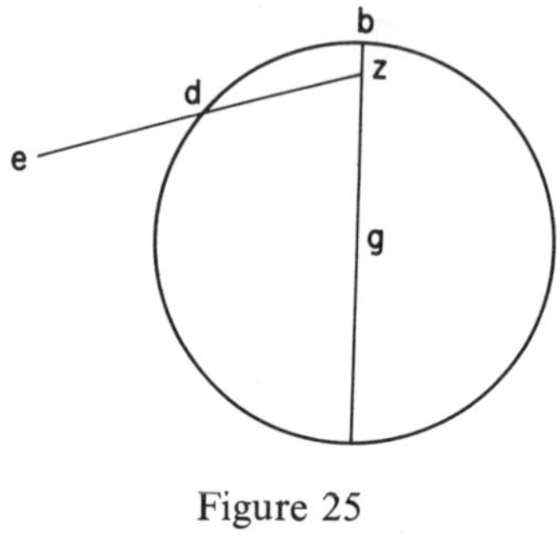

Figure 25

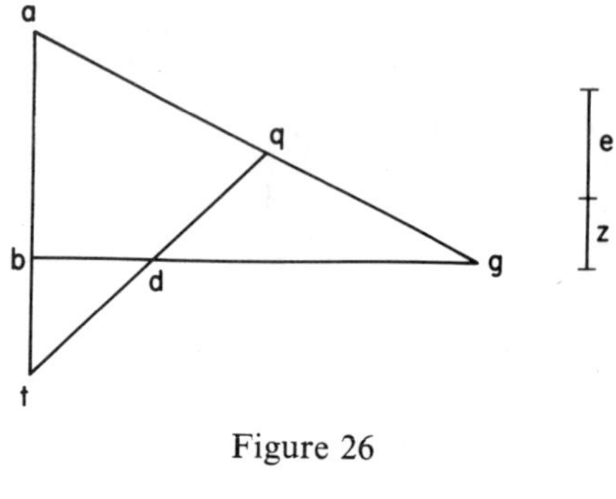

Figure 26

"A line is said to incline or verge toward a point, if it reaches the point on being extended. [It is altogether the same whether the line is said to incline toward a given point, or some point is given on it, or it passes through a given point. Indeed the term is used without distinguishing cases.][202] Now with the general problem, these are cases:

1. Two lines in given position, to place between them a given straight line segment which verges toward a given point,...
2. If a semicircle and a line perpendicular to its base, or two semicircles having bases collinear have been given, to place a given line segment between these two "lines", which verges toward the corner (literally, angle) of the (a) semicircle.
3. A rhombus given with one side produced, at an external angle, to insert a given line segment which verges toward the opposite angle.[203]
4. A circle in given position, to insert a given line segment which verges toward a given point."

Pappus then notes that these problems are distributed in Apollonius' *Neusis* as follows: in book I, the semi-circle and the line (4 cases), the circle (2 cases) and the rhombus (2 cases); in book two, the two semicircles (ten cases).[204] In other passages, for example, in book IV, he refers to neusis constructions in Archimedes' *On Spirals*.[205] These particular constructions, as well as some in Archimedes' *On the sphere and cylinder* are possible precursors of al-Haytham's Propositions 32–38,[206] in much the same way, as we have seen, that Archimedes' constructions in the latter work prefigured Near Eastern work on cubics.[207]

Before we examine these texts for parallels, it will be fruitful to list those known in Arabic in al-Haytham's time. The *Conics* of Apollonius (K. al-Maḫrūṭāt) certainly was. Recall that Hilāl ibn Abī Hilāl al-Ḥimsī had translated the first four books with the help of Aḥmad ibn Mūsā, and Thābit ibn Qurra had translated three more.[208] In fact, a copy survives in al-Haytham's handwriting.[209] Nonetheless, the *Conics* itself contains no neusis. However the *K. fī Qaṭ ͨ al-ḫūṭūṭ ͨala n-nisab* (Cutting lines in ratio), the *λόγου ἀποτομή* of Apollonius which is lost in Greek was extant in Arabic as early as the time of ibn al-Nadim.[210] The Arabic text was translated in the eighteenth century by Edmond Halley.[211] It does contain neusis, and we examine it momentarily. Similarly, the *Treatise on the sphere and cylinder* and part of Eutocius' *Commentary*, as we have already noted, were known by the time of Thābit, who is credited by Naṣir al-Dīn al-Ṭūsī with improving the existing translation.[212] Indeed, al-Haytham even wrote a short study, *Qawl fī Qismat al-Khaṭṭ alladhī ista'malahu Arshimīdis fī Kitāb al-Kura wa l'usṭuwāna* on the construction in proposition 4 of the second book.[213] On the other hand, the Archimedian treatise *On spiral lines* does not seem to have been translated into Arabic; *The Fihrist* makes no mention of it, and neither Steinschneider nor Sezgin cite manuscripts.[214] Pappus' *Collection*, except for book VIII on mechanics, which was probably translated into Arabic about the time of the Banū Musā, never entered the Arabic mathematical tradition.[215] Unguru, who edited Witelo's *Perspectiva* (1270) which is based on al-Haytham's *Optics*, has made a case for some of Witelo's propositions being from the Pappus' *Collection*,[216] namely Pappus VI, 42; VI, 44.[217] However these parallels are lacking in ibn al-Haytham's book V which served as Witelo's model for *Perspectiva*, I. Finally, it should be noted that al-Haytham apparently shared the concern of Greek geometers for problems of analysis and synthesis, as his extant treatises *Maqāla fī l-taḥlīl waʾl-tarkīb* and *Maqāla fī l-Maʿlūmāt* bear witness.[218] Consequently his mastery of the *Elements*, *Conics*, and the *Data*, and his undoubted mathematical talent might in themselves explain the *tour de force* of book V of the *Optics*.

Clagett's observation that ibn al-Haytham's neusis in proposition 32 appears (without explanation) in *On spiral lines*, proposition 4 is certainly true,[219] and there is indeed, as he claims, some similarity between ibn al-Haytham's proposition 33 and *On spiral lines*, porpositions 6–7 (neusis assumed),[220] proposition 34 and *On spiral lines*, propositions 8–9 (neusis again assumed without explanation).[221] However, since *On spiral lines* was unknown in Arabic, this does not shed any light on al-Haytham's source. Similarly, *On the sphere and cylinder* which was, contains none of the al-Haytham constructions.[222]

This leaves us with the *Cutting of lines in ratio*. As Heath observed, Apollonius treats the general problems: "Given two straight lines, parallel to one another or intersecting, and a fixed point on each line, to draw through a given point a straight line which shall cut off segments from each line (measured from the fixed points) bearing a given ratio to one another."[223] The case of parallel lines is treated in the first book, while that of intersecting lines (which Apollonius reduces to the former) is the subject of book two. One might expect the case of intersecting lines to deal with a neusis similar to ibn al-Haytham's proposition 35. However, in al-Haytham's context, Apollonius' construction allows one to get $AM/BN = r_2$ directly, while one wants $AM/MN = r_1$. (See Figure 27.)

After solving for (α, β) in terms of a, b, c, and the unknown d, one finds that:

$$r_1 = \frac{(b-a)\,d\sqrt{b^2+c^2}}{(d-c)b\sqrt{d^2+a^2}} \quad \text{and} \quad r_2 = \frac{(b-a)(d)(d-c)\sqrt{b^2+c^2}}{(bd-ac)}$$

So r_2 satisfies a quadratic equation in d, while r_1, in general, satisfies a quartic in d. Moreover r_1 and r_2 are evidently related by:

$$r_1 = \frac{r_2(bd-ac)}{(d-c)^2 b\sqrt{d^2+a^2}}$$

Still, this does not preclude al-Haytham's having found some inspiration in this

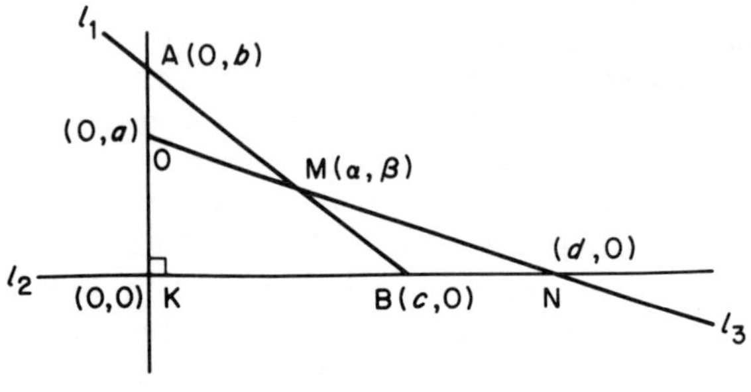

Figure 27

work of Apollonius, provided that he ever saw it. On the other hand, one must keep in mind that to one of al-Haytham's skill in the use of Apollonius' *Conics*, the circle of ideas involved in neusis constructions would not have seemed very large.

It would take us too far astray to examine in detail the possible inspiration for al-Haytham's problem (book V) itself. Suffice it to say that he was familiar with Euclid's *Optics* (Ikhtilaf al-Manāẓir) translated into Arabic by Ḥunain ibn Isḥāq and corrected by Thābit,[224] and Ptolemy's *Optics* (Kitab fi ʿl-Manāẓir), presently existing only in the Latin translation made from the (now lost) Arabic version of Eugene the Amir.[225] Euclid's *Optics* or more properly his *Catoptrics* contains only a few simple propositions dealing with reflections from plane and concave mirrors, while Ptolemy's *Optics*, which as Heath notes "suggest(s) that the author was not very proficient in geometry"[226] attempts a solution of "Alhazen's problem" for concave mirrors, based on a false assumption about the nature of reflections. The mathematics of this work seems to be based entirely on the *Elements*, the *Catoptrics* of Euclid and Theon, and Proclus' *Commentary on the Elements*. Nothing of the *Conics* is in evidence.

The four intermediate propositions 60, 62, 65, and 66 deal with locations of images in a spherical concave mirror. In order they are:

Proposition 60. In a concave spherical mirror, the image is visible at the point of reflection, in the eye, beyond, outside the mirror, between the eye and the mirror.[227]

Proposition 62. The eye placed at the centre of a concave spherical mirror sees only itself.[228]

Proposition 65. (Given) the eye and the visible point on the diameter of a concave spherical mirror at equal distances from the centre, it is possible to have a reflection from the entire periphery for which the semidiameter perpendicular to the said diameter cuts it (the periphery).[229]

Proposition 66. The eye and the visible point on different diameters of the circle, which is the section common to the surface of reflection and of the concave spherical mirror, reflect from one to the other, first from the perimeter between the semidiameters on which they are, then from the opposite arc to this, but not from the remaining two.[230]

As indicated above, these have little inherent mathematical interest, so we omit al-Haytham's proofs. Since our goal is the proof of Proposition 87, we present translations of the remaining relevant preliminary lemmas: 70–75. These are also of lesser mathematical interest than 32–36, but are offered for completeness.

Proposition 70. Amplification (see Figure 28): When the visible point and the centre of the eye are not on the same diameter, and are outside the mirror, there is no reflection from the visible point to the centre of the eye, except from just one

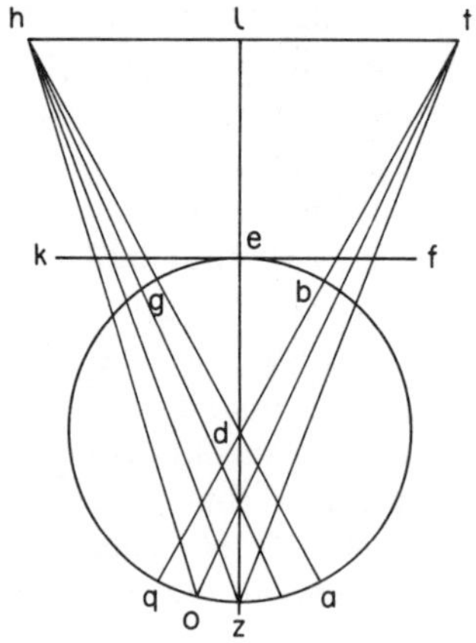

Figure 28. Figure for Proposition 70. The figure in Risner labels the intersection of ho and tq as p. However, the actual theorem makes no reference to this intersection point . (Reproduced with permission from D. C. Lindberg, ed., *Opticae Thesaurus Alhazeni Arabis...*, Johnson Reprint, N.Y., 1972, figure, p. 167 (bottom).). © by Johnson Reprint Corporation, New York, N.Y.

point of the mirror. In other words: let t be the visible point, h the centre of the eye, d the centre of the sphere, and draw lines hd, dt, ht. The plane[231] of hdt cuts the sphere in a circle [by 1th. I spher.]: this makes circle ebqg. Clearly t is not reflected to h except from some point of this circle. Therefore extend hd, td, to the circumference of the circle. Clearly, there is no reflection from arc gq, or arc ab. Reflection occurs either on arc gb or aq, as noted. [66n] Angle tdh is divided into two equal angles by line ledz [by 9 p1], and at point e a tangent is drawn, [by 17 p3] (say) kef. If points t and h be concurrent with it (kef), there is no reflection from t to h along any point of arc bg. Indeed, when a line is drawn from point t to some interior point of this arc, a line drawn from the point h to this point, falls on its exterior (lit. exterius = opposite surface) rather than on its interior. And for this reason, there will be no reflection [on a concave mirror surface]. And because from one point only, the arc aq makes a reflection; clearly, it will be from this. Indeed, draw lines tz, hz. Since angle tdh is bisected, angle tdz will be equal to angle hdz. Therefore, the lines td, hd are either equal, or not equal. If they are equal, and dz is common, triangle tzd will be equal to triangle hzd [by 4 p1], and angle tzh is bisected by line dz. And thus t is reflected to h at point z [by 12n]. Because this cannot happen from another point, the matter is settled. Point o is taken,[232] and lines to, ho are drawn, and line odm through the centre d divides that angle (hot) into two equal parts. Plainly [by 8 p3] tz is less than to, and ho less than hz, and the ratio of tz to hz is as that of tl to lh, [by 3 p6, for the angle tzh is bisected by the line zl] and it will be [for the same reason, that the ratio of to to ho is as that of tm to mh, but the ratio of ho to to is smaller than the ratio of hz to tz] because, among the four lines ho, to, hz, tz, the first is less than the third, the second greater than the fourth; the ratio of the first to the second is smaller than that of the third to the fourth as follows from 8 p3. Therefore, [by 2 p5] the ratio of hm to mt is smaller than that of hl to lt, which is impossible. [For, among the four lines hm, mt, hl, lt, the first hm is greater than the third hl; the second, mt, is certainly less than the fourth lt; so, it will follow that the ratio hm to mt is greater than that of hl to lt, as is established from 8 p5]. Clearly, therefore, if t and h are at

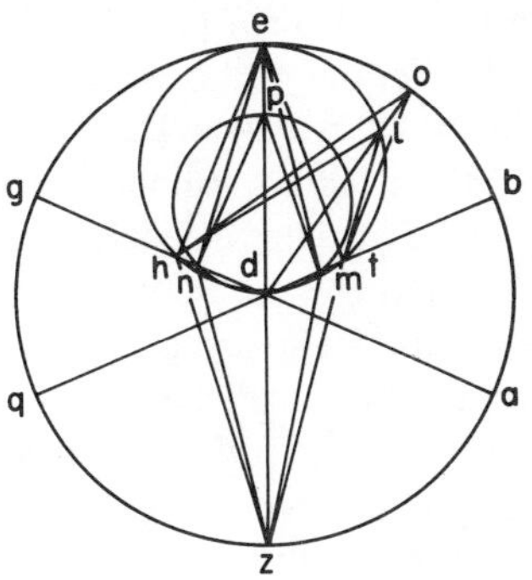

Figure 29. Figure for Proposition 71. (Reproduced with permission from D. C. Lindberg, ed., *Opticae Thesaurus Alhazeni Arabis*..., Johnson Reprint, N.Y., 1972, figure, p. 168.). © by Johnson Reprint Corporation, New York, N.Y.

equal distance from the centre, and are away from the tangent, there is no reflection from t to h, except from one single point of the mirror, and this unique point will be the site of the image. If, moreover, td, dh are unequal, let them be cut equally, and it may be demonstrated as before.

Proposition 71. Amplification (see Figure 29): Let bdq, adg be two diameters of the sphere, and diameter edz bisect angle bdg, and from point e, two perpendiculars be drawn, onto the diameters bd, dq, say et, eh. Clearly, because triangle etd is equal to triangle ehd, and angle ted equals angle hed, (the sides are equal correspondingly = *latus q'*), td equals side hd, and side et equals side eh. Since ed makes a common side, t is therefore reflected to h from point e—and, in the same manner, from point z. [Certainly, because angle bdg is bisected by line ed, angle tdz will be equal to angle hdz by 13 p1, and td is equal to hd from the (above) conclusion, and dz is a common side; therefore, angle tzd equals angle hzd by 4 p1. Therefore by 12 n4, t and h are reflected from one to the other through z.] And clearly [through 66n] it follows that t is not reflected to h, from any point of the arc ab, or the arc gq; nor is it reflected from any point of arc aq, except from the point z mentioned in the second proof [preceding number]. Indeed, because reflection is not possible from any point of arc bg, except from point e, the matter is settled. A point o is taken,[233] and lines od, ho, to are drawn; that done, (make) a circle for the line segment de, passing through the three points t, d, h. [This follows moreover from the discussion 31 p3 given in Theon in the commentaries on the three big books of constructions of Ptolemy, and in Campanus 31 p13.] The line de of this circle will be the diameter, since the angle etd, as provided for, is right. Therefore the circle will pass through e. Thus, since e is common to both circles, and lies on the diameters, the smaller circle touches (i.e. is tangent to) the larger at point e. So Euclid did prove [13 p3]. Therefore the circle will cut the line do, [otherwise it would touch the larger circle in point o, i.e. touch in two points e and o, contradicting 13 p3], say in point l. Draw lines tl, hl. Now it is clear [from above] that td is equal to hd [therefore through 28 p3 the arc (peripheria) td is equal to the arc hd]. Therefore angle tld equals angle dlh [through 27 p3] because they are on equal arcs. It follows that [through 13 p1] angle tlo equals angle hlo; and angle lot equals angle loh by hypothesis [because they are (respectively) angles of incidence and reflection] and lo is a common side. (So) [by 26 p1] triangle tlo will be equal to triangle hlo; and to will be equal to ho, which is impossible, since [by 7 p3] ho is greater than he, and to less than te. And te, as was first shown, is equal to he. [Therefore, the line ho is bigger than the line to.] It follows, therefore, that t is not reflected to h from any points, except from either e or z. Also from point e let a line be drawn through the diameter td, making em, and let a part be cut from the line hd, equal to md, say nd, and let em, en be drawn. Clearly [by 16 p1] angle emd is greater than a right angle, [because angle etd is a right angle by construction]. Let there be cut from ed[234] an angle equal to a right angle by a line pm [by 23 p1] which intersects with de [by lemma of Proclus 29 p1]. Let p be the intersection point, and let np be drawn; and let therebe a circle with respect to the segment pd, running through the three points m, d, n. Since pmd is a right angle [by construction] pd will be a diameter [as a consequence of 5 p4] and the circle will pass through p [as is plain]. It is clear, therefore, that m is reflected to n by point e, [for, since by 4 p1 triangle dmp is equal, both in angles and sides to triangle dnp, mp and np are equal, and angle dpm equals angle dpn; therefore by 13 p1. 3 ax angle mpe equals angle npe, side pe is common; therefore angle mep equals angle nep by 4 p1; whence by 12 n4, m and n are reflected from one to the other by point e], and similarly by point z; and not from any other points of the arcs ab or gq, [by 66n]. And clearly, there is no reflection from any point of arc aq, except from point z, and none from any point of arc bg, other than from point e, in the second noted manner. Indeed, with a point selected, and lines drawn from points t, d, h, and through which point the largest possible circle cuts the diameter, and with lines drawn from the cut points to the points t, h, the earlier (argument, construction) will be impossible. Clearly, therefore, it is as we have indicated; if the angle contained between two diameters is bisected by a third diameter, and from the ends of the third, perpendiculars are drawn to the other two, the points of the diameters upon which the perpendiculars fall are reflected from one to the other only from two points of the mirror. Moreover, of the points of the diameter on this side of the selected perpendiculars, that is, nearer the centre, each is reflected only from two points, and is reflected to that point which is equidistant from the centre, and each such pair is the locus of the image.

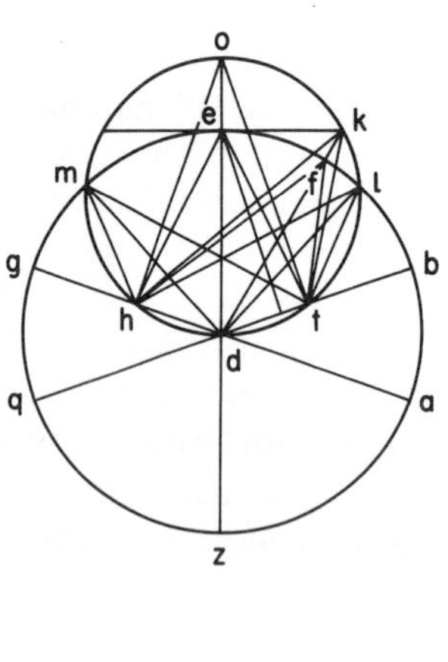

Figure 30. Figure for Proposition 72. Note that the Latin text does not explicitly place k on circle odh, until (†) in the proof, i.e. Risner's drawing is correct and f is not arbitrary. (Reproduced with permission from D. C. Lindberg, ed., *Opticae Thesaurus Alhazeni Arabis*..., Johnson Reprint, N.Y., 1972, figure, p. 169.).

© by Johnson Reprint Corporation, New York, N.Y.

Proposition 72. Amplification (see Figure 30): With the two diameters bq, ag given, and ez bisecting their common angle (gdb), a point t is taken on bd, beyond[235] the point on which the perpendicular drawn from point e falls, and on dg, dh is taken equal to dt, [by 3 p1], and te, he are drawn. At least, there is reflection from t to h by point e, and similarly by point z, but from no other point of arc aq, nor from any point of arcs ab or gq [by 66n]. Then from point t a perpendicular is drawn on td, [by 11 p1], which at least intersects with de outside the sphere, since angle bde is acute [as is demonstrated by 36n: because de and the perpendicular on td by 2 ax are concurrent, and indeed, outside circle bzg. Since this perpendicular and (ea) that which is drawn from point e onto the same semidiameter db are parallel by 28 p1, they never intersect by 35 p(d), 1. Because the perpendicular from point t (is) continued, it cuts (de extended) beyond the circle, away from e. Therefore, it intersects with the semidiameter ed outside circle bzg.] Let them therefore intersect in point o, and lines to and ho be drawn. And make a circle pass through the three points t, d, h: which will necessarily pass through point o [as was demonstrated in the preceding number] and do will be its diameter: [by the discussion of 5 p4] and let a line be drawn tangent to circle bzg at point o [by 17 p3], say ke. Clearly, since the outer circle will cut the first, that is, bzg in two points, [by 10 p3] there are two points l, m. And let lines tl, hl, ld, tm, dm, hm be drawn. Since, therefore arc td equals arc hd, [by 28 p3, because dt, dh are equal by construction], angle tld [by 27 p3] will be equal to angle dlh. And thus t is reflected to h from point l [by 12 n4]. Similarly angle tmd equals angle dhm [by 27 p3.] And thus t is reflected to h from point m. It is clear, therefore, that t is reflected to h from four points: namely, e, z, l, m, and the quadruple will be the locus of the image. And it is not possible for t to be reflected to h from any other point than one of these. Indeed, let point f be taken;[236] and lines tf, hf, df be drawn, and df extended until it intersects with the tangent ke, [it intersects, indeed, by 2 ax: because angle ked is a right angle by 18 p3, and fde acute, because it's a part of arc bde], and let the intersection be called k; and the lines tk, hk be drawn. Therefore angle tfd is equal to angle dfh by hypothesis, [and by 12 n4] it follows that angles tfk and kfh are equal. But angle tkf† is equal to angle fkh [by 27 p3][237] since they are on equal arcs: and fk is common. (So) the triangles [by 26 p1] are equal; and thus tk equals hk, which is impossible, because hk is larger than ho, and tk smaller than to [by 7 p3] and to is equal to ho. [Because the segment dt is equal to dh by construction, and angle tdo is equal to angle hdo by thesis, and line od is common, therefore by 4 p1 line to equals line ho; thus tk is less than hk.] Clearly, therefore, there is no reflection from any point other than these four. Therefore, if on different diameters two points be selected, say t, h, equidistant from the centre, and if they be over these points of the diameter, on which the perpendiculars drawn from the end point of the angle bisecting diameter fall; or lie between the centre and those points, i.e. on this side of the perpendiculars, while equally distant from the centre, reflection from t to n occurs at only two points. If, in fact, t and h be between the locus of the perpendiculars and the circle, reflection occurs from t to h through four points. If they be on the circle, or beyond, yet on this side of the tangent ke: reflection from t to h is possible at only

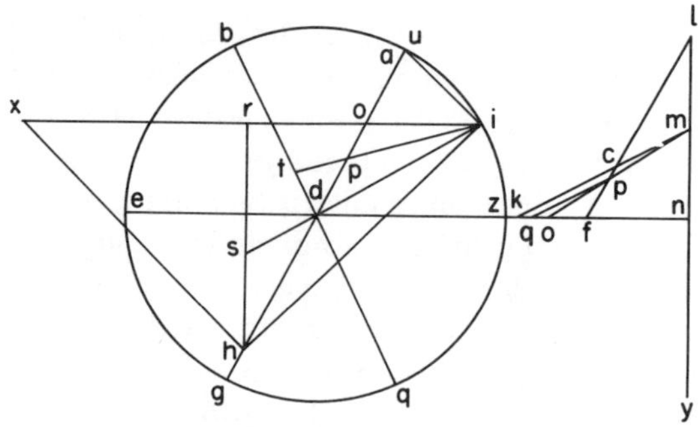

Figure 31. Figure for Proposition 73. (Reproduced with permission from D. C. Lindberg, ed., *Opticae Thesaurus Alhazeni Arabis*..., Johnson Reprint, N.Y., 1972, figure, p. 170.). © by Johnson Reprint Corporation, New York, N.Y.

two points. If, indeed, they be beyond the tangent reflection occurs between t and h from one point only. And these things happen when t and h are equidistant from the centre.

Proof of Proposition 73. Amplification (see Figure 31): Let t, h be on different diameters, and at different distances from the centre of the circle: reflection may occur from one point. In other words, let diameters adg, bdq be drawn, and let ez bisect the angle between them, and let t be closer to the centre d than h. And let line ly be taken, and [by 10 p6] let it be divided at point m, so that the ratio that ym makes to ml is that of hd to dt; and let ly be bisected at point n [by 10 p1], and from point n let perpendicular nk be drawn, [by 11 p1] and on point l make an angle equal to half angle adt with side fl. Thence angle fly will be acute [because it is equal to a smaller angle than the acute angle adt, as is evident from 36n]. Because [by 2 ax] fl intersects with nk, [because lnk is right by construction], let the intersection be in point f say, and [by 35n] from point m let a line be drawn to side fl, intersecting with side nk in a point, say k, and let that line cut side fl in point c, so that kc may make the same ratio to cl as hd to dt.[238] Then on point d make an angle equal to lcm [by 23 p1] say angle ida: and let i be a point of the circle above z, or below; and on point i make an angle equal to clm, say oid; and on this line oi [continued] let a perpendicular be drawn from point h, say hr, and let rx be extended so that it equals ri. And let lines hx, hi be drawn. Clearly as regards the second condition, it is not possible to draw a line from point m to side fl, dividing it in the manner in which mck divides the line, except by mck itself.[239] If, indeed, it be so, let mpo be such a line. Clearly po will be less than ck, because even a line pq drawn equidistant from ck will be less than ck, [since, indeed, triangle kcf and triangle qpf are equiangular by 29, 32 p1; it will (then) follow by 4 p6 that kf to qf is as kc to qp, but kf is greater than qf by 9 ax;[240] therefore, kc is greater than qp][241] and greater than po, [because it is greater than qp, which by 19 p1 is greater than op, since angle poq is obtuse by 32.13 p1] and pl is greater than cl [by 9 ax].

Therefore the ratio of po to pl will not be the same as kc to cl. [For if it be the case that kc to cl is as op to pl, by 14 p5 cl will be larger than lp, contradicting 9 ax, because kc is greater than op.] On this account po to pl will not be as hd to dt [by 11 p5]. It follows, therefore, that from point m no other line can be drawn than mck similar[242] to ei. (Apparently, ei should be di). In truth, since angle odi is equal to angle lcm, and angle oid is equal to angle clm, [by construction], triangle clm will be similar to triangle iod [by 32 p1.4 p1 d6]. Therefore angle iod will be equal to angle lmc. It follows that [by 13 p1] angle roh equals angle kmn, and angle hro, a right angle, will be equal to angle knm. Hence [by 32 p1] angle nkm equals angle rho. However, the line di, when extended, intersects with hr in point s. [It intersects by 2 ax, because angle adr is right (actually not adr, but hro), and adi is acute (rid).] Angle sdh equals angle kcf [by 15 p1.1 ax] and triangle sdh will be similar to triangle ckf [by 32 p1.4 p1 d6.] Therefore the ratio of sd to dh is as fc to ck, but [by 7 p5] hd to di [equal to dz by 15 d1] is as kc to cl [by construction]. Therefore [by 22 p5] sd to di is as fc to cl. Thus [by 18 p5] si to di is as fl to cl; but di to io is as cl to lm since triangle dio is similar to triangle clm. Therefore [by 22 p5] si to io is as fl to lm, [and by consequence of 4 p5, as io to si, so is lm to fl]. But the ratio of si to ir is as fl to ln, because triangle sir is similar to triangle fln. [Indeed, angle ris equals angle fln by construction, and sri and fnl are right angles; therefore, by 32 p1.4 p1 d6 triangles sir and fln are similar.] Therefore [by 22 p5] the ratio of io to ir is as that of lm to ln, [and as a consequence of 4 p5 as ir to io, so is ln to lm].[243] Therefore the ratio of ym to lm is as xo to io. [For indeed xi is twice ir and yl is twice ln: therefore by 15 p5, as xi is to io, so is yl to lm, and by 17 p5 as xo is to io, so is ym to lm.] (Let) a line ui be drawn through i equidistant from line hx and let line da be produced, until it intersects with ui [it will intersect, moreover, by Lemma 29 p1 of Proclus], say in u. Triangle oui will be similar to triangle hox [by 15.29.32 p1 4 p1 d6]. Therefore ho will be to ou as ym to ml. [Because triangles oui, hox are evidently similar, as ho is to ou, so xo is to io, and as xo is to io, so ym is to ml by conclusion. Therefore by 11 p5, as ho is to ou, so ym is to ml] and thus ho to ou, is as hd to dt. [For, hd is to dt as ym to ml by construction.] But since [by 4 p1] triangle hri is equal to triangle hrx, since hr be perpendicular [by construction, and xr be equal to ri, and rh be common]. Therefore angle hxr equals angle rih, and thus rih equals angle uio [because uio equals hxo on account of the similarity of triangles uio, hox]. Whence [by 3 p6] ho to ou is as hi to iu; and thus [by 11 p5] hi to iu is as hd to dt. In fact angle uid is larger than angle dih [because it is included in the equal angle oih]. Let there be cut from it (angle uid) (an angle) equal (to dih), say pid, and let line pt be drawn, and let p be a point of the diameter da. Clearly, because the ratio of hi to ui is composed of that of hi to ip and pi to ui [because the ratio of the extremes is composed of the intermediate ratios, as Theon demonstrated in 5 d6] and [by 3 p6] the ratio hi to ip, is as dh to dp, since di bisects angle pih. Therefore the ratio of hi to ui (which is hd to dt) is composed of hd to dp and dp to dt. Therefore dp to dt is as pi to ui. As a matter of fact angle oih is half angle uih [as a consequence] but angle dih is half angle pih: it follows that angle dio is half angle piu. But angle dio is half angle tdp, because it equals angle flm [which is equal to half angle adt or pdt]. Therefore angle piu is equal to angle tdp, and dp to dt is as pi to ui. Thus triangle uip is similar to triangle tpd, [by 6.4 p1 d6] and angle upi equals tpd. Therefore tpi will be a straight line [by 14 p1], since angle dpt and tpo (together) make two right angles. And thus angles opi and opt make two right angles. [In fact it actually follows from the discussion in 15 p1 demonstrated by Proclus.] And thus [by 12 n4] t is reflected to h by point i [because line tpi is the incident line, and angles tid, hid are equal by construction]. And the same will be the case, whether t be outside or inside the circle. And similarly whether the selected point h be outside or inside: as long as the points are at unequal distances from the centre.

Proof of Proposition 74. Amplification (see Figure 32): Given diameters bq, ag, and diameter ez bisecting angle bdg. I say that given any point on arc aq except point z [for, from point z there are reflected only those points, of the diameters, equidistant from the centre, as follows from previous propositions] from it (the exceptional points) there can be reflected infinitely many points. In other words, let point h be taken, and a point l on semidiameter gd. And from semidiameter bd let md be cut, equal to ld, [by 3 p1]. And let lines lm, lh, mh, dh be drawn. Let the point in which ez divides lm be f; [by 4 p1] fl will be equal to fm. And let hd extended cut lm in point n. Therefore ln is less than mn. Moreover since angle mdf is equal to fdl [by thesis] and angle qdz, [by 15 p1]; and angle mda is equal to angle ldq, [by the same]; and angle adh equals angle ndl, angle ldh will be greater than angle mdh. [For, indeed, mdn, i.e. by 15 p1, hdq is greater than ndl, that is, adh, and mda is itself equal to ldq. Therefore angle ldh is greater than angle mdh.]; therefore [by 24 p1] lh will be greater than mh, since md, dh be equal to ld, dh. Therefore angle dhl will be less than angle dhm. For indeed, let them be equal: this requires that the ratio of lh to hm be as ln to nm: [by 3 p6],[244] which is impossible. [For, indeed, the greater, lh to the smaller, hm would have the same ratio as the smaller ln to the larger nm.] If, on the other hand, it be greater, let an equal angle (i.e. cut from angle dhl an angle equal to dhm) be cut from it, and argue in the same manner. Therefore, it is less. Let there be cut, therefore, from angle mhd an angle equal to that [dhl], say thd. Thus t is reflected to l from point h [by 12n 4]. And line td is less than ld. [Since it is less than dm, which, by thesis is equal to dl.] Similarly if other points than l, m are taken on semidiameters bg, gd

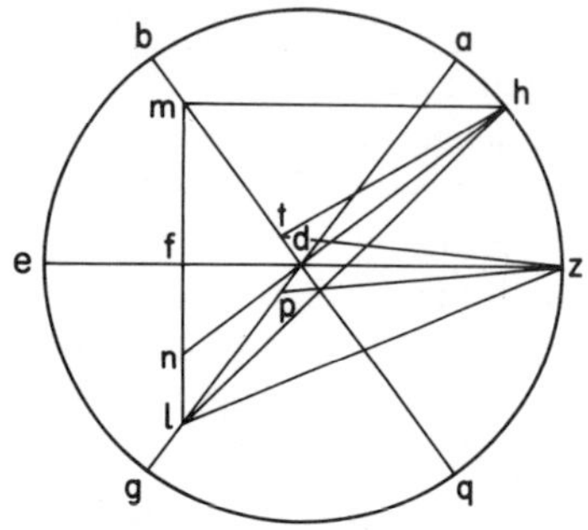

Figure 32. Figure for Proposition 74. (Reproduced with permission from D. C. Lindberg, ed., *Opticae Thesaurus Alhazeni Arabis*..., Johnson Reprint, N.Y., 1972, figure, p. 171 (top).). © by Johnson Reprint Corporation, New York, N.Y.

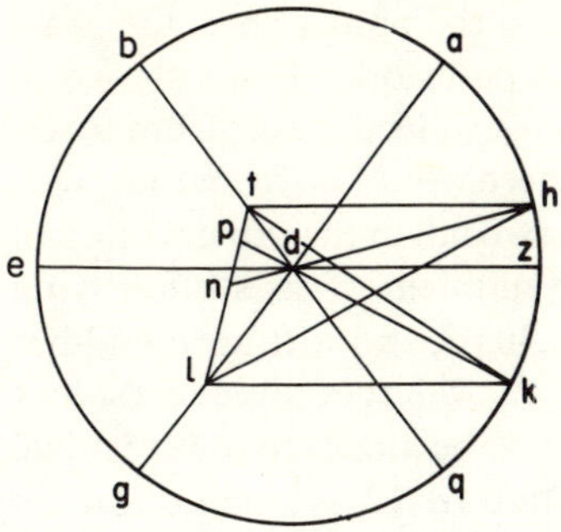

Figure 33. Figure for Proposition 75. (Reproduced with permission from D. C. Lindberg, ed., *Opticae Thesaurus Alhazeni Arabis*..., Johnson Reprint, N.Y., 1972, figure, p. 171 (bottom).). © by Johnson Reprint Corporation, New York, N.Y.

equidistant from point d, it can be shown in like manner that from point h there be mutual reflections of points, unequally distant from the centre, and thus from infinitely many points on such similarly given diameters it will be established; and from any given point on the arc aq, except from the point z.

Proof of Proposition 75. Amplification (see Figure 33): Given points t, l, on diameters, and let their distances from the centre be unequal; they are reflected mutually from point h. Reflection from t to l is not possible from any other point of the arc aq, other than from point h. If, indeed, it be possible from another, say k, let tk, lk, dk, lt, th, lh, ndh be drawn. And let dk be extended until it intersects with lt in point p. [Indeed, it does intersect, since the angle tkl subtends the base tl.] It is clear, that the ratio of lh to ht, is as ln to nt. [By 3 p6: for, indeed, by thesis t is a point of reflection; (so) by 12n 4 angle lhn is equal to angle thn.] And similarly, since angle tkp (text has tpk) be equal to angle lkp, by hypothesis, [by 3 p6] the ratio of lk to kt will be, as lp to pt. But [by 7 p3][245] lh is bigger than lk, and th less than tk; therefore, the ratio of lh to th is greater than that of lk to tk, [as follows from 8 p5]. Whence the ratio of ln to nt will be greater than that of lp to pt, which is clearly impossible. [For, indeed, ln is less than lp by 9 ax, and nt greater than pt. (So) the ratio of ln to nt will be less than the ratio of lp to pt, as follows from 8 p5.] It follows that from any part of arc aq, other than from point h, it is not possible to reflect t to l. Therefore, it is clear, what happens on arc aq.

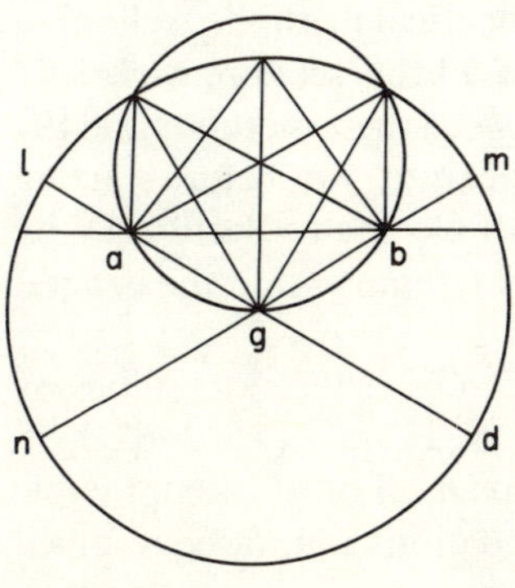

Figure 34. Figure for Proposition 87. (Reproduced with permission from D. C. Lindberg, ed., *Opticae Thesaurus Alhazeni Arabis*..., Johnson Reprint, N.T., 1972, figure, p. 181 (middle).). © Johnson Reprint Corporation, New York, N.Y.

Proof of Proposition 87 (Figure 34): Moreover when the points be at the same distance, reflection will be possible either from one point only, [as is shown in 70n], or from two, [as in 71n], or from four, [as 72n]; indeed, never from three. [Because, if there be reflection from three points, there would be from four also. For, when there is reflection from three such points two fall in the same section of the circumference, as is apparent in the preceding numbers. This section lying between these two points of reflection, cut in half by 30 p3, and with straight lines drawn from the centre, and the given points, on different diameters, but equidistant from the centre, to the point of bisection—the angles to it were made equal by 27 p3. 4 p1. Whence it is a point of reflection by 12 n4; indeed, on the periphery opposite lm, say nd, there is also a point of reflection by 73, 75n. Therefore, there is reflection from four, not from just three points.]

Abū-l-Jūd apparently also solved a fourth degree equation in the course of answering a question set by Al-Bīrūnī.[246] Given a straight line segment BC, and a point A (off BC), draw a line from A to BC, intersecting at D, say, so that $(AD)(BC) + \overline{BD}^2 = \overline{BC}^2$." The solution, by conics, uses a parabola and hyperbola. Woepcke, who presents this example with some analysis in his treatise on ʿUmar also offers another, due to an anonymous geometer: "To construct a trapezoid, all of whose sides are 10 and whose area is 90." This problem, which reduces to $x^4 + 2000x = 20x^3 + 1900$ is resolved by means of a circle and a hyperbola.[247]

However, these remain nothing but scattered examples, akin to Ibn al-Haytham's solution of a pure quintic (mentioned above),[248] and ʿUmar's proposal to do the same type of solution for a pure sextic (unresolved).[249] Yet, Rosenfeld and Youschkevitch,[250] call attention to a tantalizingly brief remark in al-Kāshī's (d. 1429) *Miftāh al-ḥisāb* to the effect that "he had discovered 'the method for the determination of unknowns in ... seventy problems (of fourth degree) which had not been touched upon either by the ancients or contemporaries.'"[251] Evidently this theory, to which he promised to devote a separate work, would have been similar to that of Abū-l-Jūd and ʿUmar for cubics. Yet nothing of the sort seems to have been written.

V. In Loco corondis

So far we have only made fleeting mention of the "numerical methods" school of solving equations, which also arose in the Near East. In this section, we briefly sketch its development, and look for relationships between these methods and the geometrical ones which are the primary concern of our study. But before starting on these two interlocking topics, we provide a proper closing to Section IV by considering the achievements of ʿUmar and Ibn al-Haytham within the general theory of geometric construction.[252]

First of all, recall that since all quartic equations can be reduced to cubics by means of the resolvent cubic (essentially a device of Ferrari [1522–65]),[253] quartic problems need not even be considered. So one need only attempt cubic constructions. The ones resulting from irreducible equations cannot be resolved by straight edge and compass, since these two instruments produce only real square roots and their iterates.[254] Interestingly enough all cubic constructions can, in turn, be reduced to two classical problems: trisecting the angle and doubling the cube.[255] These problems in turn may be solved in a number of ways: if a conic that is not a circle is given, a straight edge and compass will do;[256] if a third order curve and a fixed square are given, a straight edge is enough;[257] a marked straight edge;[258] a carpenter's square.[259] This litany should make it clear how the earlier achievements in geometric construction fit into the more recent theory. Also we should point out that the complex roots of a polynomial are graphically representable. For cubics, for example, a tangent line from the real root to the point on the curve that is the local extreme value separated from the real root by the curves inflection point determines the real and imaginary parts of the two conjugate roots.[260]

We now turn to our thumb nail sketch of the "numerical methods" school of Near Eastern mathematicians. Their methods contain such elements as iterative techniques, dating back to at least the time of Ḥabash al-Ḥāsib al-Marwazī (ninth century), who used them to solve "Kepler's equation", $f(x) = b + k \sin x$,[261] and remaining in use in al-Kāshī's, who used the trisection cubic in the form

$$f(x) = \frac{4x^3 + \sin(3°)}{x}$$

to find an approximation to sin(1°).[262] Another technique, developed and presented with great care in the *Miftaḥ al-ḥisāb* is a method for extracting *n*th roots in the manner of Horner and Ruffini. Some part of this particular technique (especially the use of the binomial theorem) may go as far back as Naṣīr al-Dīn al-Ṭūsī (*c.* 1265),[263] or even as-Samaw'al (twelfth century).[264] Another element, the rule of false position, is Egyptian.

Nonetheless, one looks in vain for a connection between the numerical methods and the geometric method in the *Miftaḥ al-ḥisāb*. However, there does exist an earlier treatise that seems to blend the two methods in an artful and totally natural way. This is the *Kitab fi'l-jabr wa'l-muqābala* of Sharaf al-Dīn al-Tūsī (d., *c.* 1213/14), which has been discussed in part by Anbouba,[265] and Rashed.[266] Unfortunately both accounts are only partial, and all too brief. Rashed promised (1974)[267] to produce a complete edition of the unique Arabic manuscript (India Office Library, Loth 767, III, fol. 35–180), which is apparently an abridgement of the original work.[268] To date, this has not appeared. Given the claims of both Anbouba and Rashed for the importance of Sharaf al-Dīn's work, an edition is urgently needed. Anbouba notes, for example, that al-Ṭusi gives ʿUmar's proofs of solvability practically unchanged, but that he takes great care in positioning the conics, and even greater care in determining the limits of solvability.[269] Both authors observe that al-Ṭūsī's classification scheme differs from ʿUmar's,[270] in that equations are grouped in classes according to "solvability", but this is far less important than a possible inference to be made from their presentations, namely: al-Ṭūsī uses the διορισμός for his groups of equations before starting his approximation scheme for determining a root of a particular equation of the group. If true, this would be an obvious and most practical link so far between the geometric and numerical theories. Of nomography there is, rightly, given the impractical nature of the constructions, not a trace.[271]

Rashed goes on to say that some of the figures and tables in the precursor of the existing manuscript of the *Algebra* are omitted in the abridgement, and that there are many scribal errors, which make the text difficult to understand.[272] On the otherhand, his own account is rather difficult to follow, and one is left wondering about the source of certain mathematical errors in it.[273] Also, there are claims made for al-Ṭūsī's use of "discriminants", and many remarks which would allow the reader to infer that Rashed claims the use of some equivalent of first derivatives of polynomials to solve certain "limits of possibility" by al-Ṭūsī.[274] Clearly, however, both Anbouba and Rashed seem bothered by al-Ṭūsī's adroit handling of the maximum problems involved in setting these limits. Anbouba acknowledges that the maximum of $x^2(a-x)$ which is central in Archimedes' work "might have been borrowed" from that author, but says the text offers no clue of how al-Ṭūsī handles other maxima, and resorts to explaining a particular problem using the calculus.[275] Since Anbouba's account is very clear and succinct, let us examine it in order to frame a small conjecture about al-Ṭūsī's method of maxima. We hope to check it against the manuscript in a future study. Al-Ṭūsī, in the example Anbouba offers,[276] is considering the equation $x^3 + b^2x + c = 3ax^2$, and shortly, the discussion hinges on the quadratic equation $x^2 + b^2/3 = 2ax$, i.e. $3x^2 + b^2 = 6ax$, the "first derivative" of the original equation. Note, however, that if one applies the simple linear transformation $x \to x + k$ to the original equation, one gets a new equation

$$x^3 + (3k - 3a)x^2 + (3k^2 - 6ak + b^2)x + (k^3 - 3ak^2 + b^2k + c) = 0$$

whose x term vanishes when k is chosen so that $3k^2 - 6ak + b^2 = 0$, i.e. when k is a root of $x^2 + b^2/3 = 2ax$. The equation then becomes

$$x^3 + 3(k - a)x^2 + (k^3 + b^2k - 3ak^2 + c) = 0$$

which is rather like Archimedes' equation

$$x^2(\alpha - x) = \beta^2\gamma.$$

It would appear, then, that the "derivatives" may be coincidental. Perhaps al-Ṭūsī recognized that his main problem in finding a *διορισμός* might be solved if he could transform it into one whose "limits of possibility" were already known. Let's look at the example more closely. Rather oddly al-Ṭūsī seems to be working with $-x^3 + 3ax^2 - b^2x$, instead of the expected $x^3 - 3ax^2 + b^2x$. In fact, if one applies the transformation $x \to x + k$ to $-x^3 + 3ax^2 - b^2x - c$ one gets

$$-x^3 + (3a - 3k)x^2 + (-3k^2 + 6ak - b^2)x + (-k^3 + 3ak^2 - b^2k - c) = 0,$$

or, denoting, as does Anbouba $k^2(3a - k) - b^2k$ by $f(k)$

$$x^2[(3a - 3k) - x] + (-3k^2 + 6ak - b^2)x + (f(k) - c) = 0.$$

Now take k to be the larger root of $-3k^2+6ak-b^2=0$, i.e. $k=a+\sqrt{a^2-b^2/3}$ as does al-Ṭūsī, who has already shown that $b<3a/2$, presumably using, as Anbouba conjectures, *Elements* II, 5.[277] Then, our transformed equation looks like Archimedes', but of course $3(a-k)<0$:

$$x^2[3(a-k)-x]=c-f(k).$$

However, since $3(a-k)$ is negative the maximum is at zero rather than at $\frac{2}{3}(3)(a-k)$, as to be expected since al-Ṭūsī has considered $-x^3+3ax^2-b^2x$ rather than $x^3-3ax^2+b^2x$. Thus Archimedes' διορισμός from *On the sphere and cylinder* II, 4 (following Eutocius' commentary), says, as does al-Ṭūsī, that the original equation, $x^3+b^2x+c=3ax^2$ has no solutions if $c-f(k)>0$, and two if $c-f(k)<0$. Of course, this is just speculation at present, but it does have the advantage of allowing us to continue to feel that derivatives are still somewhat ahead of al-Ṭūsī's time.

VI. Influence East and West

In this closing section we survey briefly the effect of the geometric theory on the Latin West, and pose a few questions concerning the development of numerical methods of solution in the Far East.

Unfortunately the traceable lines of influence of the Near Eastern developments in the geometric theory of equations are slight. Al-Khwarizmī's *Algebra*, though twice translated into Latin, and very influential in spreading knowledge of Arab algebra, is hardly an example. The work of Abū Kāmil on pentagons and decagons, which influenced Leonardo of Pisa, is also outside the geometric tradition.[278] ʿUmar's work was, of course, unknown to the Medieval West, and did not even become a part of the history of mathematics until Woepcke's translation of the *Algebra* into French in 1851. Kasir details a few earlier references to the work of ʿUmar: a notice of the manuscript on algebra by Meerman in Leiden (1742), a citation in Montucla's *Histoire des Mathematiques*, a reference in Gantz's 1823 dissertation, etc..[279] Nonetheless, the approach to cubics that ʿUmar favored did find independent expression among European mathematicians as late as the seventeenth century. For example, the English mathematician Thomas Baker (1625?–1689)[280] published his *Geometrical Key, or the Gate of Equations Unlocked*[281] in 1684. In it, he describes a rule for constructing the solutions of all equations with real coefficients and degree less than or equal to four by means of a parabola and a circle. His method differs from Descartes' in that it doesn't first require the equation to be transformed into one with its monomial of second place removed.[282] The account of his book in the *Philosophical Transactions of the Royal Society* reads very much like an earnest publisher's blurb.[283] Earlier [1670?], Isaac Newton had also offered a variety of elegant geometric constructions in a small tract, "Problems for construing equations", recently edited and published by Whitesides.[284] Some ten years after Baker's work appeared, Edmond Halley, the successor to Flamsteed as astronomer royal also published a short essay on geometric solutions.[285]

Seemingly only the work of Ibn al-Haytham, among those of all Near Eastern writers on the geometric theory, had direct influence in the West. As we noted at the end of Section V, this influence was due to Risner's 1572 edition of a Latin translation of Ibn al-Haytham's optical treatise. Baker[286] offers a bibliography of early works that make direct reference to the work, and in particular to "Alhazen's problem", from Barrow's *Lectiones*... (1669) and the correspondence of Huygens and Sluse to L'Hopital's *Traite analytique de sections coniques*, Paris, 1720.

The "Eastern" influence that forms part of the title of this section refers to the questions we pose in closing. As we have seen, the Near Eastern geometric theory of equations owes its genesis to the Greek theory of conics. What is the genesis of its rival, the "Horner–Ruffini" method which reached its full power at the hands of al-Kāshī? May it have been, as Libbrecht tentatively conjectures, a Far Eastern one, brought across central Asia by successive Mongol invasions?[287] Or, was the line of influence from Middle to Far East? Or, finally, is it possible that the two theories grew independently?

Notes

1. Gandz (53).
2. Neugebauer (117).
3. Thureau–Dangin (169).
4. Gandz (50).
5. *Ibid.*, 107–9.
6. Gandz (53).
7. Goetsch (57). See also van der Waerden (180) for a summary of Goetsch's position, and van der Waerden (182), 62–81. Two speculative papers (181) draw attention to a Chinese problem text Chiu Chang Suan Shu which contains some shared elements with Babylonian quadratic texts.
8. Bruins and Rutten (27).
9. Gandz (53), 486–7, who reduces this to a type BI.
10. Neugebauer (117); see Gandz, *supra*, for specific examples.
11. Bruins (25), 294–5.
12. Thureau–Dagin (169), especially problems 22, 35, 38.
13. Neugebauer (117) I, 76. Bruins goes on to make the interesting observation that tables of $x^3 + x^2$ and $x^3 - x^2$ would actually be sufficient to find all positive solutions of the cubic $ax^3 + bx^2 + cx + d = 0$. The trivial proof requires two transformations $x \rightarrow x' - b/3a$, followed by $x' \rightarrow 1/X$. See Bruins (25), 295. This observation is, of course, not meant to apply to Babylonian procedure, since it would be difficult to obtain using Babylonian arithmetic practice.
14. Goetsch (57), 145ff.
15. Bruins (25), 289–91, and Goetsch (57), 150.
16. For *The Rhind Mathematical Papyrus*, see the edition of Chace, A. B., Bull, L., Manning, H. P. and Archibald, R. C. 2 vols. Mathematical Association of America, Oberlin, 1927, 1929. For the Moscow papyrus, see, Struve, W. W., "Mathematischer Papyrus des Staatlichen Museums der Schönen Künste in Moskau", *Quellen und Studien zur Geschichte der Mathematik*, Abt. A. **1** (1930), 134–45. Summaries are provided in Gillings, R. J., *Mathematics in the Time of the Pharaohs*, MIT Press, Cambridge, 1972, appendices 6 and 7 respectively.
17. Gillings, *supra*, 160–5.

18. Schack-Schackenburg, H., "Der Berliner Papyrus 6619", *Zeitschrift für Ägyptische Sprache* **38** (1900), 135–40, **40** (1902), p. 65ff.
19. Griffith, F. L., ed., *The Petrie Papyri: Hieratic Papyri from Kahun and Gurob*, 2 vols. University College, London, 1897. Apparently the Cairo Demotic Mathematical Papyrus (300 B.C.) also contains, as problems 34 and 35 respectively the systems

$$\begin{cases} x^2 + y^2 = 169 \\ xy = 60 \end{cases} \qquad \begin{cases} x^2 + y^2 = 225 \\ xy = 60 \end{cases}$$

See Gillings, R. J., "The Mathematics of Ancient Egypt", DSB XV, 681–705, especially p. 698.
20. Parker, R. A., *Demotic Mathematical Papyri*, Brown University Press, Providence, 1972. See especially pp. 5–6.
21. Gandz (53), 405.
22. Heath (64), 140–1, or Tannery (165) I, 61–7.
23. Heath (64), 140. This condition, as Gandz observes, evidently comes from *Elements* II, 5, which is exactly the condition predicated by the Babylonian mode of solution.
24. Heath (64), 140.
25. Heath (65) I, 392–5, or Stamatis (157) I, 81–3.
26. Gandz (53), 468. See also Tropfke (175), **44** (1934), p. 39.
27. Gandz (53), 468.
28. For some provocative and contrary views on the "so called geometric algebra" of the Greeks and Euclid in particular, see Śzabo (164).
29. Clagett, M., "Archimedes", DSB V, 213–31.
30. Heath (66), cxxxvi.
31. Heath (66), cxxxvii.
32. Heath, *supra*, gives the νεῦσις on page cxii; in this introduction to the work of Archimedes, there is an interesting discussion of why the trisection problem, in Greek geometry, is a "solid problem". Heath's conclusion, contrary to Zeuthen's, is that the problem is viewed as such because conics ("solid loci") are used in its solution, rather than because the cubic terms were considered as volume elements.
33. Heath (66), cxxxvii.
34. Since $a:x = x:y = y:b$, $ay = x^2$ and $ab = xy$. Hence $axy = x^3$ or $a^2b = x^3$.
35. Heath (66), cxxxviii.
36. Heath, *supra*, 62.
37. Perhaps the quickest way to see that a cubic equation does arise, is to set up the problem using calculus. If the two segments are to have ratio $m < 1$, then evidently

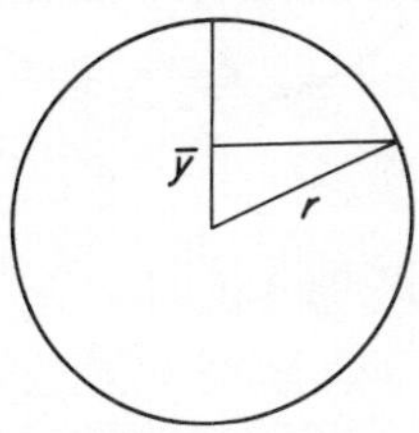

$$\frac{v_1}{v_2} = \frac{\int_{\bar{y}}^{r} \pi(r^2 - y^2)\,dy}{\int_{-r}^{\bar{y}} \pi(r^2 - y^2)\,dy} = m < 1,$$

and so

$$\frac{(r^2 y - y^3/3)|_{\bar{y}}^{r}}{(r^2 y - y^3/3)|_{-r}^{\bar{y}}} = m.$$

Thus

$$\frac{2r^3}{3} - r^2\bar{y} + \frac{\bar{y}^3}{3} = m\left(r^2\bar{y} - \frac{\bar{y}^3}{3} + \frac{2r^3}{3}\right),$$

or

$$\frac{\bar{y}^3}{3} - r^2\bar{y} + \frac{1-m}{1+m}\cdot\frac{2r^3}{3} = 0,$$

whence

$$\bar{y}^3 - 3r^2\bar{y} + \frac{1-m}{1+m}\cdot 2r^3 = 0.$$

The smaller segment, of course, has height $r - \bar{y}$, the larger, height $r + \bar{y}$.

38. See Bulmer-Thomas. I., "Eutocius of Ascalon", DSB IV, 488–91.
39. Thomas (168) II, 135ff.
40. For the full text, see Thomas, *supra*, 127ff, or Heath (66), 62ff. Mahoney gives another paraphrase in (109), especially 337–40.
41. For completeness we offer a short calculus proof of this proposition: a segment of a sphere has the same volume as a cone of height $H = h(3r - h)/(2r - h)$, where h is the height of the segment and r the radius of the sphere:

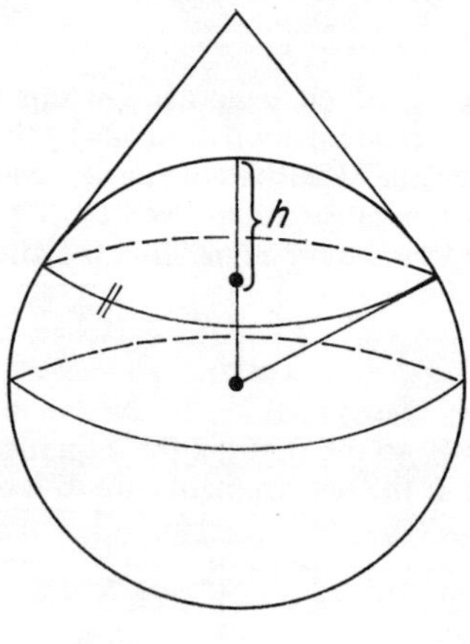

$$\text{Volume of segment} = \int_{r-h}^{r} \pi(r^2 - y^2)\,\mathrm{d}y$$

$$= \pi\left(r^2 y - \frac{y^3}{3}\right)\Bigg|_{r-h}^{r}$$

$$= \pi\left(\frac{2r^3}{3} - r^2(r-h) + \frac{(r-h)^3}{3}\right)$$

$$= \tfrac{1}{3}\pi[r^2 - (r-h)^2]H,$$

or

$$\tfrac{1}{3}\pi(2rh - h^2)H = \tfrac{1}{3}\pi(3rh^2 - h^3);$$

whence

$$H = \frac{h^2(3r-h)}{h(2r-h)} = \frac{h(3r-h)}{(2r-h)}$$

as claimed.

That the cone having $(\mathrm{K}\Delta + \Delta\mathrm{X})/\Delta\mathrm{X} = \mathrm{PX}/\mathrm{XB}$, for example, meets this condition is easily seen, for

$$\frac{\mathrm{K}\Delta + \Delta\mathrm{X}}{\Delta\mathrm{X}} = \frac{r + 2r - h_1}{2r - h_1} = \frac{\mathrm{PX}}{\mathrm{XB}} = \frac{H_1}{h_1}$$

with the H_1 being the height of the cone ΡΓΑ and h_1 the height of the segment ΑΒΓ.

42. Since

$$\frac{\mathrm{K}\Delta + \Delta\mathrm{X}}{\Delta\mathrm{X}} = \mathrm{PX}/\mathrm{XB}, \quad \frac{\mathrm{K}\Delta + \Delta\mathrm{X}}{\Delta\mathrm{X}} = \frac{\mathrm{PB} + \mathrm{XB}}{\mathrm{XB}},$$

i.e. $\mathrm{K}\Delta/\Delta\mathrm{X} = \mathrm{PB}/\mathrm{BX}$. Similarly, since

$$\frac{\mathrm{KB} + \mathrm{BX}}{\mathrm{BX}} = \frac{\Lambda\mathrm{X}}{\mathrm{X}\Lambda}, \quad \frac{\mathrm{KB}}{\mathrm{BX}} = \frac{\Lambda\Delta}{\mathrm{X}\Lambda}.$$

Consequently,

$$\frac{\Lambda\Delta}{\mathrm{X}\Lambda}\cdot\frac{\Delta\mathrm{X}}{\mathrm{K}\Delta} = \frac{\mathrm{KB}}{\mathrm{BX}}\cdot\frac{\mathrm{BX}}{\mathrm{PB}}, \quad \text{etc.}$$

43. $\mathrm{PK}/\mathrm{K}\Delta = \mathrm{K}\Lambda/\Lambda\Delta$ follows from *Elements* V, 12 or in modern notation

$$\frac{a}{b} = \frac{c}{d} \Rightarrow \frac{a+c}{b+d} = \frac{c}{d}.$$

44. Certainly that of

$$\frac{\Lambda\mathrm{X}}{\mathrm{X}\Lambda} = \frac{\mathrm{KB} + \mathrm{BX}}{\mathrm{BX}} = \frac{\mathrm{BZ} + \mathrm{BX}}{\mathrm{BX}} = \frac{\mathrm{ZX}}{\mathrm{BX}}$$

is given. But

$$\frac{\mathrm{ZX}}{\mathrm{BX}}\left(\frac{\mathrm{BX}}{\mathrm{BZ}}\right) = \frac{\Lambda\mathrm{X}}{\Delta\mathrm{X}}\left(\frac{\Delta\mathrm{X}}{\Lambda\Delta}\right)$$

since $\Lambda\Delta/\Delta\mathrm{K} = \Delta\mathrm{X}/\mathrm{BX}$, i.e.

$$\frac{\Lambda\Delta}{\mathrm{BZ}(=\Delta\mathrm{K} = \mathrm{BK})} = \frac{\Delta\mathrm{X}}{\mathrm{BX}}.$$

Hence $\mathrm{ZX}/\mathrm{BZ} = \Lambda\mathrm{X}/\Lambda\Delta$ or $\mathrm{BZ}/\mathrm{ZX} = \Lambda\Delta/\Lambda\mathrm{X}$. However, as Eutocius observes, BZ is just the radius of the sphere, and BX is assumed, meaning that $\mathrm{ZX} = \mathrm{BX} + \mathrm{BZ}$ is given. Thus $\Lambda\Delta/\Lambda\mathrm{X}$ is given. But $\Lambda\mathrm{X}/\mathrm{XP}$ is given, so since $(\mathrm{XP} + \Lambda\mathrm{X})/\Lambda\mathrm{X} = \mathrm{P}\Lambda/\Lambda\mathrm{X}$, this is also given. Finally

$$\frac{\mathrm{P}\Lambda}{\Lambda\mathrm{X}}\cdot\frac{\Lambda\mathrm{X}}{\Lambda\Delta} = \frac{\mathrm{P}\Lambda}{\Lambda\Delta}$$

is given.

45. See 44 above.

46. Thomas (168) II, 133.

47. *Loc. cit.*.
48. Or, as we shall see, possibly some translate of it.
49. Mahoney (109), *passim.*
50. *Ibid.*, 341.
51. Gulley (59) treats the logic of such analysis; Mahoney, on the other hand, deals with the problem of geometric analysis much as a mathematician would, emphasizing the role that analysis played in providing proofs as well as determining the possibility of proof.
52. Heath (65) I, 292–3, especially note 2.
53. *Ibid.*, 382–5.
54. This is possible by *Elements* I, 45: "To construct, in a given rectilineal angle, a parallelogram equal to a given rectilineal figure". Heath (65) I, 345–7. Recall, however, that this proposition is not explicitly given in the Arabic *Elements.*
55. Thomas (168) II, 141.
56. Alternatively, one might cite *Elements* I, 43 directly.
57. Archimedes (Eutocius) justifies this with a citation to Apollonius' *Conics* II, 8 converse, which does not correspond to an appropriate reference for modern editions.
58. Thomas (168) II, 141 has the obvious misprint $(BE)^2(BA)$.
59. Archimedes later proves this by a spectacular use of Apollonius' *Conics.* Using calculus, we would simply observe that one wishes to maximize $V(x)=x(a-x)^2$, $0 \leqslant x \leqslant a$, and since $V'(x)=a^2-4ax+3x^2=(3x-a)(x-a)=0$ at $x=a/3$ or $x=a$, while $V(0)=V(a)=0$, $V(a/3)>0$ and $V''(a/3)=-2a<0$, $x=a/3$ gives the required maximum.

    This, of course, is the $\delta\iota o\rho\iota\sigma\mu\acute{o}\varsigma$ for the intersection problem, i.e. $xy=ab$ and $(c^2/a)y=(x-a)^2$ intersect only if $bc^2 \leqslant 4a^3/27$, i.e. $x(x-a)^2=bc^2$ has a real solution only if $bc^2 \leqslant 4a^3/27$. In fact, the discriminant $D=4a^3bc^2-27b^2c^4 \geqslant 0$ when $bc^2 \leqslant 4a^3/27$. Consequently, by the theory of equations, e.g. Dickson (40), 47, our equation will have three distinct real roots if $D>0$, i.e. $bc^2<4a^3/27$, and at least two of these are equal real roots if $D=0$, i.e. $bc^2=4a^3/27$. Moreover, since $p(-x)=-x^3-2ax^2-a^2x-bc^2$, all three roots (by Descartes' rule of signs (Dickson (40), 76)) are positive with our choice of coordinate system. In fact, when $bc^2=4a^3/27$, the double root will be $x=a/3$, and the other, outside the range will be $x=4a/3$. This is most easily seen by writing

$$x^3-2ax^2+a^2x=bc^2$$

    as

$$x^2-2ax^2+a^2x-\tfrac{4}{27}a^3=bc^2-\tfrac{4}{27}a^3$$

    or

$$\left(x-\frac{a}{3}\right)^2\left(x-\frac{4a}{3}\right)=bc^2-\tfrac{4}{27}a^3=0.$$

    Finally, since $p(x)=x^3-2ax^2+a^2x-bc^2$ is such that $P(0)=-bc^2<0$, if $D>0$, $P(a/3)>0$, so there will be a root between zero and $a/3$, and one *bigger* than a, since $P(a)=-bc^2<0$ and $P(x)\to+\infty$ as $x\to+\infty$.
60. Thomas (168) II, 152–3, note a.
61. *Elements* I, 43 (converse) is Thomas' citation (168) II, 147. Strictly speaking, however, Euclid does not include such a theorem.
62. *Elements* VI, 4.
63. Thomas (168) II, 159, note c.
64. Thomas, "Eutocius of Ascalon", DSB IV, 488–91; see especially 490–1. Also see Heath (66), 72–9.

65. Thomas (168) II, 163, note a.
66. Thomas (168) I, 278–83.
67. See Thomas (168) I, 356ff., and Hultsch (70), iv. 43.67–44.68.
68. Thomas (168) II, 162–3, note a.
69. Heath (66), 104.
70. Compare with Thomas (168) II, 163, note a. One must, of course, take care when setting up the problem as an integral to use the mode of "integration" suggested by Archimedes method of exhaustion (Proposition 19, 20; 25, 26: Heath (66), 129–30; 136–40). If, for example, one chose, erroneously, "shells" for the first integral, the "cubic" would collapse to a simple quadratic.
71. Heath (66), cxxix.
72. For example $x^3 + bx^2 + cx + d = 0$ becomes $y^3 + py + q = 0$ under the change of variables $x = y - b/3$.
73. Heath (66), cxxvii. See the text of this paper following note 67.
74. Sezgin (154), 237.
75. Sayili (142).
76. Hartner (61), 361. Robert of Chester's translation was edited by Karpinski, and appears with translation in Karpinski (75). For citations of editions of the Gerard translation and appearance of terms from al-Khwarizmī in Fibonacci, see Hartner (61), *loc. cit.*.
77. Ruska (134); this is analysed by Gandz in (53)–(54).
78. Gandz (51), in particular, shows that numerous errors in Rosen's text and commentary on the algebra of inheritance of al-Khwarizmī are either owing to Rosen's carelessness as an editor, or simply outright misrepresentations of the actual text. See also Wieleitner (185). An earlier study (1878) of Rodet (127) argues for Indian and Greek influences on al-Khwarizmī's algebra, especially the theory of quadratics. But as we see in the main text, the Greek influence, at least, is not supported by the evidence uncovered by Gandz (53)–(54).
79. Gandz (52).
80. Gandz (53).
81. See references in 79–80.
82. Gandz (53), 510.
83. The reference is to AI–AIII in the table of Section II
84. Gandz's note at this point calls attention to the fact that al-Khwarizmī wrote his quadratics with positive coefficients. Hence $ax^2 + bx + c = 0$ offers the possibilities:

$$ax^2 = bx, \quad ax^2 = c, \quad bx = c, \quad ax^2 + bx = c,$$
$$ax^2 + c = bx, \quad ax^2 = bx + c,$$

where al-Khwarizmī includes a linear equation too, and, in fact, essentially treats it twice ($ax^2 = bx$).
85. We would also be remiss in not observing that al-Khwarizmī's work is comprehensive enough to include statements akin to διορισμός. For example, when he discusses his method for solving "squares plus units equal to so many roots", and presents as an example $x^2 + 21 = 10x$, i.e. of $x^2 + c = ax$, he remarks that when $(a/2)^2 < c$, "you have no equation". (See Grant (58), 109; or Karpinski (75).) The same sort of discussion of "logical necessity of impossibility", as ibn Turk calls it, appears in his *Kitāb al-jabr wa'l muqābala*. See Sayili (142), 163ff (English translation).
86. Levey (92), 30.
87. See Levy (94), 14, etc..
88. Levy, *supra*, 32 offers a fairly extensive bibliography. We call attention to Karpinski (74), Suter (162) Sacerdote (136), Levey (93), Schub and Levey "Book on indeterminate problems of Abū Kāmil (850–930)", *Centaurus* **13** (1968), 91–4. Levy (93) finds evidence for Abu Kāmil's being influenced by the *Elements*, particularly Book

X, since the algebra deals with problems leading to irrational roots. The Schub–Levey paper has received an important critical review by J. Sesiano, "Les Méthodes d'analyse indéterminée chez Abū Kāmil", *Centaurus* **21** (1977), 89–105.

89. Anbouba (6).
90. See Sezgin (154), 269; Luckey gave an edition and translation in "Ṯābit b. Qurra über den geometrischen Richtigkeitsnachweis der Auflösung der quadratischen Gleichungen", *Berichte über d. Verh. der Sächs. Akad. d. Wiss., Leipzig, math.-phys. Kl.* **93** (1941), 93–114.
91. Luckey, *supra*, 106. The "Arabian" Euclid and the Heath edition coincide here. See Besthorn (16), and Heath (65) I, 385–388; 388, in particular.
92. Sezgin (154), 128.
93. Sezgin (154), 254, 139.
94. Sezgin (154), 139.
95. Hartner (61), 362.
96. *Ibid.*. See also Kasir (76), 109 and 114, note 18.
97. For biographical materials see Youschkevitch and Rosenfeld (201), (131).
98. There are currently several editions in European languages. We use the published versions of Woepcke (189), Kasir (76), and Rosenfeld and Youschkevitch (133), Winter and Arafat (186). There is also a recent, critical edition by R. Rashed, but it was unavailable to us at the time of writing. Also of value is Woepcke's *Crelle* paper (193).
99. See Dold-Samplonius (45).
100. This treatise of al-Māhānī exists in a Leiden manuscript—in the form of a commentary by al-Qūhī. See Dold-Samplonius, *supra.*
101. Dold-Samplonius (44) offers biographical details on this Persian scholar from Merv in Khurāsān.
102. Hartner (61), 362.
103. Suter (163), 39.
104. Winter and Arafat (186), 31.
105. 1–1, 2–1, 3–1, 2–2, refer to the number of terms on each side of the equality.
106. This classification, at least for quadratics, follows Abū Kāmil.
107. Compare this with our remark on the Babylonian method in section II.
108. Winter and Arafat (186), 44, ¶ 4. As Kasir (76), 74 notes, this is a construction of Menaechmus.
109. If $x^2 = -4py$, $4p$ the parameter, is the length of the *latus rectum*, i.e. the focal diameter.

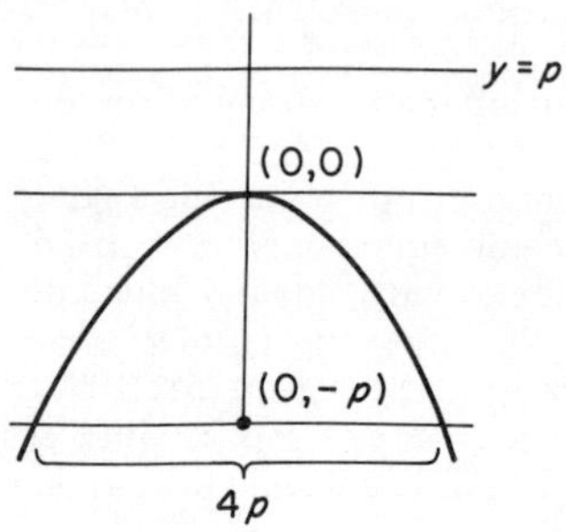

110. ʿUmar cites *Conics* I, 32 as justification, but since B is the vertex and AB ⊥ BG, the "modern" justification is clear. The citation to *Conics* is correct, according to the numbering in Heiberg's edition. This is of some interest in allowing us knowledge of the ordering of propositions in the translation of Apollonius available to ʿUmar. For example, his I, 56 is evidently Heiberg's I, 52.

111. This is just the defining property of our parabola, e.g. $y^2 = 4px$.
112. In Winter and Arafat, for example. But no such justification is provided in the manuscript used by Kasir.
113. Given m, n to find x such that $m/x = x/n$. On line l construct $AC = m$, $CD = n$. Construct a semicircle with AD as diameter. On C construct $CE \perp AD$ meeting the

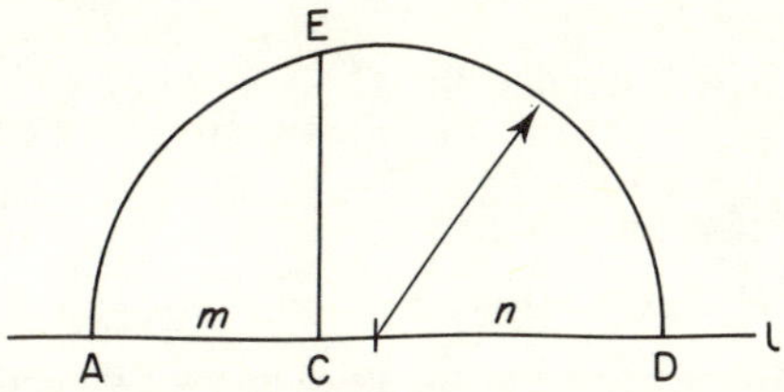

semicircle at E. CE is the required mean proportional, since $\angle AED$ is right, and so $\triangle ACE \sim \triangle ECD$. Thus $x/m = n/x$, i.e. $m/x = x/n$.
114. Winter and Arafat have AB, BG, an obvious misprint.
115. Actually, as we shall see in the text, he is solving a homogeneous form $x^3 + (a')^2 x = C$ (a cube).
116. Then AB is constructed as the mean proportional to a and l as in note 113.
117. This requires Lemma 2, starting with the solid l, l, c.
118. See note 110 above.
119. See note 111 *supra.*
120. Of course, there would be no question of his looking for the complex conjugate roots.
121. By application of Lemma 2; see note 113 for construction of AB from a.
122. This is Heiberg's I, 54; ʿUmar cites it as 58 in the Arabic text. For an illuminating account of conics in general, and their Apollonian and pre-Apollonian terminology, see Zeuthen (202), *passim.* Some of this material has been supplemented by Neugegauer in "Apollonius-Studien", *Quellen und Studien zur Geschichte der Mathematik, Astronomie und Physik*, Abt. B **2** (1933), 215–54. Van der Waerden (182), 241–63 offers highlights, while Toomer (174), 3–18 speculates that additional details may require change in light of his study of Diocles' work *On Burning Mirrors.*
123. ʿUmar cites no proposition; however, this is the defining property of the hyperbola, *Conics* I, 12.
124. He obviously does not use the other branch, whose vertex is at B and would give the one guaranteed real root of the equation, a negative one (by Descartes' rule of signs). His intersection condition corresponds to either two positive roots (equal for tangency), or unequal. For the drawing to make good geometric sense, it should be reversed from left to right; i.e. BT will be a *positive* root. It is not difficult to impart an analytic condition to ʿUmar's geometric one, but since he himself does not supply a *διορισμός*, we do not pursue the matter. It is best stated in terms of the discriminant Δ. In this case, the roots are real and distinct, i.e. one negative and two different positive roots when $\Delta > 0$, i.e. two intersections with ʿUmar's branch, and one negative and two equal positive roots when $\Delta = 0$. When $\Delta < 0$, ʿUmar's branch has no intersections, and there are two imaginary and one real negative root. See Dickson (40), 47.

More to the point is the fact that ʿUmar fails to include the second branch's intersection. To say that ʿUmar avoids it because it's "negative" is merely to beg the question. Did he have numerical evidence for not counting it as positive, or did he have some way of attaching a "sign" to points left of, say, the mid-point of BG? A likely answer is that he simply follows the Apollonian convention: hyperbola is a branch. cf. van der Waerden (182), 248.
125. Kasir (76), 77; *c.f.*, Winter and Arafat (186), 50.

126. ʿUmar cites *Conics* I, 20 which is: "If in a parabola two straight lines are dropped ordinate-wise to the diameter, the squares on them will be to each other as the straight lines cut off by them on the diameter beginning from the vertex to each other" (Heiberg edition). But this makes no sense. Arafat and Winter (186), 50 cite *Conics* I, 50 in their ʿUmar translation. Woepcke (189), 35, however identifies this with the Oxford edition's I, 21 (= Heiberg's I, 21). Here, the condition is plainly given as in ʿUmar.
127. See also Rosenfeld and Youschkevitch (133), notes 50 and 64, pages 124 and 126 respectively.
128. See note 124.
129. That is $x \to y - b/3$ changes

$$x^3 + bx^2 + cx + d = 0 \quad \text{to} \quad y^3 + py + q = 0.$$

130. *Conics* II, 4, 5; I 59 in the edition available to him see Woepcke (189), 38, note (***).
131. Again, there are (possibly) three intersections, one on his branch of the hyperbola and possibly two on the suppressed one. This equation, of course, has one positive real root ($x^3 + ax^2 - c = 0$) by Descartes' rule of signs, and either two negative ones or two imaginary conjugates.
132. Winter and Arafat (186), 52. See Kasir (76), note 34a (p. 85).
133. Our choice of coordinate system should be reflected. Then this condition is just $x > k$, in which case $x^3 > k^3 = c$ already.
134. In ʿUmar's edition. In our notation this is simply area HB $= |xy| = k^2 =$ HZ.BZ and area DB $=$ (BG)$^2 = k^2$.
135. $(BG)^2 = (HZ)(BZ)$, so $(BG)^3 = (BG)(BZ)(HZ) = (BZ)(AZ)(BZ)$.
136. $AZ = AB + BZ$.
137. Archimedes has $c \leqslant 4a^3/27$.
138. We return to this question of bounds for possibility a bit later, when we compare ʿUmar's work with that of other Near Eastern mathematicians.
139. Winter and Arafat (186), 58. ʿUmar cites *Conics* IV.
140. He has $(BH)^3 - (BD)^2(HA) + (BD)^2(AB) = (BH)^3 - (BH)^2(GH) + (BD)^2(AB) = (BH)^2(BG) + (BD)^2(AB)$, etc..
141. Standard constructions from *Elements*, or *Data*.
142. This is argued in Lemma 12, but follows directly from similar triangles:

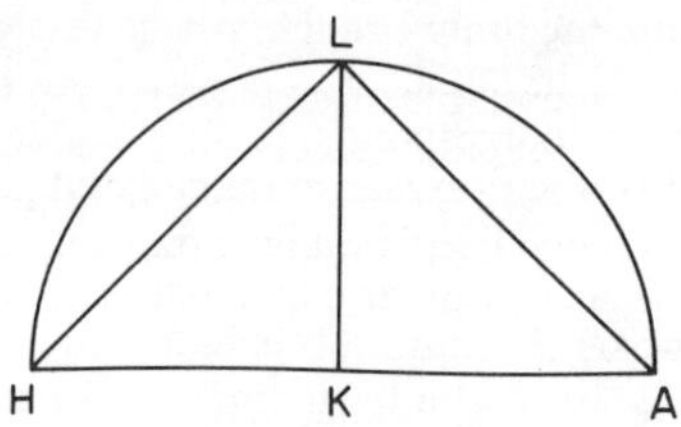

$\triangle HKL \sim \triangle LKA$, so $(LK)/(HK) = (KA)/(LK)$. $(LK)^2/(KA) = HK$; $(LK)^2/(KA)^2 = (HK)/(KA)$.

143. Kasir (76), 91.
144. The proof, which does not appear in the manuscript might run thusly: area NG = area LG = area AG. Add area MB to both, giving area NB = area DF. Thus $(NF)(MG) = (AF)(BG)$, or $(NF)/(AF) = (BG)/(MG)$ and $(NF)^2/(AF)^2 = (BG)^2/(MG)^2 = (BG)^2/(BF)^2$, but, by similar triangles $(NF)/(HF) = (AF)/(NF)$, i.e. $(NF)^2/(AF) = (HF)/1$, or $(NF)^2/(AF)^2 = (HF)/(AF)$. Thus $(BG)^2(AF) = (BF)^2(HF)$. Also $(BG)^2(AF) = (BG)^2(BF + BA) = b(BF) + c$. Thus, since $(BF)^3 + (BG)^2(AF) = (BF)^2(BF + HF) = (BF)^2 BH$, $x = BF$ gives $x^3 + bx + c = ax^2$.

145. Compare Woepcke (189), 51–2, note (**).
146. For example, $a = 1$, $b = 2$, $c = 1$.
147. Kasir (76), 91.
148. Recall, too, that Archimedes solved L.10 in terms of a parabola and hyperbola, while ʿUmar does also, but in an awkward three case decomposition.
149. Sesiano (153), especially 190–4.
150. Woepcke (189). See "Additions §B, etc.", pp. 96–127.
151. Woepcke (189), "Additions, §B, §C". The first example however is a manuscript, citing at second hand (possibly by al-Qūhī), al-Māhānī's work on the Eutocius' commentary of Archimedes' *Sphere and Cylinder*. Of course, it is Archimedes' διορισμός which is cited. The second manuscript is evidently an extract of a longer treatise by al-Qūhī on problems either stated in *Sphere and Cylinder*, or related to them. However, the crucial passage is paraphrased by Woepcke, so it's not possible to comment with certainty on the exact degree of improvement. See also Section V of the present study for a discussion of the work of Sharaf al-Dīn al-Tūsī (d. *c.* 1213/14).
152. Some of this literature has been treated by Sayili (143) and Schoy (145).
153. Hogendijk (68).
154. *Ibid.*, 31–3. Al-Shannī certainly doesn't seem to be an unbiased observer, if the quotes from Hogendijk are representative.
155. *Loc. cit.*.
156. See Bruins (25), van der Waerden (180).
157. The numbering is Sacerdote's (136). Since $\sqrt{51200000} \approx 7155.4$, while $\sqrt{2048000} \approx 1431.1$ these are well defined. We are citing the equations with Suter's corrections, however. See Suter (162).
158. Sabra (135), 200. See also the appendix of our paper for a translation of some of the statements of the propositions of Book V of al-Haytham's *Optics.*
159. Lohne (99).
160. Baker (12), 328.
161. Cited by Lohne (99), 8.
162. Naẓif (115).
163. Sabra (135), 200–1. He informs us that he is also preparing editions of the Latin and Arabic texts of *The Optics* with detailed analysis.
164. Bruins (26).
165. This is just the simple analytic geometry condition for equality of the angles.
166. Lindberg (98), 181. See also Lindberg's introduction, v–xxxiv, which gives an excellent account of the Latin edition upon which we base our translation.
167. This, unfortunately, can only be established through careful reading of the text, since references to previous propositions are often omitted. In fact, the division into propositions seems to be a Latin addition, rather than an integral part of al-Haytham's text.
168. This follows note (198) in the text.
169. Lindberg (98), 132.
170. Risner's "Candido Lectori" states: There are two things we like to say a bit about here for the benefit of the would-be reader. The first of these is that whatever commentaries we've made on Alhazen, we've included between brackets, thus [ ]. The second . . . (the use of abbreviations, e.g.), 29 p1, 1 ax., 4p, 1 d6: are proposition 29, book 1; axiom 1, proposition 4; definition 1 book 6 of the *Elements* . . . , (while) 19 n1; 14 n2, 13n are references to number 19 book 1 of Alhazen's *Optics*, number 14 book 2, and number 13 of the book in which the claim is made."
171. See note 170. This is Heath (65), II, 73–4, i.e. Euclid's proposition 36 of Book 3 of the elements. From now on, we offer a note for such justification only if its numbering differs from Heath's edition of the *Elements*.
172. *totali.*

173. *partiali.*
174. *quod fit ex ductu.*
175. Clagett's remark (*Archimedes in the Middle Ages*, vol. 4), 20 should be noted here. Risner has, indeed, arbitrarily divided a single proposition which includes both 32 and 33 into two parts. Clagett, in the work cited here, 23–6, also offers translations of propositions 33 and 34. Ours was done independently and checked against his. The inelegance is partially our fault, and mostly the fault of the Arabic to Latin translators, who chose to preserve the omnipresent Arabic "ands" instead of adopting short sentences!
176. *Sectio pyramidalis.*
177. This is Heath's *Apollonius*, p. 56; i.e., II, 4 is the construction of a branch of a hyperbola, given asymptotes and one point.
178. This is guaranteed by compass construction.
179. Heath, *ibid.*, 57.
180. This is the place in the text to which note 182 refers.
181. since $(ad)^2 = (bd)(dg) = (dg)^2 + (bg)(dg)$, and $(ad)^2 = (ad)(dn) + (ad)(na)$, $(ad)(na) = (bg)(dg)$.
182. See note 180 above.
183. *quae quidem sectio non continget aliquam linearum zn, hl, inter quas iacet.*
184. Heath, *Apollonius*, p. 56.
185. Risner offers no citation here, but Clagett (204), 25 reasonably suggests *Conics*, I, 14 and II, 15. See Heath, *Apollonius*, p. 13 and 53.
186. *constat enim e duobus angulis per fabricationem equalibus angulis agb, abg rectum equantibus.*
187. *Apollonius*, p. 58.
188. That is, if c_1 and c_2 are the two points of intersection, the two parallels will give rise to angles hl_1z, hl_2z say. These are evidently to be used as hlz had been before.
189. The text has dnm.
190. That is, on the extension of a side containing the right angle.
191. The parenthesis here is apparently Risner's.
192. Our reprint seems to have em rather than cm.
193. It would be better to refer to triangle tea here.
194. Compare this with al-Haytham's earlier approach to similar construction; there he chose to show that u, a, d are collinear.
195. $\angle gal = \frac{1}{2}\angle dgu$, $\angle gal + \angle lah = 90°$, $\angle dgu + \angle dge = 180°$. $\therefore \frac{1}{2}\angle dgu + \frac{1}{2}\angle dge = \angle gal + \angle lah$, or $\frac{1}{2}\angle dge = \angle lah$.
196. For $\angle tau + \angle tad = 180°$, $\angle tau = \angle dgu$, and $\angle dgu + \angle dge = 180°$.
197. Proofs of propositions 37 and 38 appear in Appendix II. We include them so we can discuss all the special constructions of al-Haytham.
198. Gulley (59) and Mahoney (109).
199. Heath (66), c–cxxii; Zeuthen (202), 271ff.
200. Heath (66), c. The first constructions of this type are attributed to Hippocrates of Chios (second half of the fifth century, B.C.) in a passage on the quadrature of lunes preserved in Simplicius' *Commentary on Aristotle's Physics.* The particular construction amounts to a quadratic equation, easily resolved by *Elements* II, 6. However, the general opinion seems to be that Hippocrates solved his problem by mechanical means. For the relevant passages and references to further historical commentaries, see Thomas (168) I, 235–45. See also van der Waerden (182), 134, and Heath (62) I, 183–200.
201. Hultsch II, 670–3.
202. The bracketed passage is probably an interpolation.
203–4. Hultsch II, 671, 673 refers to an analysis (and reconstruction) of Horsley who challenged the number of cases necessary and noted that a variant of Pappus' 3, due to Apollonius, appears in the *Collection*, VII, proposition 73.

205. Hultsch I, 273, 298, etc..
206. Clagett (204) IV, pt. I, 26–9.
207. For a recent summary and analysis of the trisection neusis in Greek mathematics, see Bruins, E. M., "Duplication and trisection", *Janus* **67** (1980), 63–89.
208. Dodge (205) II, 637.
209. Sezgin V, 139.
210. Dodge (205), II 637. See also Sezgin V, 142.
211. Halley (206).
212. See Sezgin V, 128. Al-Nadīm mentions a commentary on the first part (Dodge (205) II, 638), while Sezgin V, 188 remarks that part of Eutocius' commentary on the second part is preserved in al-Ṭūsī's *Taḥrīr*.
213. Sabra (135), 207. Woepcke (189), 91–3 gives a translation. See also Sezgin V, 130 for manuscript information.
214. Steinschneider (158), *passim*. Sezgin V, 128–36. See also Clagett (204) I, 3–4. For the *Fihrist*, see Dodge (205) II, 636.
215. See the interesting paper of Berggren, J. L., "A coincidence of Pappus' Book VIII with al-Bīrūnī's Taḥdīd", *J. History of Arabic Science* **2** (1978), 137–42. The Arabic manuscript of this book is discussed in Jackson, D. E. P., "The Arabic translation of a Greek manual of mechanics", *Islamic Quarterly* **16** (1972), 96–103. [Ahmet III, no. 3457].
216. Unguru, S., "Pappus in the thirteenth century Latin West", *Arch. History of the Exact Sciences* **13** (1974), 307–24.
217. *Op. cit.*, 311–15. He notes that Risner cites Pappus VI, 42 to support Witelo I, 22; that Witelo I, 38 clearly parallels Pappus VI, 42, and Witelo I, 39 is Pappus VI, 44.
218. Sabra (135), 202. There seems to be no evidence that al-Haytham had a direct knowledge of Greek, though he seems to have written summaries of Aristotle and Galen. Again, see Sabra (135), 190 and Suter (163), 91–5.
219. Clagett (204) IV (1), 29. See Heath (66), 156–7.
220. *Loc. cit.*. See Heath (66), 157–8.
221. *Loc. cit.*. Again see Heath (66), 159–62, to see that the similarity is only that.
222. Clagett's claim (204) IV (1), 29 that "the neusis of Alhazen of v. 34 was employed for the same purpose in proposition XVIII of the *Verba filiorum* of the Banū Mūsā, but with a mechanical solution given" seems unfounded. In that proposition of the *Verba filiorum* one has essentially the following construction:

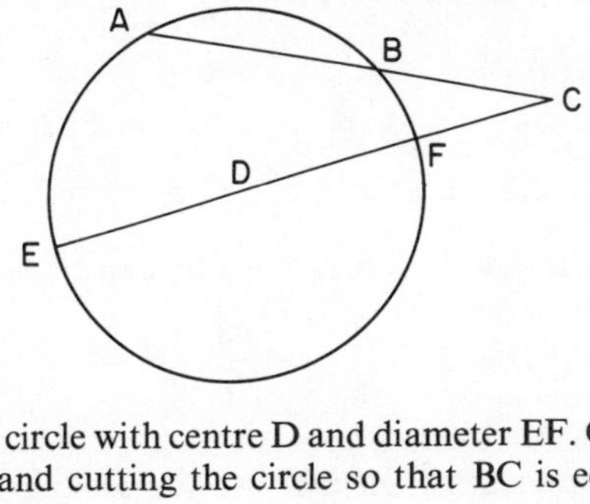

Arc AE is given on a circle with centre D and diameter EF. One is to draw a line from A to EF extended, and cutting the circle so that BC is equal to DF. See van der Waerden (182), 225–6.
223. Heath (62) II, 175–6.
224. Steinschneider (158), 171. A convenient source for acquiring some notion of al-Haytham's indebtedness to Euclid can be gleaned from Paul ver Eecke's translation: *Euclide: L'optique et la Catoptrique.* (new printing) Libraire Scientifique et Technique Albert Blanchard, Paris, 1959. See especially the part "Catoptrique", pp. 99–123, which contains Euclid's work on reflection.

225. Steinschneider (158), 216. See A. Lejeune, ed., *L'Optique de Claude Ptolémée dans la version latine d'après l'arabe de l'emir Eugène de Sicile*, Publications Universitaires de Louvain, 1956. Of special note are books III and IV, pp. 87–221. A convenient summary of the propositions contained in the work appears between pp. 123*–31*.
226. Heath (62) II, 293.
227. Lindberg (98), 162.
228. *Ibid.*, 163.
229. *Ibid.*, 164.
230. *Ibid.*, 165.
231. Literally, "*Vercies quid hdt secat...*"
232. The supposition is that 'o' is another such point.
233. Here, 'o' is assumed to be a reflection point.
234. The printed text of Risner seems to have eo instead.
235. *Supra.*
236. Here, 'f' is assumed to be a reflection point.
237. See the caption to Figure 30. This is a minor matter, since f can be taken arbitrarily if one works with the intersection of fd produced and circle ohd. The tangent through e plays no essential role in the proof.
238. Text has dz.
239. Uniqueness.
240. This curious reference to a ninth axiom is repeated several times.
241. This is a parallel argument to that for po > ck.
242. This rather vague statement is clarified in what follows.
243. This extended piece of arithmetic goes as follows (starting from)

$$\frac{sd}{dh}=\frac{fc}{ck}\cdot\frac{hd}{di}=\frac{hd}{dt}=\frac{kc}{cl},$$

so

$$\frac{sd}{dh}\cdot\frac{hd}{di}=\frac{fc}{ck}\cdot\frac{kc}{cl},$$

i.e. sd/di = fc/cl. But

$$\frac{sd+di}{di}=\frac{fc+cl}{cl};\quad \frac{si}{di}=\frac{fl}{cl}.$$

And di/io = cl/ml, giving

$$\frac{si}{di}\cdot\frac{di}{io}=\frac{fl}{cl}\cdot\frac{cl}{ml},$$

or si/io = fl/ml, or io/si = ml/fl. But si/ir = fl/ln, and

$$\frac{io}{si}\cdot\frac{si}{ir}=\frac{ml}{fl}\cdot\frac{fl}{ln},$$

or io/ir = ml/ln, or ir/io = ln/lm.
244. Applied to triangle hml since angle mhl is assumed to be "bisected".
245. This is a doubtful application on the part of the Latin commentator.
246. Woepcke (189), 114–15.
247. Woepcke (189), 115–16. The complete text appears in Woepcke (194).
248. ʿUmar refers to it in the section of his *Algebra* on "Equations with Fractional Terms". See Kasir (76), 109.
249. Kasir (76), 34.

250. Rosenfeld and Youschkevitch (200), 257.
251. *Loc. cit.*. Rosenfeld and Youschkevitch count them (132), p. 357, note 171 and find only 65.
252. See Breidenbach and Süss (23), especially 229–37 or Bieberbach (17), 72–4.
253. For example, see Dickson (40), 51–2 for the (nearly) trivial proof.
254. For a proof, see Hudson (69).
255. Bieberbach (17), 72ff.
256. This is a result of Kortum and Smith; see Breidenbach and Süss (23), 231.
257. London (100)–(101).
258. The trisection problem with marked straight edge was solved by Pappus and by Archimedes, and Vieta provided a solution of the duplication problem with this instrument. (See Breidenbach and Süss (17), 232–3 for the actual constructions.)
259. Breidenbach and Süss (17), 235–6.
260. For cubics, see Henriquez, G., "The graphical interpretation of the complex roots of cubic equations", *American Mathematical Monthly* **42** (1953), 383–4. The general situation was described in Gehman, H. M., "Complex roots of a polynomial equation", *American Mathematical Monthly* **48** (1941), 237–9.
261. Kennedy (79), 248.
262. *Ibid.*, 250.
263. Youschkevitch and Rosenfeld (200), 257. See also Luckey (103)–(104).
264. Rashed (124).
265. Anbouba (4), 151–16.
266. Rashed (126).
267. *Ibid.*, 249, continuation of note 5 of the preceding page.
268. Loth, O., *A Catalogue of Arabic Manuscripts in the Library of the India Office*, Biblio Verlag, Osnabrüch, 1975 [1877], 222–223. See also Sezgin V, 399 (Nachträge). For the little that is known of Sharaf al-Dīn's life, see Anbouba (4), and Suter (163), 134. Both mention another manuscript of possible importance for the geometric theory: *Jawāb ʿalā suʾāl liʾ amīr al-umarāʾ Shams al-Dīn* (Leiden 1027; Smith, Or. 45, 2), on the division of a square into a parallelogram and three trapezoids with a prescribed ratio. Such work is also reminiscent of Euclid's work on divided figures; see Archibald (208).
269. Anbouba (4), 515.
270. Anbouba (4), 515 and Rashed (126), 272.
271. Still this is, so far, only negative evidence. See Karpova, L. M. and Sergeeva, N. D., "Graphs of functions by Al-Marakeshi," *Istoriko mat.-issled.* (1978), 231–4 for a thirteenth century instance of function graphs.
272. Rashed (126), 249, end of note 5.
273. For example, in Rashed (126), p. 277 it is stated that the equation $x(b-x^2)=2(b/3)^{3/2}$ has a unique (presumably positive) solution, and that "la démonstration donnée est une vérification". But it is readily checked that $x=(b/3)^{1/2}$ is a *double* root. Similar statements dot the text. Some arguments are reversed, and gaps appear in the reasoning. But most disturbing of all is Rashed's lack of historical reference. The obvious dependence of Sharaf al-Dīn and his predecessors on Archimedes is never mentioned.
274. Rashed (126), 280: "Le rôle du discriminant $D=b^3/27-a^2/4$ est ainsi établi et algébriquement élaboré pour l'étude de l'equation cubique". Then, on page 288: "On conclut donc que dans ce cas AL-ṬŪSĪ est amené tout d'abord à trouver le maximum de l'expression bx^2+cx-x^3. Pour déterminer ce maximum, il annule $\frac{2}{3}bx+\frac{1}{3}c=x^2$, c'est à-dire dans un autre langage la dérivée première de cette équation". Similar remarks appear in §V of the same paper, starting on page 289.
275. Anbouba (4), 515.
276. *Loc. cit.*.

277. Anbouba (4), 516, note 16. It is enough, of course, that $|b| \leqslant \sqrt{3}|a|$.
278. See Sacerdote (136) and Suter (162). One might also expect to find an Arabic influence in Leonardo's work on cubics. Indeed, Woepcke (195) published a study of Leonardo's solution of $x^3 + 2x^2 + 10x = 20$, and found elements in common with Arab procedures, namely, only positive roots are treated, and the demonstrations are geometric. However, no conics are involved, and Leonardo spends much of his study in proving that the solution cannot be formed of quadratic surds before resorting to a numerical solution of remarkable accuracy.
279. Kasir (76), 6–7.
280. DNB I, 937.
281. *Loc. cit.*; Baker (13).
282. DNB I, 937.
283. *Phil. Trans. Royal Soc., London* **14** (part 157), 549.
284. Whitesides, D. T., (ed. and tr.), *The Mathematical Papers of Isaac Newton*, vol. II (1667–1670), Cambridge University Press, 1968, especially pp. 450–517.
285. We are relying on C. A. Ronan's brief description of the essay, p. 71 of his study, "Halley, Edmond", in DSB VI, 67–72. The essay itself, "Methodus... inveniendi radices aequationum...", appears in the *Philosophical Transactions of the Royal Society* **18** (1694), 136–48.
286. Baker (12), 327–8.
287. Libbrecht (97), especially his Chapter 13, "Numerical Equations of Higher Degree" (in the Shu-shu chiu-chang). See also Needham (116) *passim*, and Mikami (113). Rashed (126) offers evidence that numerical methods arose in the Near East earlier than they did in China. This is also an appropriate place to remark on our silence on Indian influences. As far as the geometric theory of equations is concerned, there simply are none. In the numerical theory, there may have been some early ones. See, for example, Datta and Singh, *History of Hindu Mathematics* (2 parts), Lahore, 1935 and Kaye (77).

Appendix I

125

Summary of ibn al-Haytham's *Optics* (Kitāb al-Manāẓar in Risner's Latin edition of 1572), Book V.

Book V, 125[1]

The book is divided into two parts. The first is a preface. The second is about reflections (images).

Preface (Proem) of the book. Chapter I.

1. A visible image is formed upon reflection from a polished surface.

It is clear from book IV [2n][2] that the forms of visible things are reflected from polished bodies, and that the eye can identify them on the polished body on account of reflection, and how it may be open to the acquisition of things by reflection of their forms. And the eye comprehends a visible thing in a fixed place. And, firstly, since the location of the visible thing does not change with respect to the eye. And the form seen on the polished body is called the image. And we shall explain in this book the location of the image on the polished body, and tell how knowledge of the location is acquired, and how it is found logically.

Concerning the locations of images (reflections). Chapter II.

2. In a plane mirror the image appears at the intersection of the perpendicular of incidence and of the line (ray) of reflection.

126

3. In a spherical convex mirror, the image appears at the intersection of the perpendicular of incidence and of the line (ray) of reflection.

127

4. In a cylindrical or conical convex mirror, the image appears at the intersection of the perpendicular of incidence and of the line (ray) of reflection.

128

5. Of all straight lines falling from a visible point onto a plane or convex mirror, the minimum is perpendicular.

128

6. In a concave spherical mirror, the image appears at the intersection of the incident perpendicular and of the reflected ray.

131

7. In a cylindrical or conical concave mirror, the image appears at the intersection of the incident perpendicular and the line of reflection.

131

8. The image, in all these cases, appears at intersection of the incident perpendicular and the line of reflection.

130

9. The image in a plane mirror appears on the perpendicular of incidence.

130

10. The image in a spherical or cylindrical concave or convex mirror appears on the perpendicular of incidence.

131

11. The object and the image with respect to the surface of a plane mirror appear on opposite sides, at equidistance.

132

12. The eye and the visible object given, to find the point of reflection in a plane mirror.

132

13. If a straight line from one eye makes a perpendicular with a plane mirror, just one point, in which the eye cuts the surface at the one point of the mirror at which it (the perpendicular) falls, is reflected to that eye.

133

14. From one point of a plane mirror, exactly one visible point is reflected to an eye.

133

15. In a plane mirror, the image of one and only one point is visible to the eye in a given position.

134

16. In a spherical convex mirror lines of reflection and incidence perpendiculars are concurrent, and the image is visible in precisely this locus.[3]

134

17. A limit of contingency on a spherical mirror is a point on the tangent line to the mirror at the reflection point determined by its intersection with a line through the mirror centre and the image. Moreover the line from the centre of a convex spherical mirror to the image is greater than that drawn from the image to the reflection point.[4]

135

18. If in a convex spherical mirror the incidence perpendicular is cut by the line of reflection, and the mirror is tangent at the point of reflection, it will follow that entire perpendicular segment to the lesser segment is as the larger to the radius. And the part of the perpendicular between the point of contingency and the periphery, . . . , will be less than the semidiameter.

135

19. If a straight line from one eye makes a perpendicular with a convex spherical mirror, only one point, in which it cuts the surface is reflected to the eye.

135

20. If the part of the line of reflection (which is the common section between the reflecting surface and the convex spherical mirror) continued through the periphery of the circle is equal to the radius, the image appears inside the mirror.

136

21. If the "reflection" (which is the common section between the point of reflection and the convex spherical mirror) is equal to the interior part of the segment connecting the eye to the centre of the mirror (i.e. both equal the radius of mirror), the image appears inside the mirror.

137

22. If the reflection be made from the (part of the) circumference of a circle (which is common to the reflecting surface and a convex spherical mirror), lying between a line from the eye to the centre of the circle, and the line of reflection, its part inside the periphery being equal to a radius, the image is visible inside the mirror.

137

23. If the line of reflection cutting the diameter of a convex sphere makes a segment between the boundary of the mirror and the said diameter equal to the segment of that diameter coterminus with the centre of the mirror, this segment will represent the image.

138

24. If, on the diameter of a convex spherical mirror, drawn outward from its centre, an image is marked, the image of the said diameter will be visible between the "mark" and the spherical surface.

138

25. If the line of reflection cutting the convex spherical mirror makes a segment (between the sphere's boundaries) equal to a radius, and the radius through the end of the line of reflection intersects a line drawn tangent to the mirror from the eye, the images of the intersection of the radius are visible between the point of intersection and the surface of the mirror.

139

26. If the line of reflection, equaling in its inscribed portion a semidiameter is cut (away from the boundary of the circle), the incidence perpendicular, cutting along this line of reflection and a line from the eye tangent to the mirror, has its image either inside, outside, or on the mirror.

139

27. If the line of reflection is cut off on the circumference (which is the surface common to reflection and to a convex spherical mirror) with its inscribed part equal to a semidiameter, the perpendicular of incidence cutting the circumference between the ends of the line of reflection and a quarter circle, measured beginning from the point of tangency of a line from eye to mirror, will have its image outside the mirror.

140

28. The incidence perpendicular cutting the periphery of the circle between the end points of a line through the eye-centre and the mirror, passing through a point of the line from the eye tangent to the circle, has no image.

140

29. From one point on a convex spherical mirror, only one visible point is reflected to the eye. Also, a single image of a point is seen.

141

30. If two points of an incidence perpendicular are reflected from a spherical convex mirror to the eye, this is the situation with regard to the position of the image and reflection: the point closer to the centre of the mirror will be the more remote; the image, from centre; the reflection, from the eye.

142

31. Object and eye are equally distant from the centre of a convex spherical mirror, to find the point of reflection.

142

32. From the mid-point of a semicircle to draw a line so that its segment coterminus with the diameter extended equals a given line.

143

33. (Same as 32, with "*non medio*", i.e. away from the mid-point.)

144

34. From a given point on the circumference of a circle and outside a given diameter, to draw a line so that the segment between the diameter and the point on the circumference opposite to the given point is equal to a given line, smaller than the diameter.

145

35. From a point on one of the sides of a right triangle containing the right angle to draw a line to the side opposite the right angle (such that its) segment coterminus at its beginning with the remaining side has to the segment coterminus with the first side, a given ratio of two segments.

146–7

36. Given two points outside the circle, or one outside and the other inside, to find a point

on the circumference, through which the tangent to it bisects the angle contained by the two lines drawn from the said points to the required point.

148

37. From a given point outside a circle, to draw to a given diameter, a straight line whose parts, between the circumference and the given diameter equal the part of the diameter coterminus with the centre of the circle.

149

38. From a given point on one of the sides of a right triangle containing the right angle, to draw to the side opposite the right angle a line intersecting the other side extended in such a way that the whole to the segment opposite the right angle and coterminus with the first side, has the ratio of two given line segments.

150

39. Given the eye and object at unequal distances from a spherical convex mirror, to find the point of reflection.

151

40. If the incident radius from the visible object to the convex spherical mirror forms, with the semidiameter, an angle less than a right angle, there is no reflection to the eye from the point of incidence.

152

41. A visible point (object) reflected to the eyes from two points on a convex spherical mirror has a single image.

152

42. In a convex spherical mirror the image points correspond in order and location (in the eyes) with those of the visible points.

153

43. If the reflecting surface of a mirror is a circular cylinder, then the reflections and images occur in the same manner as in a plane and a spherical convex mirror.

153

44. If the incident perpendicular is cut by lines, the line of reflection, inside the ellipse (which is the common surface section of reflection and of the convex cylindrical mirror), and tangent in the point of reflection, then it will follow that the entire perpendicular (is) to the “lower segment” as “the upper” is to the “intermediate”. And the largest of the “lower” segments will be on the line of reflection.

154

45. If the common surface of reflection, and of a convex cylindrical mirror is an ellipse, the visible image of reflection may be oblique, on the surface of the mirror, or visible inside or outside the mirror.

154

46. If the common surface section of reflection and of a cylindrical convex mirror is the side of a cylinder or circle parallel to the base, from one point one visible point is reflected to the eye.

155

47. If the common surface section of reflection and of a convex cylindrical mirror is an ellipse, from one point, one visible point is reflected to the eye.

156

48. If the common surface section of reflection and of a convex cylindrical mirror is an ellipse, the eye and visible point given, to find the point of reflection.

156

49. If the common surface section of reflection, and of a convex conical mirror is the side (latus) of a cone, the locations of both reflection and image can be found in the same manner as in a plane mirror.

156

50. The common surface section of reflection and a convex conical mirror is not a circle.

156

51. If the common surface section of reflection and of a convex conical mirror is an ellipse, the image of an obliquely reflected visible point is visible either on the surface of the mirror, inside, or outside it.

156

52. If from a given point on the surface section common to the reflection and to a convex conical mirror, reflection occurs; the eye and the visible point are so located, with respect to such a point, so that from a point of a circle parallel to the base, reflection reaches the eye.

157

53. If the surface section common to the reflection and the convex conical mirror is the side (latus) of a cone, from one point, one visible point is reflected to the eye.

158

54. The eye and visible object are between the base of a convex conical mirror and a plane drawn through the vertex and parallel to the base, to find the point of reflection.

159

55. The eye and the visible point in a plane through the vertex of a convex conical mirror and parallel to the base, to find the point of reflection.

160

56. The eye and the visible point beyond (above) a plane drawn through the vertex of a convex conical mirror and parallel to the base, to find the point of reflection.

160

57. The eye in a plane drawn through the vertex of a convex conical mirror and parallel to the base, the object below the plane, to find the point of reflection.

161

58. The eye in a plane drawn through the vertex of a convex conical mirror and parallel to the base, the object above the plane, to find the point of reflection.

161

59. The eye below a plane through the vertex of a convex conical mirror, and parallel to the base, the visible point above the plane, or vice versa, to find the point of reflection.

162

60. In a concave spherical mirror, the image is visible, at the point of reflection, in the eye, beyond, outside the mirror, between the eye and the mirror.

162

61. In a concave spherical mirror the image appears in various orders and locations.

163

62. The eye placed at the centre of a concave spherical mirror only sees itself.

163

63. (Given) a semidiameter of a concave spherical mirror, on which the eye is away from the centre; none of its points incident obliquely to the mirror is reflected to the eye, but those on the extension of the semidiameter are.

163

64. In a concave spherical mirror, the perpendicular of incidence and line of reflection intersect; it follows that the perpendicular of incidence is concurrent with the line joining the centre and the image, the line between the visible point and the finite contingency is concurrent with the line between the finite contingency and the image.

164

65. (Given) the eye and the visible point on the diameter of a concave spherical mirror at equal distances from the centre, it is possible to have a reflection from the entire periphery for which the semidiameter perpendicular to the said diameter cuts it.

165

66. The eye and the visible point on different diameters of the circle (which is the section common to the surface of reflection and of the concave spherical mirror) reflect from one to the other, first from the perimeter between the semidiameters on which they are, then from the opposite arc to this, but not from the remaining two.

165

67. If the eye and the visible point are located on different diameters of a circle (which is the section common to the surface of reflection and to the concave spherical mirror, a line parallel to the diameter containing the visible object and containing the eye cuts the circumference of the said circle. The image reflected from the circumference between the parallel and the diameter containing the visible, appears outside the mirror; from the circumference between the parallel and the diameter containing the eye (appears) beyond the eye; from the opposite periphery, between the eye and the mirror.

166

68. On any point of the diameter of the circle (which is the section common to the surface of reflection and of the concave spherical mirror) however far extended, it is possible to see the image.

166

69. If the eye and the visible point are situated on the same diameter (which is the section common to the surface of reflection and of the concave spherical mirror) the image is visible at the eye, on a semicircle, or, secondly, a reflection to the eye is possible from any point on the fixed circle.

167

70. The eye and the visible point are situated outside the circle (which is the section common to the surface of reflection and of the concave spherical mirror) and on different diameters (extended). Then reflection occurs from one point, and one image is visible.

168

71. If the angle contained by two diameters, in the centre of the circle (which is the section common to the surface of reflection and of the concave spherical mirror) cuts a third line (the angle bisector), and from the end of it on the periphery subtended by the said angle, perpendiculars are drawn to the said diameters, the points of the diameters both on which the perpendiculars fall, and those below, equidistant from the centre of the mirror, are reflected among themselves from the cut points of the diameter, and have two images.

168–9

72. If the angle contained by two diameters in the centre of a circle (which is the section common to the surface of reflection and of a concave spherical mirror) cuts a third line (angle bisector), and from the end of it on the periphery subtended by the said angle, perpendiculars are dropped on the said diameters, the points of the diameters between the periphery and the ends of the perpendiculars equidistant from the centre are reflected among themselves from four points of the periphery, and have four images.

169

73. The eye and visible point on different diameters of a circle (which is a section common to the surface of reflection and of a concave spherical mirror), at unequal distances from the centre: from a point of the periphery between the semidiameters, beyond the eye and visible point, reflection is possible.

170–1

74. If the angle contained by two diameters in the centre of a circle (which is the section common to the surface of reflection and of a concave spherical mirror), cuts a third line (angle bisector), the points on the said diameters at unequal distances from the centre, are reflected from any point of the periphery between the semidiameters, inclusively, with the exception of the points in which the diameters terminate.

171

75. If the eye and visible point are on different diameters of a circle (which is the section common to the surface of reflection and of a concave spherical mirror) at unequal distances from the centre, they are reflected among themselves from any point of the periphery between the semidiameters; (moreover) reflection is one to one.

171

76. The eye placed on the diameter of a circle (which is the section common to the surface of reflection and of a concave spherical mirror) inside the circle, the visible point with the eye, at an arbitrary distance from the centre: reflection is possible from any point of the semicircle.

172

77. If from the eye two lines tangent to the circle (which is the section of the common surface of reflection and of a concave spherical mirror) are cut by a third through the

centre, the visible point and the eye equidistant from the centre, reflection is possible from any point of the periphery between the points of tangency with the exception of the points of tangency and the ends of the (cutting = secant) diameter.

172

78. If the eye and visible point inside a circle (which is the section of the common surface of reflection and of a concave spherical mirror) at unequal distance from the centre are reflected into each other, the angle exterior to the diameter of the eye and that of the visible point is either greater or less than the angle of incidence and (likewise) the angle of reflection.

173

79. If the eye and visible point on different diameters of a circle (which is the section of the common surface of reflection and of a concave spherical mirror) at unequal distances from the centre are reflected one to the other, the angle exterior to the diameter of the eye and that of the visible point is unequal to the incidence angle (similarly) reflection angle.

174

80. If the eye and visible point on different diameters of a circle (which is the section of the common surface of reflection and of a concave spherical mirror) at unequal distances from the centre are reflected one to the other from two points of the periphery between the semidiameters on which the eye and visible point are, neither the angle of incidence nor reflection will be less than the angle exterior to the said diameters.

175

81. Two points on different diameters of a circle (which is the common section of the surface of reflection and of a concave spherical mirror, at unequal distances from the centre can be reflected from one to the other from two points of the periphery contained between the semidiameters on which they are located.

176

82. (As 81, but with the addition: from no other point of the periphery is reflection possible.)

178

83. Given two points on different diameters of a circle (which is the section common to the surface of reflection and of a concave spherical mirror) at unequal distances from the centre, to find on the periphery between their semidiameters, two points of reflection.

179

84. If two points outside a circle (which is the section common to the surface of reflection and of a concave spherical mirror), or one inside, the other outside, on different diameters,

at different distances from the centre, are reflected from the periphery contained between the semidiameters beyond which they are (*extra quas ipsa sunt*), they are reflected from only one point.

179

85. If a straight line connecting two points on different (extended) diameters of a circle (which is the section common to the surface of reflection and of a concave spherical mirror), at unequal distances from the centre, is tangent to the said circle, or is beyond the tangent, reflection occurs from one point only.

179

86. If a straight line connecting two points on different diameters of a circle (which is the section common to the surface of reflection and of a concave spherical mirror), at unequal distances from the centre, extended, cuts the circle, the points can be reflected from one to the other through one, two, three, or four points.

181

87. If a straight line connecting two points on different diameters of a circle (which is the section common to the surface of reflection and of a concave spherical mirror), at equal distances from the centre, extended, cuts the circle, reflection between the said points is possible from one, two or four points of the mirror, but never from three, only.

181

88. In a concave spherical mirror the image of each visible point is visible to the eye either once or twice.

182

89. The section common to the surface of reflection and of a concave cylindrical mirror is either a side (generator) of a cylinder, a circle or an ellipse.

182

90. If the section common to the surface of reflection and of a concave cylindrical mirror is an ellipse, the image is visible either beyond the mirror, or on the surface, or below (citra) the eye, in the eye, between the eye and the mirror.

182

91. If the eye and the visible point are on a straight line, perpendicular to the plane tangent to a concave cylindrical mirror (at a point of tangency), reflection occurs from one, or two points of the mirror, and the image appears at the centre of the eye.

183

92. If the eye is at the centre of a circle in a concave cylindrical mirror, it is reflected from every point of the circle's circumference; similarly, the periphery of the circle is reflected through the centre of the eye, and the image appears at the eye's centre.

183

93. If the section common to the surface of reflection and of a concave cylindrical mirror is an ellipse, a visible point is reflected to the eye from many points.

183

94. If two points lie on the axis of a concave cylindrical mirror, they can be reflected from one to the other by an entire circle of points, and the image is visible on the periphery of a circle lying beyond the surface of the mirror.

184

95. If the section common to the surface of reflection and of a concave cylindrical mirror is a circle, or ellipse, there is reflection from one, two, three, or four points of the mirror, and in every case the images are visible.

185

96. The eye and visible point given, on a concave cylindrical mirror to find the point of reflection.

185

97. The section common to the surface of reflection and of a concave conical mirror is the side (generator) of the cone, or an ellipse.

185

98. If the eye is on a section common to the axis and to a line perpendicular to a plane tangent to a conical concave mirror, it is reflected from a full circle (whose centre is common to the said section) through perpendicular lines, and the image appears at the centre of the eye.

186

99. If the eye and the visible point are on the axis of a concave conical mirror, they can be reflected from one to the other on a full circle, and the image is visible on the periphery of a circle, outside the circle of the described mirror.

186

100. If the section common to the surface of reflection and of a concave conical mirror is an ellipse, the eye and the visible point, outside the axis in the base, or in a plane parallel to

it, are reflected between themselves, either from one, two, three, or four points of the mirror, and there are as many images as reflections.

187

101. As 100, but eye and visible point inside mirror away from axis, base or plane parallel to base,....

188

102. The eye and visible point given, to find the point of reflection in a concave conical mirror.

Notes to Appendix I

1. 125, etc., refer to the page numbers upon which the propositions of book V appear in Risner's edition.
2. As in our translation of the Latin in the main part of this paper, brackets preserve the additions of the Latin editor, while parenthesis contain our own additions, except in propositions 20–22, 44, 66–87, where the common phrase "which is common..." is a parenthetical insertion in the Latin text.
3. Literally, intersection (concurrence).
4. An attempt has been made to translate proposition 17 as the amplius demands. The literal reading obscures the sense of the passage, whose first aim is a definition of "*finis contingentiae*" (limit of contingency, end of tangent).

Appendix II

Proof of Proposition 37. Amplification: given a circle with centre g, and given on it a diameter bg,[1] and a point e outside the circle, there is to be drawn from point e to the diameter bg, a line cutting the circle, so that the part of it from the circle to the point (at which it cuts the diameter) be equal to the part of the diameter lying between it (the line) and the centre. In other words, let there be drawn a perpendicular from point e onto the diameter, say ec; draw line eg, and construct a line qt equal to ec. Then [by 33 p3] a part of

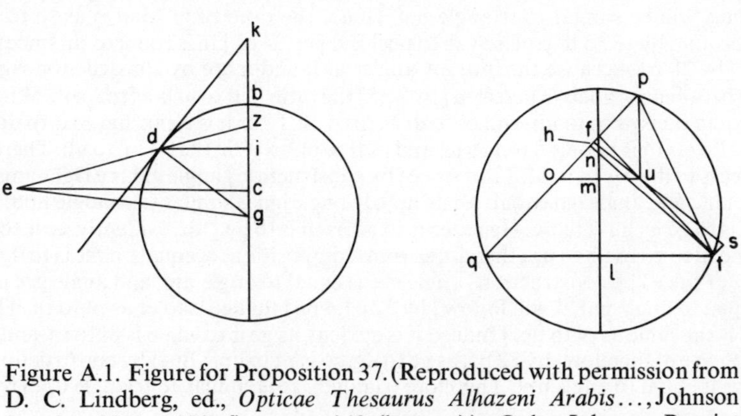

Figure A.1. Figure for Proposition 37. (Reproduced with permission from D. C. Lindberg, ed., *Opticae Thesaurus Alhazeni Arabis*..., Johnson Reprint, N.Y., 1972, figure, p. 148 (bottom).). © by Johnson Reprint Corporation, New York, N.Y.

a circle is constructed, so that the angle contained in this part is equal to angle egb; and let the circle be completed [by 25 p3], and from the mid-point of qt let there be drawn in both directions a portion of the perpendicular (extended) to the circle; it will be [by the porism to 1 p3] the diameter of this circle. And, from the point q let a line be drawn to this diameter, cutting it in a point f, and let it be extended upto a point p on the circle, so that fp be equal to one half gb [by 34n] and let the lines pt and tf be drawn. (Now) from point p a line, say pu, is drawn parallel to the diameter. It intersects the line tf in point u. [Moreover, it will intersect by Proclus' lemma to 29 p1.][2] and from point u a line is drawn parallel to tq, say uo. (Then) from point t, a perpendicular is drawn on pq, say tn, and from point t a parallel is drawn to pq, say ts, and from u, a perpendicular on pq, say uh. Then from angle bge there is cut an angle equal to angle qpu [by 23 p1]. [That is possible, since the whole angle qpt is equal to the angle bge, so that part can be taken from the whole.] Let it be bgd, and let line edz be drawn. I say that dz is equal to zg,[3] and from point d a perpendicular, say di on bg, and at d [by 17 p3] a tangent to the circle, say dk. Clearly, since the diameter fl is perpendicular to qt [by construction] and ou [by 29 p1], and pu is parallel to fl,[4] it will follow that [by 29 p1] angle oup is right. And since ou has the diameter as its perpendicular bisector[5] [for by construction, 29, 32 p1 triangles fql, fom are equiangular[6] as are triangles flt, fmu; thus by 4 p6, as ql is to lf so is om to mf, and as fl is to lt, so is fm to mu; therefore, by 22 p5, as ql is to lt, so om is to mu; and also by construction ql is equal to lt itself; whence om equals mu][7] it will follow [by 4 p1] that fo equals fu, and angle fou is equal to angle fuo. But since the two angles pou, opu sum to a right angle [by 32 p1, since angle oup (text: adu) is clearly a right angle] angle tup will be equal to angle fpu. [Indeed, angle fou equals angle fuo by argument, and angle pou and opu together equal a right angle. Thus angles fuo, fpu sum to a right angle, and angle oup is right. Therefore with the removal of the common angle fuo, the remaining angle fup is equal to the angle fpu by 3 ax.] Thus [by 6 p1] fp is equal to fu; and so [by 1 ax.] is equal to fo. Thus po equals bg [because fb is equal, by construction, to half of bg, and from our argument so is fo; their sum po, therefore, equals bg] and gd, [by 15 d1]. Thus [by 7 p5] ec is to gd as tq to po.[8] But, since the right angle kdg [by 18 p3] is equal to angle gid, and angle igd is common, triangle igd will be similar to triangle kdg [by 32 p1. 4 p.1 d6] and the ratio of gd to di as gk to kd. But [by construction] angle kgd is equal to angle opu, and angle kdg is right, equal to angle oup. So triangle kgd is similar to triangle oup, and [by 1 d6] the ratio of gk is to kd as op to ou. Therefore [by 2 p5] dg is to di as op to ou. Therefore, the ratio of ec to di is as tq to ou [for it was demonstrated that ec is to gd as tq to po, and gd to di as po to ou; therefore by 22 p5, as ec is to di, so is tq to ou]. But the ratio of qt to ou is as tf to fu, since triangle tfq is similar to ofu [by 29 p1.4 p.1 d6]. Moreover[9] [by construction and 29 p1] angle uts equals angle hfu because they are alternate angles, and the right angle ust[10] equals the right angle fhu, (so) triangle ust will be similar to triangle huf. Hence the ratio of tu to uf is as su to uh, [on which account, by 18 p5 tf to uf is as sh to uh.] But [by 34 p1] tn is equal to sh, since they are parallel [by 28 p1, because the interior angles at h and n are by construction right] and they are between parallels. Therefore [by 7 p5] the ratio of tf to uf is as tn to uh. Whence the ratio of qt to ou is as tn to uh, and ec to di as tn to uh. [For it is clear that ec is to di as tq to ou, and also, as tq is to ou, so is tf to uf, and as tf to uf, so is sh, that is tn, to uh. Therefore by 2 p5, as ec is to di so is tn to uh.] But since [by construction] angle gid is a right angle equal to angle phu, and angle igd equals angle hpu, triangle igd is similar to triangle hpu, and [by 1 d6] id is to gd as hu is to pu. Hence ec is to gd as tn is to pu [for, evidently, ec is to di as tn to uh, and di to gd as uh to up; therefore, from (composition) of equals, as ec is to dg, so is tn to up.] But since [by construction] angle cge is equal to angle npt, and angle gce is a right angle equal to angle pnt, it will follow [by 32 p1.4 p6] that ge is to ec as pt to tn. Therefore ge to gd is the same as pt to up. [Indeed it is evident, as ge is to ec, so is pt to tn, and as ec to dg so is tn to up; therefore, by 22 p5, as ge to dg, so is pt to up.] But [by construction, 3 ax.] angle dge is equal to angle upt. Therefore triangle dge is similar to triangle upt [by 6.4 p.1

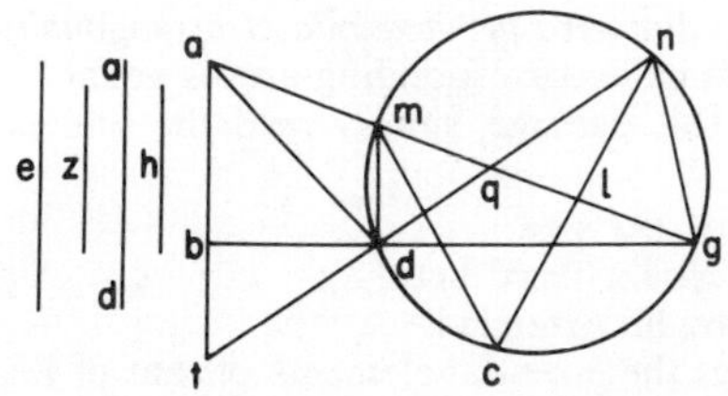

Figure A.2. Figure for Proposition 38.[11] (Reproduced with permission from D. C. Lindberg, ed., *Opticae Thesaurus Alhazeni Arabis*..., Johnson Reprint, N.Y., 1972, figure, p. 149.). © by Johnson Reprint Corporation, New York, N.Y.

d6]. Hence angle gde is equal to angle put. It follows, consequently, [by 13 p1.3 ax.] that angle gdz equals angle fup [and by construction angle fpu equals angle zgd] on which account the third angles are equal [by 32 p1, and for that reason triangles zgd, fpu have equal angles] and [by 4 p6] dz is to zg as uf to fp. But uf equals fp. Therefore dz equals zg. That is the proposition.

Proof of Proposition 38. Amplification: with a given right triangle abg whose right angle is abg, and with a point d given on bg or ab, a line is to be drawn from point d to the side ag, intersecting in a point, say q, and in the other direction intersecting with the other side (extended), so that its whole is to gq as is e to z (where e and z are given segments). In other words, a parallel to ab is drawn through d, say dm, and [by 5 p4] let there be a circle passing through the three points d, m, g. mg will be the diameter [by consequence of 5 p4.] Line ad is drawn, and let there be a segment h [by 12 p6] which itself is to ad, as e to z. And since [by 29 p1] angle dmg is equal to angle bag, let there be cut from it an angle equal to angle dag, say angle cmd, and mc drawn, which cuts the circle in c, from which (i.e., from c) [by 34n] a line is drawn to the diameter mg and beyond[12] to the circle, so that ln is equal to segment h. (Now) line ng is drawn, and (also) line dn, intersecting with ag in point q and ab in point t. Since, therefore, [by 27 p3] angle dmc equals angle dnc, because they are on the same arc, it will follow [by 1 ax.] that angle qnl equals angle daq, and [by 15 p1] nql equals angle dqa. Whence [by 32 p1.4 p.1 d6] triangle nql is similar to triangle dqa. Therefore aq is to qn as ad to nl. In fact, since angle dmg is equal to angle dng [by 27 p3], qng will equal [by 1 ax.] taq. Let t be the point in which dn intersects with ab, and [by 15 p1] (note that) angle tqa equals to angle nqg. Therefore triangle tqa is similar to triangle nqg, and [by 1 d6. 16 p5] the ratio of aq to qn is as tq to qg. Therefore [by 2 p5] the ratio of tq to qg is as that of ad to nl. But [by construction] nl equals h, and ad is to h as e to z. Thus [by 2 p5] tq to qg is as e to z. That is the proposition. It will also happen, in the case that two lines will be drawn from point c, like cln, and then two lines will be drawn from point d, like tq, that as each is to the part which it cuts from ag, so will e be to z, and this follows by the same proof.

Notes to Appendix II

1. Actually, bg is only a radius, but it is called a diameter in the original text. This should cause no particular confusion.
2. For all Proclus' references, the reader may consult G. R. Morrow's edition of the *Commentary*, Princeton University Press, 1970.
3. The proof of this claims takes up the remainder of the text of proposition 37.
4. The Risner text appears to have ei here instead of fl.

5. Literally, "*dividatur à diametro per aequalia et orthogonaliter*".
6. Triangles fql and fom have corresponding angles equal.
7. To get the sense of this passage, simply read the sentence through, omitting the bracketed material.
8. Recall that ec = qt and po = gd.
9. Actually, the word "*verũ*" is used here.
10. Angle ust is formed by hu extended.
11. Our figure reproduces the essential elements of that of Risner's edition.
12. *usq.*

Bibliography

(*) after a title indicates that it was unavailable to the authors. [= 1, 84] after a bibliographic entry means that the entry repeats an earlier one, e.g., no. 84 of essay 1.

Essay 1

1. Adamson, J. W., "Education", 255–85 in C. G. Crump and E. F. Jacob, eds., *The Legacy of the Middle Ages*, Oxford University Press, 1926 [1969].
2. Alonso, M. A., "Juan Sevillano, sus obras propias y sus traducciones", *Al-Andalus* **18** (1953), 17–49.
3. Altaner, B., "Zur Kenntnis des Arabischen im 13. und 14. Jahrhundert", *Orientalia Christiana Perodica* **2** (1936), 437ff.
4. Alverny, M. T. d'. and Vajda, G., "Marc de Tolède, traducteur d'Ibn Tūmart", *Al-Andalus* **16** (1951), 99–140; 259–307; **17** (1952), 1–56.
5. Anawati, G. C., "Ḥunayn ibn Isḥāq", DSB XV, 230–4.
6. Arnold, T. and Guillaume, A., eds., *The Legacy of Islam*, Oxford University Press, 1931. [1968]
7. Berthelot, M., *Die Chemie im Alterum und im Mittelalter*, G. Olms, Hildesheim, 1970. [Reprint of 1909 ed.]
8. Berthelot, M., *La Chemie du Moyen Âge*, 3 volumes, Philo Press, Amsterdam, 1967. [Reprint of 1893 ed.]
9. Bischoff, B., "Das griechische Element in der abendländischen Bildung des Mittelalters", *Byzantinische Zeitschrift* **44** (1951), 27–55.
10. Bischoff, B., "The Study of Foreign Languages in the Middle Ages", *Speculum* **36** (1961), 209–23.
11. Bossong, G., *Probleme der Übersetzung wissenschaftlicher Werke aus dem Arabischen in das Altspanische zur Zeit Alfons des Weisen*, Max Niemeyer Verlag, Tubingen, 1979.
12. Bouyges, M., "Roger Bacon a-t-il lu des livres arabes?", *Arch. his. litt. Moyen Âge* **5** (1930), 311–15.
13. Brooke, C., *Europe in the Central Middle Ages*: 962–1154. Longman, London, 1975.
14. Bullough, V. L., "Alderotti", DSB I, 107.
15. Chejne, A. G., *Muslim Spain : Its History and Culture*, University of Minnesota Press, Minneapolis, 1974.
16. Clagett, M., "John of Palermo", DSB VII, 133–4.
17. Clagett, M., *Studies in Medieval Physics and Mathematics*, Variorum Press, London, 1979.
18. Cook, M. A., ed., *Studies in the Economic History of the Middle East*, Oxford University Press, London, 1970.
19. Coulton, G. G., *Life in the Middle Ages*, 4 volumes in 2, Cambridge University Press, 1967. [1910]
20. Crombie, A. C., Augustine to Galileo: *The History of Science, A.D.* 400–1650, Falcon Press, London, 1952.

21. Crombie, A. C., *Robert Grosseteste*, Oxford University Press, 1953. [1971]
22. Crombie, A. C. and North, J. D., "Roger Bacon", DSB I, 377–85.
23. Dales, R. C., *The Scientific Achievement of the Middle Ages*, University of Pennsylvania Press, Philadelphia, 1973.
24. Daniel, N., *The Arabs and Mediaeval Europe*, Longman, London, 1975.
25. Daniel, N., *The Cultural Barrier*, University Press, Edinburgh, 1975.
26. Daumas, M., ed., *A History of Technology and Invention*, Tr., E. B. Hennessy, Volume 1, Crown Pub., Inc., N.Y. 1969 [1962]
27. Destombes, M., "La diffusion des instruments scientifique du haut moyen âge au XVe siècle", *Cahier d'Histoire Mondiale* **10** (1966/67), 31–51.
28. Dodge, B., *Al-Azhar : A Millenium of Muslim Learning*, The Middle East Institute, Washington, D.C., 1961.
29. Dodge, B., *Muslim Education in Medieval Times*, Washington, D.C., 1962.
30. Douglas, D. C., *The Norman Fate* : 1100–1154, Eyre Methuen, London, 1976.
31. Duhem, P., *Le Système du Monde*, 10 volumes. Hermann, Paris, 1956. [Especially volume 5.]
32. Dulieu, L., "L'arabisme médical à Montpellier du XIIe au XIVe siècle", *Cahiers de Tunisie* **3** (1955), 86–95.
33. Dunlop, D. M., *Arab Civilization to A.D.* 1500, Longman, London, 1971.
34. Dunlop, D. M., "Arabic Science in the West", *J. Pakistan Historical Soc.* **5** (1957), 1–22, 77–99, 129–50, 200–19.
35. Dunlop, D. M., "The Work of Translation at Toledo", *Babel* **6** (1960), 55–9.
36. Feldhaus, F. M., *Die Technik der Antike und des Mittelalters*, G. Olms, Hildesheim, 1971. [Reprint of 1930 ed.]
37. Forbes, R. J., *Man the Maker : A History of Technology and Engineering*, Constable and Company, Ltd, London, 1958.
38. Forbes, R. J., "Metallurgy and Technology in the Middle Ages", *Centaurus* **3** (1953), 49–57.
39. Forbes, R. J., *Studies in Ancient Technology*, Volume 2, E. J. Brill, Leiden, 1965.
40. Gabrieli, F., *Arab Historians of the Crusades*, Tr., E. J. Costello, Routledge and Kegan Paul, London, 1969 [G. Einaudi edition (Italian), 1957]
41. Gille, B., "Technological developments in Europe: 1100–1400," in G. S. Métraux and F. Crouzet, eds, *The Evolution of Science*, N.Y., 1963, pp. 168–219.
42. Gillispie, C. C., ed., *Dictionary of Scientific Biography* (DSB), 16 volumes, Charles Scribner's Sons, N.Y., 1970–1980.
43. Glick, T. F., "Éloge: José María Millás Vallicrosa", *Isis* **68** (1977), 276–83.
44. Glick, T. F., *Islamic and Christian Spain in the Middle Ages*, Princeton University Press, 1979.
45. Goiten, S. D., *A Mediterranean Society*, 2 volumes, University of California Press, Berkeley, 1967, 1971.
46. Goiten, S. D., *Studies in Islamic History and Institutions*, E. J. Brill, Leiden, 1968.
47. Goldstein, B. R., "The Role of Science in the Jewish Community in Fourteenth-Century France", *Annals of the New York Academy of Sciences* **314** (1978), 39–49.
48. Goldziher, I., "Education (Muslim)", in J. Hastings, ed., *The Encyclopaedia of Religion and Ethics*, vol. 5, T. and T. Clarke, Edinburgh, 1912 pp. 198–207. [1967]
49. Grabar, O., *The Alhambra*, Harvard University Press. Cambridge, 1978.
50. Grant, E., ed., *A Source Book in Medieval Science*, Harvard University Press, Cambridge, 1974.
51. Grant, E., *Physical Science in the Middle Ages*, Cambridge University Press, London, 1977. [John Wiley, 1971]
52. von Grunebaum, G. E., *Classical Islam*, Tr., K. Watson, George Allen and Unwin, Ltd, London, 1970. [German edition, 1963.]
53. von Grunebaum, G. E., *Islam and Medieval Hellenism : Social and Cultural Perspectives*, Variorum Reprints, 1976.

54. Gwatkin, H. M., Whitney, J. P., *et al.*, eds., *The Cambridge Medieval History*, 8 volumes in 9, Cambridge University Press, 1964.
55. Harant, H. and Vidal, Y., "Les influences de la médecine arabe sur l'École de Montpellier", *Cahiers de Tunisie* **3** (1955), 60–85.
56. Haskins, C. H., "Arabic Science in Western Europe", *Isis* **7** (1925), 478–85.
57. Haskins, C. H., *Studies in the History of Mediaeval Science*, Harvard University Press, 1927. [Ungar Reprint, 1967.]
58. Haskins, C. H., *The Renaissance of the Twelfth Century*. Harvard University Press, 1927. [pp. 282–91, 310–14 which are relevant to our study are reprinted as pp. 79–88 in A. R. Lewis, ed., *The Islamic World and the West : A.D.* 622–1492, John Wiley and Sons, N.Y., 1970.]
59. Haskins, C. H., *The Rise of Universities*, Brown University Press, 1923. [Cornell Reprint, 1972.]
60. Al-Hassan, A. Y., *et al.* ed., *Proceedings of the First International Symposium for the History of Arabic Science*, Volume 2, IHAS, Aleppo, 1978.
61. Hill, D. R., ed. and tr., *Al-Jazarī's Book of Knowledge of Ingenious Mechanical Devices* (*Kitāb fī ma ʿrifat al-ḥiyal al-handasiyya*), D. Reidel, Dordrecht, 1974.
62. Hill, D. R., ed. and tr., *Banu Musa : The Book of Ingenious Devices* (*Kitāb al-Ḥiyal*), D. Reidel, Dordrecht, 1979.
63. Hill, D. R., "The Banū Mūsà and their *Book of Ingenious Devices*", *J. History of Technology* (1977), 39–76.
64. Holmes, U. T., "Life among the Europeans in Palestine and Syria in the Twelfth and Thirteenth Centuries", pp. 3–35 in H. W. Hazard, ed., *The Art and Architecture of the Crusader States*, volume 4 of K. M. Setton's, *A History of the Crusades*, University of Wisconsin Press, Madison, 1977.
65. Holt, P. M., *et al.*, eds, *The Cambridge History of Islam*, Volume 2, Cambridge University Press, 1970.
66. Hourani, G. F., "The Early Growth of the Secular Sciences in Andalusia", *Studia Islamica* **32** (1970).
67. Hourani, G. F., "The Medieval Translations from Arabic to Latin Made in Spain", *The Muslim World* **62** (1972), 97–114.
68. Hunt, R. W., "English Learning in the Late Twelfth Century", in R. W. Southern, ed., *Essays in Medieval History*, Macmillan, N.Y., 1968, pp. 106–28.
69. Huygens, R. B. C., "Guillaume de Tyr étudiant. Un chapitre de son *Histoire* retrouvé", *Latomus* **21** (1962), 811–29.
70. Iskandar, A. Z., "Ḥunayn the Translator", DSB XV, 234–49.
71. Kimble, G. H. T., *Geography in the Middle Ages*, Russell and Russell, N.Y., 1968. [1938]
72. King, D. A., "Review of Hill's Edition of al-Jazarī's The Book of Knowledge of Ingenious Mechanical Devices", *Hist. Sci.* **13** (1975), 284–9.
73. Kish, G., ed., *A Source Book in Geography*, Harvard University Press, Cambridge, 1978.
74. Knowles, D., *The Evolution of Medieval Thought*, Longman, London, 1962.
75. Kren, C., "Dominicus Gundissalinus", DSB V, 591–3.
76. Kren, C., "Roger of Hereford", DSB XI, 503–4.
77. Krey, A. C., "William of Tyre", *Speculum* **16** (1941), 149–66.
78. Kristeller, P. O., *Renaissance Thought and Its Sources*, ed., M. Mooney, Columbia University Press, N.Y., 1979.
79. Kunitzsch, P., *Der Almagest*, Otto Harrassowitz, Wiesbaden, 1974.
80. Kurz, O., *European Clocks and Watches in the Near East*, E. J. Brill, Leiden, 1975.
81. Laistner, M. L. W., *Thought and Letters in Western Europe*, Cornell University Press, N.Y., 1931. [1966]
82. Laver, H. H., "Zur Beurteilung des Arabismus in der Medizin des mittelalterlichen England", *Sudhoffs Arch.* **51** (1967), 326–48.

83. Lemay, R., "Dans l'Espagne du XIIe siecle: Les traductions de l'Arabe au Latin", *Annales* **18** (1963), 639–65.
84. Lemay, R., "Gerard of Cremona", DSB XV, 173–92.
85. Levey, M., "Ibn Ezra", DSB IV, 502–3.
86. Levey, M., "Savasorda", DSB I, 22–3.
87. Levi-Provençal, E., *Histoire de L'Espagne Musulmane*, 3 volumes, Éditions G.-P. Maisonneuve et Cie, Paris, 1950, 1950, 1967.
88. Lewis, A. R., *Naval Power and Trade in the Mediterranean : A.D.* 500–1100, Princeton University Press, 1951.
89. Lindberg, D. C., ed., *Science in the Middle Ages*, University of Chicago Press, 1978.
90. von Lippmann, E. O., *Entstehung und Ausbreitung der Alchemie*, G. Olms, Hildesheim, 1978. [Reprint of 1919 ed.]
91. McVaugh, M., "Arnald of Villanova", DSB I, 289–91.
92. McVaugh, M., "Constantine the African", DSB III, 393–5.
93. Makdisi, G., "Muslim Institutions of Learning in Eleventh Century Baghdad", *Bulletin of the School of Oriental and African Studies* **24** (1961), 1–56.
94. Metlitzki, D., *The Matter of Araby in Medieval England*, Yale University Press, New Haven, 1977.
95. Mez, A., *The Renaissance of Islam*, Tr., S. Khuda Bukhsh and D. S. Margoliouth, United Publishers, Beirut, 1973.
96. Mieli, A., *La Science Arabe*, E. J. Brill, Leiden, 1966.
97. Millás Vallicrosa, J. M., "El literalismo de los traductores de la corte de Alfonso el Sabio", *Al-Andalus* **1** (1933), 155–87.
98. Millás Vallicrosa, J. M., "Translations of Oriental Scientific Works", in Métraux, G. S. and Crouzet, F., eds, *The Evolution of Science*, New York, 1963, pp. 128–67.
99. Minio-Paluello, L., "Michael Scot", DSB IX, 361–5.
100. Minio-Paluello, L., "Plato of Tivoli", DSB XI, 31–3.
101. Nakosteen, M., *History of Islamic Origins of Western Education*, University of Colorado Press, Boulder, 1964.
102. Needham, J., *Science and Civilisation in China*, especially volumes 3–5, Cambridge University Press, 1959–1974.
103. Otte, J. K., "The Life and Writings of Alfredus Anglicus", *Viator* **3** (1972), 275–91.
104. Parry, V. J. and Yapp, M. E., eds, *War, Technology and Society in the Middle East*, Oxford University Press, London, 1975.
105. Pedersen, O. and Pihl, M., *Early Physics and Astronomy*, American Elsevier, N.Y., 1974.
106. Peters, F. E., *Allah's Commonwealth : A History of Islam in the Near East*—600–1100 *A.D*, Simon and Schuster, N.Y., 1970.
107. Peters, F. E., *Aristotle and the Arabs*, New York University Press, N.Y., 1968.
108. Pidal, G. M., "Cómo trabajaron las escuelas alfonsíes", *Neuva Revista de Filología Hispánica* **5** (1951) 363–80.
109. Pidal, G. M., "Los llamados numerales árabes en occidente", *Boletín de la Real Academia de Historia* **145** (1959), 179–208.
110. Poulle, E., "Henry Bate of Malines", DSB VI, 272–5.
111. Pring-Mill, R. D. F., "Ramon Lull", DSB VIII, 547–51.
112. Procter, E. S., "The Scientific Works of the Court of Alfonso X of Castille", *MLR* **40** (1945), 12–29.
113. Rashdall, H., *The Universities of Europe in the Middle Ages*, 3 volumes, New ed. of Powicke, F. M., Emden, A. B. Oxford University Press, 1936. [1969; Original ed., 1895]
114. Rivera, J. F., "Nuevos datos sobre los traductores Gundisalvo y Juan Hispano", *Al-Andalus* **31** (1966), 267–80.
115. Rose, V., "Ptolemaeus und die Schule von Toledo", *Hermes* **8** (1874), 327–49.

116. Rosenthal, F., tr., *Ibn Khaldun : The Muqaddimah*, 3 volumes, Routledge & Kegan Paul, London, 1958.
117. Runciman, S., *A History of the Crusades*, 3 volumes, Cambridge University Press, 1968, 1962, 1966. [First edition, 1951–52]
118. Ruska, J., *Arabische Alchemisten*, 2 parts, Martin Sändig GmbH, Wiesbaden, 1977. [Reprint of 1924 ed.]
119. Russell, J. C., "Hereford and Arabic Science in England about 1175–1200", *Isis* **18** (1932), 14–25.
120. Sa'di, L. M., "Reflection of Arabian Medicine at Salerno and Montpellier", *Annals of Medical History* (*n.s.*) **5** (1933), 215–25.
121. Sánchez-Albornoz, "Observaciones a unas páginas de Lemay sobre los traductores toledanos", *Cuadernos de Historia de Espana* **41/42** (1965), 313–24.
122. Sarton, G., *Appreciation of Ancient and Medieval Science During the Renaissance*, University of Pennsylvania Press, Philadelphia, 1955.
123. Sarton, G., *Introduction to the History of Science*, 3 volumes in 5, Carnegie Institution, Washington, D.C., 1927–1948.
124. Schipperges, H., *Arabische Medizin im lateinischen Mittelalter*, Springer-Verlag, Berlin, 1976.
125. Schipperges, H., "Assimilations-Zentren arabischer Wissenschaft im 12. Jahrhundert", *Centaurus* **4** (1956), 325–50.
126. Schipperges, H., *Die Assimilation der Arabischen Medizin durch das lateinische Mittelalter*, Franz Steiner Verlag, Wiesbaden, 1964.
127. Schipperges, H., "Die Schulen von Chartres unter dem Einfluss des Arabismus", *Sudhoffs Arch.* **40** (1956), 193–210.
128. Sezgin, F., *Geschichte des arabischen Schrifttums*, Volumes 3, 4, 5, 6, E. J. Brill, Leiden, 1970, 1971, 1974, 1978.
129. Silverstein, T., "Daniel of Morley, English Cosmologist and Student of Arabic Science", *Medieval Studies* **10** (1948), 179–96.
130. Singer, C., *et al.*, eds, *A History of Technology*, Voluumes 2 and 3, Oxford University Press, 1957.
131. Singer, C., *A Short History of Scientific Ideas to* 1900, Oxford University Press, 1959. [1966]
132. Smail, R. C., *The Crusaders*, Thames and Hudson, London, 1973.
133. Smalley, B., *Historians in the Middle Ages*, Thames and Hudson, London, 1974.
134. Smith, N., *A History of Dams*, Peter Davies, London, 1971.
135. Spuler, B., *The Muslim World : Part I — The Age of the Caliphs*, tr., F. R. C. Bagley, E. J. Brill, Leiden, 1960.
136. Stahl, W. H., *Roman Science*, University of Wisconsin Press, Madison, 1962. [Greenwood Reprint, 1978.]
137. Steele, R., "Practical Chemistry in the Twelfth Century", *Isis* **12** (1929), 10–46.
138. Steele, R., "Roger Bacon and the State of Science in the Thirteenth Century", pp. 121–50 in C. Singer, ed., *Studies in the History and Method of Science*, Volume 2, Oxford University Press, 1921. [Arno Press Reprint, 2 volumes in one, 1975.]
139. Steinschneider, M., *Die arabischen Übersetzungen aus dem Griechischen*, Akademische Druck-U. Verlagsanstalt, Graz, 1960.
140. Steinschneider, M., *Die europäischen Übersetzungen aus dem Arabischen bis Mitte des 17. Jahrhunderts*, Akademische Druck-U. Verlagsanstalt, Graz, 1956.
141. Steindschneider, M., *Die hebräischen Übersetzungen des Mittelalters und die Juden als Dolmetscher*, Berlin, 1893.
142. Talbot, C. H., "Stephen of Antioch", DSB XIII, 38–9.
143. Taylor, H. O., *The Mediaeval Mind*, 2 volumes, Harvard University Press, Cambridge, 1925. [Reprint of 4th edition.]
144. Thomas, P. E., "Alfonso el Sabio", DSB I, 122.

145. Thompson, J. W., "The Introduction of Arabic Science into Lorraine in the Tenth Century", *Isis* **12** (1929), 184–93.
146. Thompson, J. W., *The Literacy of the Laity in the Middle Ages*, University of California Press, Berkeley, 1939. [Johnson Reprint, 1966.]
147. Thorndike, L., "John of Seville", *Speculum* **34** (1959), 20–38.
148. Tibbetts, G. R., *Arab Navigation in the Indian Ocean Before the Coming of the Portuguese*, Royal Asiatic Society of Great Britain and Ireland, London, 1971.
149. Totah, K. A., *The Contribution of the Arabs to Education*, Columbia University, New York, 1926. [AMS Press reprint, 1972.]
150. Ullman, M., *Die Medizin im Islam*, E. J. Brill, Leiden, 1970.
151. Ullman, M., *Die Natur- und Geheim- Wissenschaften im Islam*, E. J. Brill, Leiden, 1972.
152. Usher, A. P., *A History of Mechanical Inventions*, McGraw-Hill, N.Y., 1929.
153. Vernet, J., "Ibn Tibbon, Jacob ben Machir", DSB XIII, 400–1.
154. Vernet, J., "Les traductions scientifiques dans l'Espagne du Xe siècle", *Cahiers de Tunisie* **18** (1970), 47–59.
155. Vernet, J., "Moses ibn Tibbon", DSB XIII, 401–2.
156. Vernet, J., "Una bibliografía de la historia de las ciencias matemáticas y astronómicas entre los arábes", *Al-Andalus* **21** (1956), 431–440; **23** (1958), 215–36, 456–95.
157. Walzer, R., "Arabic Transmission of Greek Thought to Medieval Europe", *Bulletin of the John Rylands Library* **29** (1945/46), 160–83.
158. Walzer, R., "Arabische Übersetzungen aus dem Griechischen", *Antike u. Orient im Mittelalter* (1962), 179–95.
159. Weiss, R., "England and the Decree of the Council of Vienne on the Teaching of Greek, Arabic, Hebrew, and Syriac", *Bibliothèque d'Humanisme et Renaissance* **14** (1952), 1–9.
160. Welborn, M. C., "Lotharingia as a Center of Arabic and Scientific Influence in the Eleventh Century", *Isis* **16** (1931), 188–99.
161. White, L., Jr, *Medieval Technology and Social Change*, Oxford University Press, 1962.
162. Wiedemann, E., *Aufsätze zur arabischen Wissenschaftsgeschichte*, 2 volumes, G. Olms, Hildesheim, 1970.
163. Wright, J. K., *The Geographical Lore of the Time of the Crusades*, Dover, N.Y., 1965. [1925]

Essay 2

1. Bond, J. D., "The Development of Trigonometric Methods Down to the Close of the XVth Century", *Isis* **4** (1921/22), 295–323.
2. Boyer, C. B., *A History of Mathematics*, J. Wiley & Sons, N.Y., 1968.
3. von Braunmühl, A., "Beitrag zur Geschichte der prosthapäretischen Methode in der Trigonometrie", *Bibliotheca mathematica* (1896), 105–8.
4. von Braunmühl, A., "Beiträgezur Geschichte der Trigonometrie", *Abh. der Kaiser. L.-P. Deutschen Akad. d. Naturforscher* (*Halle*) **LXXI** (1897), no. 1.
5. von Braunmühl, A., *Vorlesungen über Geschichte der Trigonometrie*, Leipzig, 1900.
6. Brockelmann, C., *Geschichte der arabischen Litteratur*, 2 volumes, and 3 supplements, E. J. Brill, Leiden, 1943–1949. Especially vol. II, 536.
7. Bruins, E. M. and Rutten, M., *Textes mathématiques de Suse*, Paris, 1961.
8. Cantor, M., *Vorlesungen über Geschichte der Mathematik*, Bd II, Johnson Reprint, N.Y., 1965. [Teubner, 1900.]
9. Carra de Vaux, B., "L'Almageste d'Abu'l-Wéfa Albuzdjani", *J. Asiatique*, (*ser.* 8) **10** (1892), 408–71.

10. Caussin de Perceval, "Le livre de grande table Hakémite...", *Notices et Extraits des Manuscrits de la Bibliothèque Nationale* **7** (1804), 16–240.
11. Delambre, M., *Histoire de L'Astronomie du Moyen Âge*, Johnson Reprint, N.Y., 1965. [Paris, 1819]
12. Efendi, S. Z., "Notation algébrique chez les orientaux", *J. Asiatique* (*ser.* 9) **11** (1898), 35–52.
13. Flügel, G., ed., *Hajjī Khalīfa-Kashf al-Ẓunūn*, vol. II. (1837), Johnson Reprint, N.Y., 1964.
14. Goetsch, H., "Die Algebra der Babylonier", *Arch. Hist. Exact Sci.* **5** (1968), 79–153.
15. Hairetdinova, N. G., "Trigonometry in the works of al-Fārābi and ibn Sīnā", (Russian), *Voprosy Estest. i Tekh.* **28** (1969), 29–31.
16. Heath, T. L., ed., *The Works of Archimedes with the Method of Archimedes*, Cambridge University Press, 1912.
17. Juschkewitsch, A. P., *Geschichte der Mathematik im Mittelalter*, Pfalz-Verlag, Basel, 1964. (= Youschkevitch) [see 4, 101].
18. Kennedy, E. S., "The History of Trigonometry: An Overview", 31*st Yearbook*, National Council of Teachers of Mathematics, pp. 333–50, Washington, D.C., 1969.
19. King, D. A., "A Double-Argument Table for the Lunar Equation Attributed to Ibn Yūnus", *Centaurus* **18** (1974), 129–46.
20. King, D. A., "Islamic Mathematics and Astronomy", *J. History of Astronomy* **9** (1978), 212–19.
21. King, D. A., "The Astronomical Works of Ibn Yūnus", *Ph.D. Dissertation*, Yale, 1972.
22. King, D. A., "Ibn Yūnus", DSB XIV, 574–80.
23. King, D. A., "Ibn Yūnus' Very Useful Tables for Reckoning Time by the Sun", *Archive for the History of the Exact Sciences* **10** (1973), 342–94.
24. Kline, M., *Mathematical Thought from Ancient to Modern Times*, Oxford University Press, N.Y., 1972.
25. Nasr, S. H., *Islamic Science: An Illustrated Study*, World of Islam Festival Publishing Co., Ltd., London, 1976.
26. Neugebauer, O., *Mathematische Keilschrift-Texte*, 3 volumes, Springer-Verlag, N.Y. 1973. [J. Springer, 1935.]
27. Neugebauer, O., *The Exact Sciences in Antiquity*, 2nd edition, Dover, N.Y., 1969. [Brown, 1957.]
28. Neugebauer, O. and Sachs, A., *Mathematical Cuneiform Texts*, American Oriental Society, New Haven, 1945.
29. Schoy, C., "Beiträge zur arabischen Trigonometrie", *Isis* **5** (1923), 364–99.
30. Schoy, C., "Das 20 Kapitel der grossen Hâkemitischen Tafeln...", *Annalen der Hydrographie und maritimen Meterologie* **48** (1920), 97–111.
31. Schoy, C., "Die Bestimmung der geographischen Breite eines Ortes durch Beobachtung...", *Annalen der Hydrographie und maritimen Meterologie* **50** (1922), 3–20.
32. [= 1, 128].
33. Smith, D. E., *A Source Book in Mathematics*, volume I, Dover, N.Y., 1959.
34. [= 1, 139].
35. Suter, H., *Die Mathematiker und Astronomen der Araber und Ihre Werke*, Johnson Reprint, 1972. [Teubner, 1900]
36. Tekli, S., "Ḥabash al-Ḥāsib", DSB V, 612–20.
37. Thureau-Dangin, F., *Textes mathématiques babyloniens*, Leiden, 1938.
38. Youschkevitch, A. P., "Abū'l-Wafā'", DSB I, 39–43. (= Juschkewitsch)

Essay 3

1. Bose, D. M., Sen, S. N., and Subbarayappa, B. V., *A Concise History of Science in India*, Indian National Academy of Science, New Delhi, 1971.

2. Boyer, C. B., "Clairaut le Cadet and a Theorem of Thābit ibn Qurra", *Isis* **55** (1964), 68–70.
3. Brockelmann, C., *Geschichte der arabischen Literatur, Supplement I*, E. J. Brill, Leiden, 1937.
4. Bruins, E. M., "On Interpretation in the History of Mathematics", *Janus* **66** (1979), 83–129.
5. Bruins, E. M., "On Plimpton 322: Pythagorean numbers in Babylonian Mathematics", *Kon. Neder. Akad. Wetensch. Proc.* **52** (1949), 629–32.
6. Bruins, E. M., "Pythagorean triads in Babylonian Mathematics", *Mathematical Gazette* **41** (1957), 25–8.
7. Bruins, E. M., "Quelques textes mathématiques de la mission de Suse", *Proc. K. Neder. Akad. van Wetensch.* **53** (1950), 1025–33.
8. Bruins, E. M., "Reciprocals and Pythagorean Triads", *Physis* **9** (1967), 373–92.
9. Bruins, E. M. and Rutten, M., *Textes mathématiques de Suse, Memories de la Mission Archeologique en Iran*, Paris, 1961.
10. Buck, R. C., "Sherlock Holmes in Babylon", *American Math. Monthly* **87** (1980), 335–145.
11. Bulmer-Thomas, I., "Pappus", DSB X, 293–304.
12. Al-Daffaʾ, Ali A., *Al-Mūjaz fī al-turāth al-'ilmī al-Arabī al-Islāmī*, John Wiley & Sons, N.Y., 1979.
13. Al-Daffaʾ, Ali A., *Nawābigh 'ulamā' al-'Arab wa al-Muslimīn fī al-riyādīyāt*, John Wiley & Sons, N.Y., 1978.
14. Datta, B. and Singh, A. N., *History of Hindu Mathematics*, 2 volumes, London and Bombay, 1962.
15. Datta, B., *The Science of the Śulba*, University of Calcutta, 1932.
16. Dodge, B., *The Fihrist of al-Nadīm*, 2 volumes, Columbia University Press, N.Y., 1970.
17. Gandz, S., "The Origin and Development of Quadratic Equations in Babylonian, Greek and Early Arab Algebra", *Osiris* **3** (1937), 405–543.
18. Gillings, R. J., *Mathematics in the Time of the Pharaohs*, MIT Press, Cambridge, 1972.
19. Gillings, R. J., "The Mathematics of Ancient Egypt", DSB XV, 681–705.
20. Goetsch, H., "Die Algebra der Babylonier", *Arch. Hist. Exact Sciences* **5** (1968), 79–153.
21. Heath, T. L., *A History of Greek Mathematics I*, The Clarendon Press, Oxford, 1921.
22. Heath, T. L., *The Thirteen Books of Euclid's Elements*, Volumes 1–3, Dover, N.Y., 1956. [Reprint of Cambridge University Press (1908) 1925.]
23. Huber, P., "Bemerkungen über mathematische Keilschrifttexte", *Enseignement mathematique* (*ser.* 2) **3** (1957), 19–27.
24. Huber, P., "Zu einem mathematischen Keilschrifttext (VAT 8512)", *Isis* **46** (1955), 104–6.
25. Hultsch, F., *Pappi Alexandrini—Collections*, volume I, Verlag Adolf M. Hakkert, Amsterdam, 1965. [Reprint of Berlin, 1875].
26. Kaye, G. R., "Indian Mathematics", *Isis* **2** (1914), 326–56.
27. Mikami, Y., *The Development of Mathematics in China and Japan*, 2nd ed., Chelsea Publishing Co., N.Y., 1974. [Reprint of 1913 Leipzig edition.]
28. [= 1, 102, vol. 3].
29. [= 2, 26].
30. [= 2, 27].
31. [= 2, 28].
32. Parker, R. A., *Demotic Mathematical Papyri, Brown Egyptological Studies, VII*, Providence, 1972.
33. Parker, R. A., [P. Dem. Heidelberg 663], *J. Egyptian Archaeology* **61** (1975), 189–96.
34. Plato, *Meno*. tr. and edited by W. R. M. Lamb, Loeb Classical Library, Harvard University Press, Cambridge, 1924 (1967).

35. Price, D. J. de S., "The Babylonian 'Pythagorean triangle' tablet", *Centaurus* **10** (1964), 219–31.
36. Rosenfeld, B. A. and Grigorian, A. T., "Thābit ibn Qurra", DSB XIII, 288–95.
37. [= 1, 123, vol. I].
38. Sayili, A., "Sābit ibn Kurra'nin Pitagor Teoremini Tamimi", *Belleten* **22** (1958), 527–49.
39. Sayili, A., "Thābit ibn Qurra's Generalization of the Pythagorean Theorem", *Isis* **51** (1960), 35–7.
40. Scriba, C. J., "John Wallis' 'Treatise of Angular Sections' and Thābit ibn Qurra's Generalization of the Pythagorean Theorem", *Isis* **57** (1966), 55–6.
41. [= 1, 128, vol. V].
42. [= 1, 139].
43. [= 1, 140].
44. Steinschneider, M., "Thabit (Thebit) ben Korra: Bibliographische Notiz", *Zeitschrift f. Math. u. Physik* **18** (1873), 331–8.
45. [= 2, 35].
46. [= 2, 37].
47. Unguru, S., "History of Ancient Mathematics: Some Reflections on the State of the Art", *Isis* **70** (1979), 555–65.
48. van der Waerden, B. L., "Mathematics and Astronomy in Mesopotamia", DSB XV, 667–80.
49. van der Waerden, B. L., *Science Awakening*, 2 volumes, P. Noordhoff Ltd, Groningen 1961 (vol. 1); Noordhoff International, Leyden, 1974 (vol. 2).

Essay 4

1. Ahadova, M. A., "Application of motion in geometry in the works of mathematicians of the Middle East", (Russian), *Taškent Gos. Univ.-Buharsk. Ped. Inst. Nauck, Trudy Vyp.* 3 (**17**) 1967, 76–81. (*)
2. Akhmedov, A., "The treatise of al-Samarqandī—The Fundamental Propositions", in *History of the Exact Sciences in the Near and Middle East in the Middle Ages*, Tashkent, 1972. (*)
3. Amir-Moéz, A. R. "Discussion of Difficulties in Euclid", *Scripta Math.* **24** (1959) 275–303. (*)
4. Anawati, G. C., "Ibn Sīnā", DSB XV, 494–8.
5. Barnes, J., ed., *Aristotle's Posterior Analytics*, Clarendon Press, Oxford, 1973.
6. Besthorn, R. C. and Heiberg, J. L., eds., *Euclidiis Elementa ex interpretatione al-Hadschdschadschii cum commentariis al-Narizii*, Copenhagen, 1893–1932. [See 5, 11].
7. Bonola, R., *Non-Euclidean Geometry*, Dover, New York, 1955.
8. Bulmer-Thomas, I., "Euclid", DSB IV, 414–37.
9. Busard, H. L. L., "Clavius", DSB III, 311–12.
10. [= 2, 8, vols I–IV].
11. Carruccio, E., "Cataldi", DSB III, 125–9.
12. Clagett, M., "The Medieval Latin Translations from the Arabic of the *Elements* of Euclid, with Special Emphasis on the Versions of Adelard of Bath", *Isis* **44** (1953), 16–42.
13. [= 3, 13].
14. Al-Dahir, M. W., "Concerning the parallel postulate", (Arabic), *Bulletin College of Arts and Sciences, Baghdad* **3** (1958), 60–5. (4 plates).
15. Dicks, D. R., "Geminus", DSB V, 344–7.
16. Dilgan, H., "Démonstration du Ve postulat d'Euclide par Schams-ed-Din Samarkand. Traduction de l'ouvrage Aschkâl-üt-teessis de Samarkandi", *Rev. Hist. Sci. Appl.* **13** (1960), 191–6.

17. Dilgan, H., "Qādī Zādā al-Rūmī", DSB XI, 227–9.
18. Dilgan, H., "Al-Samarqandī", DSB XII, 91.
19. Dold-Samplonius, Y. "Al-Sijzī", DSB XII, 431–2.
20. Frankland, W. B., *Theories of Parallelism*, Books for Libraries Press, Freeport, 1972 (1910).
21. Gans, D., *Non-Euclidean Geometry*, Academic Press, New York, 1973.
22. Al-Gasri, Mohammed Fayiz, *Mazhr al-Thagafat ul-Islamiah wa atharuhal fī al-Hazarah*, Muṭba ͑at Zaid ibn Thābit, Damascus, 1979.
23. [= 1, 42].
24. Gray, J., "Non-Euclidean Geometry—A Re-interpretation", *Historia Mathematica* **6** (1979), 236–58.
25. Hājjī Khalifah, *Kesf el-Zunun*, Mothana's Library, Beirut, 1941.
26. Hatipov, A. E.-A., "The first book of Omar Khayyām's treatise on geometry", (Russian), *Trudy Samarkand Gos. Univ.* (*N.S.*) **107** (1960), 9–16. See also: *Trudy Uzbek. Gos. Univ.* (*N.S.*) **78** (1958), 131–6. (*)
27. [= 3, 21, vols I–II].
28. Heath, T. L., *A Manual of Greek Mathematics*, Oxford University Press, London, 1931.
29. [= 3, 22].
30. Jaouiche, K., "Le Traité de Naṣīr al-Dīn al-Ṭūsī sur la theorie des parallèles", *Proceedings of the First International Symposium for the History of Arabic Science, Aleppo* (*April* 5–12, 1976), 1978. volume II, 62–6.
31. Jolivet, J. and Rashed, R., "Al-Kindī", DSB XV, 261–7.
32. Kagan, V. F., *The Foundations of Geometry*, (Russian), Moscow, Leningrad, 1949. (*).
33. Kahalah, ͑Amar Riza, *Al-Alūme al-Bahtah fī al-͑Asoor al-Islamiyah*, Muṭba ͑at al-Turgi, Damascus, 1972.
34. Kennedy, E. S., "The Exact Sciences in Iran under the Saljuqs and Monguls", in *The Cambridge History of Islam*, J. A. Boyle, ed., Volume 5, 659–79. See especially 664.
35. [= 1, 84].
36. Mamedbeili, G. D., "Muhammed Nasirèddin Tusi on the Theory of Parallel Lines and the Theory of Ratios. (Azerbaijani)", *Izdat. Akad. Nauk Azerbaĭdzansk, SSR, Baku*, 1959 (*).
37. Mamedov, K. M., "On the Commentaries of Nasireddin Tusi in Books I–III of 'Tahir Èglidis' in the *Elements* of Euclid", (Azerbaijani. Russian Summary), *Akad. Nauk Azerbaĭdžan, SSR Trudy Inst. Mat. Meh. ser* 2, **10** (1963), 147–58.
38. Martin, G. E., *The Foundations of Geometry and the Non-Euclidean Plane*, Intext Educational Publishers, New York, 1975.
39. Mazhar, Jalal, *Athar al-͑Arab fī al-Hazarah al-Europiyah*, Dar al-Rayeed, Beirut, 1967.
40. Meschkowski, H., *Noneuclidean Geometry*, Academic Press, New York, 1964.
41. Morrow, G. R., "Proclus", DSB XI, 160–2.
42. Morrow, G. R., ed., *Proclus. Commentary on the First Book of Euclid's Elements*, Princeton University Press, 1970 (*).
43. Al-Mūlla, Ahmed A., *Athar al-͑Alama al-Muslimeen fī al-Hazarah al-Europiyah*, Dar al-Fikr, Cairo, 1978.
44. Murdoch, J., "Euclid: Transmission of the *Elements*", DSB IV, 437–59.
45. Nasr, S. H., *Science and Civilization in Islam*, Havard University Press, Cambridge, 1968.
46. Nasr, S. H., "Al-Ṭūsī, Naṣir al-Dīn", DSB XIII, 508–14.
47. Owen, G. E. L., "Aristotle", DSB I, 250–8.
48. Page, T. E. *et al.*, eds., *Corpus Aristotelicum* (Loeb Classical Library), Harvard University Press, Cambridge, 1960.
49. Peters, F. E., *Aristotle and the Arabs*, University of London Press, London, 1968.

50. Petrosian, G. B. and Rosenfeld, B. A., "Aganis' Proof of Euclid's Fifth Postulate", (Russian, Armenian summary), *Izad. Akad, Nauk Armjan. SSR Ser. Fiz-Mat. Nauk* **13** (1960), 153–64 (*).
51. Rosenfeld, B. A., "Introduction, Text and Notes on Ibn al-Haytham and Levi ben Gerson, and the Parallel Postulate", (Russian) *Istoriko-mat. issled.* **11** (1958) 743–82. (*).
52. Rosenfeld, B. A., "New research on the prehistory of non-Euclidean Geometry", in Kagan, V. F. (ed.) *The Foundations of Geometry.* (*)
53. Rosenfeld, B. A. "(Thābit's) Book on the fact that two lines, drawn at angles less than two right angles (to a given line) intersects", (Russian), *Istoriko-mat. issled,* **15** (1963), 363–80. (*)
54. Rosenfeld, B. A. "The Proofs of Euclid's Fifth Postulate by the Medieval Mathematicians Ibn al-Haytham and Levi ben Gerson", (Russian) *Istoriko-mat. issled* **11** (1958), 733–82. (*)
55. Rosenfeld B. A. "The theory of parallel lines in the Medieval East", in *Actes du Onzième Congrès International d'Histoire des Sciences* (1965, *Cracow*). *Sect. III : Histoire des Sciences Exactes,* Wroclaw, 1968, 175–178. (*)
56. Rosenfeld, B. A., Kubesov, A. K., and Sabirov, G. S. "Who was the Author of the Rome Edition Recension of Euclid by Naṣir al-Din al-Ṭūsī?", (Russian), *Voprosy istorii estestvoznaniya i tekhniki* **20** (1966), 51–3. (*)
57. [= 3, 36].
58. Rosenfeld, B. A. and Hairetdinova, N. G., "Husam ad-Dīn as-Salar, Premises for the proof of the parallel postulate that was introduced by Euclid at the beginning of his first book", (Russian), *Istoriko-mat. issled.* **19** (1974), 285–93. (*)
59. Rosenfeld, B. A. and Youschkevitch, A. P., "Demonstrations of the Fifth Postulate of Euclid by Thābit ibn Qurra and Shams al-Dīn al-Samarqandī", (Russian) *Istorikomat. issled.* **14** (1961), 587–92.
60. Rosenfeld, B. A. and Youschkevitch, A. P., "Die Mathematiker der Lander des Ostens im Mittlelalter", in *Sowjetische Beitrage zur Geschichte der Naturwissenschaft,* Berlin, 1960. (*)
61. Rosenfeld, B. A., and Youschkevitch, A. P., "Al-Khayyāmī", DSB VII, 323–34.
62. Rosenfeld, B. A. and Youschkevitch, A. P. "Notes on the Treatise of Naṣir al-Dīn al-Ṭūsī on Parallel Lines", (Russian), *Istoriko-mat. issled.* **13** (1960), 525–32.
63. Rosenfeld, B. A. and Youschkevitch, A. P. "The Mathematical Treatises of Omar Khayyam", (Russian), *Istoriko-mat. issled* **6** (1953), 11–172. See especially 67–84.
64. Rosenfeld, B. A. and Youschkevitch, A. P. "The Treatise of Naṣir al-Dīn al-Ṭūsī on the Theory of Parallels", (Russian), *Istor.-Mat. issled.* **13** (1960), 475–82.
65. Ross, W. D., ed., *The Works of Aristotle,* Oxford University Press, 1908.
66. Sabra, A. I. "Burhān Naṣīr al-Dīn al-Ṭūsī ʿalā muṣādarat Uqlīdis al-Khāmisa", *Bulletin Fac. Arts Alexandria Univ.* **13** (1959), 133–70. (*)
67. Sabra, A. I., "Ibn al-Haytham", DSB VI, 189–210.
68. Sabra, A. I., "Al-Jawharī", DSB VII, 79–80.
69. Sabra, A. I. "Al-Nayrīzī," DSB XI, 5–7.
70. Sabra, A. I., ed., *Omar Khayyam, Risalah fi sharh ma ashkala min musadarat kitab Uqlidis,* Al-Maaret Est., Alexandria, 1961. (*)
71. Sabra, A. I., "Simplicius's Proof of Euclid's Parallels Postulate", *Journal of the Warburg and Courtland Institutes* **32** (1969), 1–24. (*)
72. Sabra, A. I. "Thābit ibn Qurra on Euclid's Parallels Postulate", *Journal of the Warburg and Courtland Institutes* **31** (1968), 12–32. (*)
73. Samso, J., "Levi ben Gerson", DSB VII, 279–82.
74. [= 1, 123, vol. II, pt. II].
75. Scriba, C. J., "Lambert", DSB VII, 595–600.
76. Struik, D. J., "Saccheri", DSB XII, 55–7.
77. Scriba, C. J., "Wallis", DSB XIV, 146–55.

78. Sezgin, F., *Geschichte des arabischen Schrifttums, Band V (Mathematik)*, E. J. Brill, Leiden, 1974.
79. Sezgin, F., *Geschichte des arabischen Schrifttums, Band VI (Astronomie)*, E. J. Brill, Leiden, 1978.
80. Sharif, M. M., ed., *A History of Muslim Philosophy*, 2 Volumes, Otto Harrassowitz, Wiesbaden, 1963.
81. Siddiqi, B. H., "Naṣīr al-Dīn Ṭūsī", in *A History of Muslim Philosophy*, **1**, 564–80.
82. Ibn Sīnā (Avicenne), *Le Livre de Science* (Dānīsh-Nāma), 2 volumes, Especially volume 2: *Physique, Mathematiques*, Translation by M. Achena and H. Massé, Societe d'Edition "Les Belles Lettres", Paris, 1958.
83. Smith, D. E., "Euclid, Omar Khayyam and Saccheri", *Scripta Mathematica* **3** (1935), 5–10. (*)
84. Sommerville, D. M. Y., *Bibliography of Non-Euclidean* Geometry, Chelsea, New York, 1970. (1911)
85. [= 2, 35].
86. Szabó, Á., *The Beginnings of Greek Mathematics*, D. Reidel, Dordrecht, 1978.
87. Al-Ṭayar, Hashim and Sa ᶜid, Yahya, *Mujaz Tarikh al-Riyadiyāt*, Dar al-Kutub, Mosul, 1977.
88. Tekeli, S., "Al-Maghribī", DSB IX, 555–7.
89. Thomas, I., *Greek Mathematics*, 2 volumes, Harvard University Press, Cambridge, 1941. Especially vol. 2.
90. Toogan, Qadri Hafiz, *Turath al-ᶜArab al-Alomi fī al-Riyadiyāt wa falak*, Dar-ul-Qalm, Cairo, 1963.
91. Torretti, R., *Philosophy of Geometry from Riemann to Poincare*, D. Reidel, Dordrecht, 1978.
92. Toth, I., "Das Parallelproblem im Corpus Aristotelicum", *Arch. Hist. Exact Sciences* **3** (1966/67), 249–422.
93. Al-Ṭūsī, Naṣīr al-Dīn, "al-Risāla ʾl-shāfiya ᶜan al-shakk fi al-khutut al-mutawāziya", in *Majmūᶜ Rasāʾil al-Ṭūsī*, II (17–26), Hyderabad, 1359 AH.
94. Al-Ṭūsī, Naṣīr al-Dīn "al-Risāla", (Russian tr. by Rosenfeld, B. A. and Youschkevitch, A. P.), *Istoriko-mat. issled.* **13** (1960), 475–532.
95. Al-Ṭūsī, Naṣīr al-Dīn "The Treatise which settles doubts on the matter of parallel lines", (Russian tr. Rosenfeld, Youschkevitch), *Istoriko-mat. issled.* **13** (1960), 483–524.
96. Verbeke, G. "Simplicius", DSB XII, 440–3.
97. Wallis, J. *Opera Mathematica II*, 665–678, "De postulato quinto; et definitione quinta. Lib. 6 Euclidis; disceptatio geometrica" (Oxford, 1693) Reprint, G. Olms, Hildesheim, 1972.
98. Warmington, E. H., "Posidonius", DSB XI, 103–6.
99. Wiedemann, E., "Beiträge zur Geschichte der Naturwissenschaften 78—Naṣīr al-Dīn al-Ṭūsī" in *Aufsätze zur Arabischen Wissenschaftsgeschichte*, II, 701–28, Georg Olms Verlag, Hildesheim, 1970.
100. Wolfe, H. E., *Non-Euclidean Geometry*, Holt-Rinehart and Winston, New York, 1945.
101. Youschkevitch, A. P., *Geschichte der Mathematik im Mittelalter*, Pfalz-Verlag, Basel, 1964, pp. 277–88. See also: *Les mathématiques Arabes*, a French tr. by M. Cazenave and K. Jaoviche of a part of the above work. VRIN, Paris, 1976, pp. 112–23.
102. Dehn, M., "Die Legendre'sche Satze über die Winkelsumme in Dreiecke", *Math. Annalen* **53** (1900), 405–39.

According to the IHAS Newsletter (June, 1979), the following publications are "forthcoming" from the presses of the Institute for the History of Arabic Science:

103. Rosenfeld, B. A. and Youschkevitch, A. P., *The Theory of Parallels in the Arabic Literature of the 9th to the 14th Centuries.* (*)

104. Jaouiche, K., *The Book of Arabic Parallel Lines* (Arabic). This author also expects to offer a French translation. (*)

Essay 5

1. Afnan, S. M., *Avicenna : His Life and Works*, George Allen and Unwin, London, 1958.
2. Afnan, S. M., "The Commentary of Avicenna on Aristotle's Poetics", *J. Royal Asiatic Society* (1947), 188–90.
3. Ahadova, M. A., *Arifmeticheskaya chast' "Kniga Znaniya" Ibn Sino*, Uch. Zap. Buharsk. Gos. Vyp 12, Buharsk, 1964. (*)
4. Ahadova, M. A., "The part of Ibn Sina's "Book of Knowledge" devoted to geometry", (Russian), *Buharsk. Gos. Ped. Inst. Učen. Zap. Ser. Fiz.-Mat. Nauk Vyp.* **1** (13) (1964), 143–205. (*)
5. Ali, J., tr., *Al-Bīrūnī's The Determination of the Coordinates of Cities*, American University Press, Beirut, 1967.
6. Alverny, M. T. d.', "Notes sur les traductions médiévales d'Avicenna," *Arch. hist. doctr. litt.* **27** (1952), 337–58.
7. Anawati, G. C., "Bibliographie de la philosophie médiévale en terre d'Islam pour les années 1959–1969", *Bulletin de philosophie médiévale* **10–12** (1968/70), 343–9.
8. Anawati, G. C., "Essai de bibliographie avicennienne", *Revue Thomiste* **51** (1951), 407–40.
9. [= 4, 4].
10. Anbouba, A., "L'algèbre arabe aux IXe et X[e] siecles. Aperçu général", *J. History Arabic Sci.* **2** (1978), 66–100.
11. Besthorn, R. O. and Heiberg, J. L., eds, *Codex Leidensis 339 I, Euclids Elementa ex interpretatione Al-Hadschdshadschii cum commentariis Al-Narizii*, F. Hegel & Fil, Hauniae, 1893. Partis I, fasc. I; 1897 Partis I, fasc. II; 1932, Partis III, fasc. II (eds G. Junge, J. Raeder, and W. Thomson).
12. Boilot, D. J., "Al-Bīrūnī", *Encyclopaedia of Islam*, (*New Edition*), eds. Gibb, H. A. R. *et al.*, E. J. Brill, Leiden, 1960, Vol. I, pp. 1236–8.
13. Bosworth, C. E., *The Later Ghaznavids: Splendour and Decay*, The University Press, Edinburgh, 1977.
14. [= 2, 6, Band I, Suppl. I].
15. Brown, P., "A Medieval Analysis of Infinity", *J. History of Philosophy* **3** (1965), 242–3.
16. Browne, E. G., *A Literary History of Persia*, 4 volumes, Cambridge University Press, London, 1959–64.
17. [= 4, 8].
18. [= 2, 8, vol. I].
19. Carra de Vaux, B., *Avicenna*, Philo Press, Amsterdam, 1974. [Paris, 1900]
20. Casiri, M., *Bibliotheca Arabico-Hispana Escurialensis*, T.I. Biblio Verlag, Osnabrück, 1969. [Reprint of 1760 edition]
21. Chelkowski, P. J., ed., *The Scholar and the Saint: Studies in Commemoration of ... al-Bīrūnī*, New York University Press, 1975.
22. Chistyakov, V. D., *Rasskazi o matematikah* (*razdel Avitsenna*), Izd., 2-e. Minsk, 1966. (*)
23. Collison, R., *Encyclopaedias: Their History Throughout the Ages*, Hafner, N.Y., 1966.
24. [= 1, 20].
25. Crombie, A. C., "Avicenna's Influence on the Medieval Scientific Tradition", in Wickens, G. M., ed., *Avicenna: Scientist and Philosopher*, London, 1952, pp. 84–107.
26. [= 1, 21].
27. [= 1, 22].
28. Dickson, L. E., *History of the Theory of Numbers I*, Carnegie Institute, Washington, 1919. [Reprint: Chelsea, N.Y., 1971]

29. Dieterici, F., *Die Philosophie bei den Arabern im X Jahrhundert n. Chr.* III, E. S. Mittler und Sohn, Berlin, 1865 [Reprint, G. Olms, Hildesheim, 1969]
30. Drake, S., tr., *Galileo Galilei—Two Sciences*, University of Wisconsin Press, Madison, 1974.
31. Elkana, Y., ed., *The Interaction Between Science and Philosophy*, Humanities Press, Highlands, N.J., 1974.
32. Eneström, H., (title not available), *Bibliotheca Math.* **3** ser 7., 81. (*)
33. Farmer, H. G., "Arabian Music", vol. I, pp. 179–187 of *Grove's Dictionary of Music and Musicians*, ed., Blom, E., St. Martin's Press, New York, 1961 (1973).
34. Farmer, H. G., "The Music of Islam", pp. 421–77, 500–3 in Ancient and Oriental Music, ed., Wellesz, E., Oxford University Press, London, 1957 (1975).
35. Farmer, H. G., "Ibn Sīnā", vol. 4, pp. 437–8 of *Grove's Dictionary of Music and Musicians*, ed., Blom, E., St. Martin's Press, New York, 1961 (1973).
36. Frye, R. N., ed., *The Cambridge History of Iran*, Volume 4, Cambridge University Press, London, 1975.
37. [=1, 42].
38. Gohlman, W. E., *The Life of ibn Sīnā: A critical edition and annotated translation*, State University of New York Press, Albany, 1974.
39. Goichon, A.-M., "Ibn Sīnā", *Encyclopaedia of Islam* (*New Edition*), ed. Gibb, H. A. R. *et al.*, E. J. Brill, Leiden, 1960, Vol. III, pp. 941–7.
40. Goldstein, B. R., "A Treatise on Number Theory from a Tenth Century Arabic Source", *Centaurus* **10** (1964), 129–160.
41. [=2, 15].
42. [=1, 57].
43. [=4, 28].
44. [=3, 21, vol. I–II].
45. Heath, T. L., *Mathematics in Aristotle*, Clarendon Press, Oxford, 1949.
46. [=3, 22].
47. Heiberg, J. L., "Die arabische Tradition der Elemente Euklid's", *Hist.-lit. Abthlg. d. Zeitschrift für Mathematik und Physik* **29** (1884), 1–22.
48. Heiberg, J. L. and Menge, H., eds., *Euclidis Opera Omnia*, B. G. Teubner, Leipzig. Bd V. Elementorum qui feruntur libri XIV–XV et scholia in Elementa [graece] cum prolegomensis criticis et appendicibus edidit J. L. Heiberg, 1888.
49. Heinen, A. M., "'A Priori' Positions Determining al-Bīrūnī's Scientific Work", in *The Scholar and the Saint*, ed. Chelkowski, P. J., NYU Press, 1975, pp. 44–66.
50. [=1, 65].
51. [=3, 25; vols I–III].
52. Huxley, G. L., "Theon of Smyrna", DSB XIII, 325–6.
53. Iskandar, A. Z., "Ibn Sīnā: Medicine", DSB XV, 498–501.
54. [=2, 17].
55. Kahhorov, A., and Hodžiev, I., "Ibn Sīnā-mathematician", (Russian), *Izv. Akad. Nauk Tadžik. SSR Otdel. Fiz.-Mat. i Geolog.-Him. Nauk* **3** (65) (1977), 121–4.
56. Kapp, A. G., "Arabische Übersetzer und Kommentatoren Euklids", *Isis* **22** (1934), 150–72; **23** (1935), 55–9; **24** (1935), 37–79.
57. Kennedy, E. S., *A Commentary Upon Bīrūnī's Kitāb Taḥdīd al-Amākin*, American University of Beirut, 1973.
58. [=2, 18].
59. Ibn Khaldūn, *The Muqaddimah*, tr. by F. Rosenthal, 3 volumes, Routledge & Kegan Paul, London, 1958 [1967].
60. King, D. A., "On Medieval Islamic Multiplication Tables", *Historia Math.* **1** (1974), 317–23.
61. King, D. A., "Supplementary Notes on Medieval Islamic Multiplication Tables", *Historia Math.* **6** (1979), 405–17.
62. Klamroth, M., "Ueber den arabischen Euklid", *Zeitschrift der Deutschen Morgenlandischen Gesellschaft* **35** (1881), 270–326, 788.

63. *Kūshyār ibn Labbān, Kitab fī usūl hisāb al-hind* (*Principles of Hindu Reckoning*), tr. Levey, M. and Petruck, M., University of Wisconsin Press, Madison, 1965.
64. Kutsch, Wilhelm, *Ṯābit ibn Qurra's Arabische Übersetzung der* ᵓ*Αριθμητιχη Εισαγωγή des Nikomachos von Gerasa zum Ersten mal Herausgegeben*, Imprimerie Catholique, Beirut, 1958.
65. [= 1, 89].
66. Lloyd, A. C., "Iamblichus", DSB VII, 1.
67. Lokotsch, K., *Avicenna als Mathematiker, besonders die planimetrischen Bucher seiner Euklidübersetzung*, Inaugural-Dissertation, Rheinischen Friedrich-Wilhelms-Universität zu Bonn, Erfurt, 1912.
68. Lokotsch, K., Review of Codex Leidensis 399, 1: Part I, fasc 1–2; Part II, fasc. 1–2; Part III, fasc. 1, *Orientalistische Literaturzeitung* **15** (1912), Col. 19–23.
69. Mahdi, M., "Al-Fārābī", DSB IV, 523–6.
70. Marmura, M. E., "Avicenna", in Edwards, P., ed., *The Encyclopedia of Philosophy*, Macmillan Co., and The Free Press, N.Y., 1967, Vol. I., pp. 226–9.
71. Marmura, M. E., "Avicenna and the problem of the infinite number of souls", *Mediaeval Studies* **22** (1960), 232–9.
72. Matvievskaya, G. P., *Uchenie o chisle na Srednevekovom Vostoki*, Tashkent, 1967. (*)
73. Maziarz, E. A. and Greenwood, T., *Greek Mathematical Philosophy*, Frederick Ungar Publishing Co., New York, 1968.
74. Minio-Paluello, L., "Aristotle: Tradition and Influence", DSB I, 267–81.
75. Minio-Paluello, L., "Boethius", DSB II, 228–36.
76. Montferrier, A. S., ed., *Dictionnaire des sciences mathématiques*, Paris, 1835. (*).
77. Morewedge, P., tr. and commentary, *The Metaphysica of Avicenna* (*Ibn Sīnā*), Routledge and Kegan Paul, London, 1973.
78. [= 4, 44].
79. [= 1, 101].
80. Nallino, C. A., ed., *Al-Battānī sive Albatenii: Opus Astronomicum*, G. Olms, Hildesheim, 1977. [Reprint 1899–1907 edition].
81. Nasr, S. H., *An Introduction to Islamic Cosmological Doctrine*, Revised edition, Thames and Hudson, London, 1978.
82. [= 4, 45].
83. *Nicomachus of Gerasa, Introduction to Arithmetic*, tr., D'Ooge, M. L. MacMillan, New York, 1926.
84. Owen, G. E. L., "Aristotle", DSB I, 250–8.
85. Pines, S., "Philosophy, Mathematics and Concepts of Space in the Middle Ages", in *The Interaction Between Science and Philosophy*, Elkana, Y., ed., Humanities Press, Highland, N. J., 1974, pp. 75–90.
86. [= 1, 113].
87. Rosen, F., tr., *The Algebra of Mohammed Ben Musa*, John Murray, London, 1831. *See* also: S. Gandz, "The Algebra of Inheritance: a rehabilitation of Al-Khuwārizmī", *Osiris* **5** (1938), 319–91.
88. Ross, W. D., ed., *The Works of Aristotle, Volume I: Categories,..., De Sophisticis Elenchis*, Oxford University Press, London, 1928 [1971].
89. Ross, W. D., ed., *The Works of Aristotle, Volume VIII: Metaphysica*, Clarendon Press, Oxford, 1928. (Second Edition).
90. Rypka, J. (ed. Jahn, K.), *History of Iranian Literature*, D. Reidel, Dordrecht, 1968.
91. Said, H. M., ed., *Al-Bīrūnī Commemorative Volume*, Hamdard National Foundation, Karachi, 1979.
92. Saidan, A. S., tr., *The Arithmetic of Al-Uqlīdisī*, D. Reidel, Dordrecht, 1978.
93. Saidan, A. S., "The Trigonometry of Al-Bīrūnī", in *Al-Bīrūnī Commemorative Volume*, H. M. Said, ed., Hamdard National Foundation, Karachi, 1979, pp. 681–90.
94. [= 1, 123, vol. I].
95. [= 3, 39].
96. Sayili, A., "Was Ibn Sīnā an Iranian or a Turk", *Isis* **31** (1939), 8–24.

97. Sellheim, R., "Review of Osman Ergin: İbni Sina Bibliografiyasi", *Oriens* **11** (1958), 231–9.
99. [= 1, 128, vol. V–VI].
99. Sharipova, M., "Voprosi teorii chisel v 'Knige Istseleniia' Ibn Siny", *Uch. Zap. DGPI*, **71**, Vyp. 6, 3–30, Dushanbi, 1970.
100. Ibn Sīnā, *Le Livre de Science II* (tr. of *Dânesh-nâmè*), Achena, M. and Massé, H., trs., Société d'Éditions "Les Belles Lettres", Paris, 1958.
101. *Ibn Sīnā, Al-Shifā': al-Hisāb*, Preface by Madkour, I., edited by Lotfi, A. H., L'Organisation Egyptienne Generale du Livre, Cairo, 1975.
102. *Ibn Sīnā, Al-Shifā' : Usūl al-handasah*, Preface by Madkour, I; edited by Sabra, A. H. and Lotfi, A. H., L'Organisation Egyptienne Generale du Livre, Cairo, 1970.
103. Smith, D. E., *Rara Arithmetica*, 4th edition, Chelsea, N.Y., 1970. [Reprint with corrections and additions of 1908, 1910, 1939 edns].
104. [= 1, 140].
105. Steinschneider, M., "Euklid bei den Arabern", *Hist.-lit. Abthlg. d. Zeitschrift für Mathematik und Physik* **31** (1886), 81–110.
106. Steinschneider, M., *Al-Fārābī*, Philo Press, Amsterdam, 1966 [Reprint of St. Petersburg, 1869].
107. Stock, B., "Science, Technology, and Economic Progress in the Early Middle Ages", in *Science in the Middle Ages*, ed. D.C. Lindberg, University of Chicago Press, 1978, pp. 1–51.
108. Struik, D. J., ed., *A Source Book in Mathematics*, 1200–1800, Harvard University Press, Cambridge, 1969.
109. [= 2, 35].
110. Tannery, P., "Sur l'invention de la preuve par neuf", *Bull. des sciences mathématiques* **6** (1882), 142–4.
111. Tarán, L., "Nicomachus of Gerasa", DSB X, 112–4.
112. Theon of Smyrna, *Mathematics Useful for Understanding Plato*, English tr. by R. & D. Lawlor from the French–Greek edn. of J. Dupuis (1892, Paris), Wizards Bookshelf, San Diego, 1979.
113. Thomas, I., "A 12th Century Paradox of the Infinite", *J. of Symbolic Logic* **23** (1958), 133–4.
114. [= 4, 89].
115. Thornton, J. L. and Tully, R. I. J., *Scientific Books, Libraries, and Collections*, 3rd rev. edn., Library Association, London, 1971.
116. Ünver, A. S., "Avicenna's praise of Euclid", *J. History of Medicine* **2** (1947), 198–200.
117. Van den Bergh, S., tr., *Averroes' Tahafut al-Tahafut*, 2 volumes, Luzac and Co., London, 1969.
118. Vernet, J., "Al-Khuwārizmī, Abū 'Abd Allāh Muḥammad ibn Aḥmad ibn Aḥmad Ibn Yūsuf", DSB VII, 357–8.
119. Wallace, W. A., "Albertus Magnus", DSB I, 99–103.
120. Wiedemann, E., Aufsätze zur arabischen Wissenschaftsgeschichte, 2 volumes, G. Olms, Hildesheim, 1970.
121. Wiedmann, E., "Ueber ein von Ibn Sīnā (Avicenna) hergestelltes Beobachtungsinstrument", *Zeitschrift für Instrumentenkunde* **45** (1925), 269–75.
122. Woepcke, F., *Journal asiatique* **1** (1863), 502–4. (*)
123. Woepcke, F., "Recherches ... les travaux mathématiques des arabes. I: Traduction d'un Chapitre des Prolégomènes d'Ibn Khaldoûn relatif aux sciences mathématiques", *Atti dell' Accademia Pontificia dei Nuovi Lincei* [*Rome*] **10** (1856/57), 236–48.
124. Wolfson, H. A., *Crescas' Critique of Aristotle*, Havard University Press, Cambridge, 1957 (1929).
125. [See 4, 101].
126. Zabeeh, F., ed., *Avicenna's Treatise on Logic: Part One of Danesh-Name Alai*, Martinus Nijhoff, The Hague, 1971.

Essay 6

1. [= 2, 27].
2. [= 1, 102].
3. Libbrecht, U., *Chinese Mathematics in the Thirteenth Century: The Shu-shu chiu-chang of Ch'in Chiu-shao*, The MIT Press, Cambridge, 1973.
4. [= 3, 1].
5. [= 3, 49].
6. [= 3, 53].
7. Goldstine, H. H., *A History of Numerical Analysis from the 16th Through the 19th Century*, Springer-Verlag, New York, 1977.
8. Suter, H., "Abhandlung über die Ausmessung der Parabel von Ibrahim ben Sinan ben Thabit ben Kurra", *Vierteljahrsschrift der Naturforschenden Gesellschaft in Zurich* **63** (1918), 214–28.
9. Suter, H., "Die Abhandlung Qosṭa ben Lūqās und zwei andere anonyme über die Rechnung mit zwei Fehlern und mit der angenommen Zahl", *Bibliotheca mathematica*, (*Ser.* 3) **9** (1908), 111–22.
10. Suter, H., "Die Abhandlungen Thâbit b. Ḳurras und Abû Sahl al-Kûhîs über die Ausmessung der Paraboloide", *Sitzungsberichte der Physikalisch-medizinischen Societät in Erlangen* **48–49** (1918), 186–227.
11. Luckey, P., "Die Ausziehung der n-ten Wurzel und der binomische Lehrsatz in der islamischen Mathematik", *Mathematische Annalen* **120** (1948), 217–74.
12. Luckey, P., *Die Rechenkunst bei Ǧamšīd b. Masʿūd al-Kāšī mit Rückblichen auf die ältere Geschichte des Rechnens*, Deutsche Morgenlandische Gesellschaft, Kommissions-verlag Franz Steiner GmbH, Wiesbaden, 1951.
13. [= 5, 57].
14. Kennedy, E. S., "A Fifteenth Century Lunar Eclipse Computer", *Scripta Mathematica* **17** (1951), 91–7.
15. Kennedy, E. S., "A Medieval Interpolation Scheme Using Second Order Differences", in *A Locust's Leg: Studies in Honour of S. H. Taqizadeh*, ed. by W. B. Henning, *et al.*, Percy Lund, Humphries & Co., Ltd, London, 1962, pp. 117–20.
16. Kennedy, E. S. and Transue, W. R., "A Medieval Iterative Algorism", *American Mathematical Monthly* **63** (1956), 80–3.
17. Kennedy, E. S., "An Early Method of Successive Approximation", *Centaurus* **13** (1969), 248–50.
18. Kennedy, E. S., "Parallax Theory in Islamic Astronomy", *Isis* **47** (1956), 33–53.
19. [= 2, 13].
20. Youschkevitch, A. P., "Note sur les determinations infinitesimales chez Thabit ibn Qurra", *Archives Internationales d'Histoires des Sciences* **17** (1964), 37–45.
21. [= 5, 92].
22. Saidan, A. S., "The Earliest Extant Arabic Arithmetic", *Isis* **57** (1966), 475–90.
23. Rashed, R., "L'Extraction de la Racine n-ieme et l'Invention des Fractions Decimales (XIe–XIIe Siecles)", *Archive for History of Exact Sciences* **18** (1978), 191–243.
24. Rashed, R., "Résolution des Équations Numériques et Algèbre: Šaraf-al-Din al-Tūsī, Viète", *Archive for History of Exact Sciences* **12** (1974), 244–90.
25. Henrici, P., *Elements of Numerical Analysis*, John Wiley & Sons, New York, 1964.
26. [= 5, 63].
27. [= 2, 24].
28. [= 4, 61].
29. Dold-Samplonius, Y., "Al-Māhānī", DSB 9, 21–22.
30. Rosenfeld, B. A. and Youschkevitch, A. P., "Al-Kāshī", DSB 7, 255–62.
31. Saidan, A. S., "Al-Nasawī", DSB 9, 614–15.
32. Noble, B., *Numerical Methods*, 2 volumes, Oliver and Boyd, Edinburgh, 1970.
33. Chandrasekhar, S., and Swerdlow, N., "Review of Neugebauer's A History of Ancient

Mathematical Astronomy", *Bulletin of the American Mathematical Society* **84** (1978), 428–44.
34. Saidan, A. S., "Kūshyār Ibn Labbān", DSB 7, 531–3.
35. [=4, 69].
36. [=4, 46].
37. Dold-Samplonius, Y. and Hermelink, H., "Al-Jayyānī", DSB 7, 82–3.
38. Schrader, D. V., "Aḥmad Ibn Yūsuf", DSB 1, 82–3.
39. Toomer, G. J., "Al-Khwārizmī", DSB 7, 358–65.
40. Saidan, A. S., "Al-Uqlīdīsī", DSB 13, 544–6.
41. Saidan, A. S., "Al-Nasawī", DSB 9, 614–15.
42. Pingree, D., "Al-Fazārī", DSB 4, 555–6.
43. Saidan, A. S., "Al-Umawi", DSB 13, 539–40.
44. [=2, 2].
45. Harvey, E. R., "Qusṭa Ibn Lūqā Al-Baᶜlabakkī", DSB 11, 244–6.
46. Clagett, M., *Archimedes in the Middle Ages. Volume I, The Arabo-Latin Tradition*, University of Wisconsin Press, Madison, 1964.
47. Anbouba, A., "Al-Samawʾal", DSB 12, 91–5.
48. Anbouba, A., "Al-Ṭusi, Sharaf Al-Din", DSB 13, 514–17.
49. [=4, 17].
50. Rosenfeld, B. A. and Youschkevitch, A. P., "On the treatise of Qāḍī Zāda al-Rumī for the calculation of the sine of one degree", *Istoriko-matematicheskie Issledovanija* **13** (1960), 533–538. [Russian]
51. Kari-Niazov, T. N., "Ulugh Beg", DSB 13, 535–7.
52. Schoy, C., "Al-Bīrūnī's Method of Approximation of Chord 40°", *American Mathematical Monthly* **33** (1926), 95–6.
53. Al-Dabbagh, J., "Banū Mūsā", DSB 1, 443–6.
54. [=3, 22].
55. Toeplitz, O., *Calculus : A Genetic Approach*, University of Chicago Press, Chicago, 1963.
56. [=3, 36].
57. Rashed, R., "Ibrāhīm Ibn Sinān", DSB 7, 2–3.
58. Dold-Samplonius, Y. "Al-Qūhī", DSB 11, 239–41.
59. Suter, H., "Die Abhandlung über die Ausmessung des Paraboloides von Ibn el-Haitham", *Bibliotheca mathematica* (1912), 289–332.
60. [=1, 128, Vol. V].
61. [=4, 67].
62. [=3, 27].
63. [=2, 3].
64. [=1, 42].
65. [=2, 11].

Essay 7

1. Aaboe, A., "Al-Kāshī's iteration method for the determination of sin 1°", *Scripta Math.* **20** (1954), 24–9.
2. Amir-Moéz, A. R., "A Paper of Omar Khayyam", *Scripta Mathematica* **26** (1963), 323–37.
3. [=6, 47].
4. [=6, 48].
5. Anbouba, A., "Construction de l'heptagone régulier par les Arabes au 4ᵉ siècle de l'hégire", *J. History of Arabic Science* **2** (1978), 264–9.
6. [=5, 10].
7. Anbouba, A., "Un algébriste arabe, Abū Kāmil", *Horizons Techniques du Moyen Orient* **2** (1963), 6–15.

8. Anbouba, A., "Un Traité d'Abū Jaʿfar [al-Khazin] sur les triangles rectangles numeriques", *J. History of Arabic Science* **3** (1979), 134ff.
9. Archbold, J. W., *Algebra*, 4th edn., Pitman and Sons, London, 1970.
10. Archibald, R. C., ed., *Euclid's Book on Divisions of Figures*, Cambridge University Press, 1915.
11. Archibald, R. C., *et al.*, eds., *Jacob Steiner's Geometrical Constructions with a Ruler*, Scripta Mathematica, Yeshiva University, N.Y., 1950.
12. Baker, M., "Alhazen's Problem", *American J. of Mathematics* **4** (1881), 327–31.
13. Baker, T., *The Geometrical Key;* or *the Gate of Equations Unlock'd : A New discovery of the Construction of all Equations, howsoever affected, not exceeding the fourth Degree*, J. Playford for R. Clavel, London, 1684.
14. Bashmakova, I. G., *et al.*, "An Arabic Version of Diophantus' 'Arithmetic'", (Russian) *Istoriko-mat. issled.* **23** (1978), 192–225.
15. Becker, O., *Das mathematische Denken der Antike*, 2nd edn, Vandenhoeck and Ruprecht, Göttingen, 1966.
16. [= 5, 11].
17. Bieberbach, L., *Theoire der Geometrischen Konstruktionen*, Birkhäuser Verlag, Basel, 1952.
18. *Al-Bīrūnī, Qānūn-i-Masʿūdī. Dāiratuʾl-Maʿārif-il-Oṣmania*, Hyderabad, 1955 [1374 A. H.]. 3 volumes.
19. Bode, P., "Die Alhazensche Spiegel-Aufgabe", *Jahresbericht des physikalischen vereins zu Frankfurt am Main*, 1893, 63–107.
20. [= 2, 1].
21. Boyle, J. A., "ʿUmar Khayyam. Astronomer, Mathematician and Poet", Chapter 20(b) (pp. 658–65) in R. N. Frye, ed., *Cambridge History of Iran*, vol. 4, Cambridge University Press, London, 1975.
22. Breidenbach, W., "Der Rechte Winkel und das Einschiebe lineal", *Zeitschrift für math.-nat. Unterricht* **56** (1925), 4–13.
23. Breidenbach, W. and Süss, W., "Geometric Constructions", pp. 198–237 in volume 2 of H. Behnke, *et al.*, eds., *Fundamentals of Mathematics*, MIT Press, Cambridge, 1974.
24. [= 2, 6 all parts].
25. Bruins, E. M., "Interpretation in Cuneiform Mathematics", *Janus* (1978), 277–341.
26. Bruins, E. M., "Problema Alhaseni at the Tercentenary of Huygens' Solution", *Centaurus* **13** (1968/69), 269–77.
27. [= 3, 9].
28. [= 3, 11].
29. Busard, H. L. L., "Der Traktat ʿDe Isoperimetris,ʾ der unmittelbar aus dem Griechischen ins Lateinische übersetzt worden ist", *Medieval Studies* **42** (1980), 61–88.
30. Busard, H. L. L., "Lʾ Algèbre au Moyen Âge: Le "Liber Mensurationum" dʾ Abû Bekr", *Journal des Savants* (1968). 65–124.
31. [= 2, 8, vols. I–II].
32. Cassina, U., "Sulle equazioni cubiche di al-Bīrūnī", *Period. Mat.* **21** (1941), 3–20.
33. Cesàro, G., "Les Problèmes de Géometrie résolus par le compas, sans la régle", *Memoires de la Société royale des Sciences de Liège* (*ser.* 3) **1** (1899), 1–11.
34. Coolidge, J. L., *The Mathematics of the Great Amateurs*, Dover, N.Y., 1963. [Oxford, 1949]
35. Curtze, M., *Urkunde zur Geschichte der Mathematik im Mittelalter und der Renaissance*, 2 parts, Johnson Reprint, N.Y., 1968. [Leipzig, 1902].
36. Czwalina, A., ed., *Archimedes Werke*, Wissenschaftliche Buchgesellschaft, Darmstadt, 1972.
37. [= 6, 53].
38. [= 3, 15].
39. Delambre, J. B. J., *Histoire de l'Astronomie Ancienne*, vol. 2, Johnson Reprint, N.Y., 1965 [Sources of Science, no. 23]. [Paris, 1817].

40. Dickson, L. E., *New First Course in the Theory of Equations*, John Wiley & Sons, N.Y., 1967. [1939].
41. Dickson, L. E., "On the Trisection of an Angle and the Construction of Regular Polygons of 7 and 9 Sides", *American Mathematical Monthly* **21** (1914), 259–62.
42. [= 4, 18].
43. [= 4, 17].
44. Dold-Samplonius, Y., "Al-Khāzin", DSB VII, 334–5.
45. Dold-Samplonius, Y., "Al-Māhānī", DSB IX, 21–2.
46. Dold-Samplonius, Y., "Al-Qūhī", DSB XI, 239–41.
47. [= 4, 19].
48. Eves, H., *A Survey of Geometry*, rev. edn., Allyn and Bacon, Boston, 1972.
49. Gandz, S., Review of B. Datta and A. N. Singh: *History of Hindu Mathematics*, I, *Isis* **25** (1936), 478–88.
50. Gandz, S., "Studies in Babylonian Mathematics III: Isoperimetric Problems and the Origin of the Quadratic Equations", *Isis* **32** (1940), 103–15.
51. Gandz, S., "The Algebra of Inheritance", *Osiris* **5** (1938), 319–91.
52. Gandz, S., "The 'Mishnat Ha Middot'", *Quellen und Studien zur Geschichte der Mathematik, Astronomie und Physik, Abt. A*, **2** (1932), 1–96, 4 plates.
53. Gandz, S., "The Origin and Development of the quadratic equation in Babylonian, Greek and early Arabic Algebra", *Osiris* **3** (1938), 405–557.
54. Gandz, S., "The Sources of al-Khowārizmī's Algebra", *Osiris* **1** (1936), 263–77.
55. Gerretsen, J. and Verdenduin, P., "Polygons and Polyhedra", pp. 238–90 in vol. 2 of H. Behnke, *et al.*, eds., *Fundamentals of Mathematics*, MIT Press, Cambridge, 1974.
56. [= 1, 42].
57. Goetsch, H., "Die Algebra der Babylonier", *Archive for History of the Exact Sciences* **5** (1968), 79–153.
58. [= 1, 50].
59. Gulley, N., "Greek Geometrical Analysis", *Phronesis* **3** (1958), 1–14.
60. Hankel, H., *Zur Geschichte der Mathematik in Altertum und Mittelalter*, 2nd edn., with forward by J. E. Hofmann, G. Olms, Hildesheim, 1965. [Leipzig, 1874]
61. Hartner, W., "Al-Djabr waʾl-Muḳābala", *Encyclopaedia of Islam*, 2nd edn., **2**, 360–2.
62. [= 3, 21, vols. I–II].
63. [= 4, 28].
64. Heath, T. L., *Diophantus of Alexandria*, Cambridge University Press, 1910. [Dover, N.Y., 1964]
65. [= 3, 22].
66. Heath, T. L., ed., *The Works of Archimedes with the method of Archimedes*, Dover, N.Y., n.d. [Cambridge, 1912]
67. Heiberg, J. L., *Apollonius Pergaeus*, 2 volumes, Teubner, Stuttgart, 1974. [1891]
68. Hogendijk, J. P. "On the trisection of an angle and the construction of a regular nonagon by means of conic sections in medieval Islamic geometry", Univ. of Utrecht. Dept. of Mathematics, Preprint No. 113, March, 1979. 52pp.
69. Hudson, H. P., *Ruler and Compass*, in *Squaring the Circle and other Monographs*, Chelsea Publishing Co., N.Y., 1969.
70. [= 3, 25].
71. [= 4, 31].
72. [= 2, 17].
73. [= 6, 51].
74. Karpinski, L. C., "The algebra of Abu Kamil Shojaᶜ ben Aslam", *Bibliotheca Mathematica* **12** (1911/12), 40–55.
75. Karpinski, L. C. and Winter, J. G., *Contributions to the History of Science*, University of Michigan Press, Ann Arbor, 1930.
76. Kasir, D. S., *The Algebra of Omar Khayyam*, AMS Press, N.Y., 1972. [Teachers College, Columiba University, 1931].

77. [= 3, 26].
78. Kazim, M. A., "Al-Bīrūnī and Trigonometry", pp. 161–70 in V. Courtois, ed., *Al-Bīrūnī Commemoration Volume*, Iran Society, Calcutta, 1951.
79. [= 6, 17].
80. Kennedy, E. S., "Al-Bīrūnī", DSB II, 147–58.
81. Kennedy, E. S., Review of Rosenfeld, *et al.*, eds., *Al-Kāshī's Key to Arithmetic*, in *Isis* **50** (1959), 72–5.
82. Kennedy, E. S., "The Exact Sciences", pp. 378–95 (Chapter 11) in R. N. Frye, ed., *The Cambridge History of Iran*, vol. IV, Cambridge University Press, 1975.
83. [= 4, 34].
84. [= 2, 18].
85. Kohl, K., "Zur Geschichte der Dreiteilung des Winkels", *Sitzungsberichte der Phys.-Med. Soz. Erlangen* **54** (1925), 180–9.
86. Krasnova, S. A., "Abū᾿ l-Wafā᾿ al-Būzjānī: 'Geometrical Constructions for the Artisan'", (Russian), *Fiziko-matematicheskie nauki v stranakh vostoka I* (1966), 42–140.
87. Krasnova, S. A. and Rosenfeld, B. A., "First treatise on algebra: Omar Khyyam", (Russian) *Istoriko-mat. issled.* **15** (1963), 445–72.
88. Kubesov, A., ed., *The Mathematical Treatises of Al-Fārābī*, (Russian), Izdat. "Nauka" Kazah. SSR, Alma-Ata, 1972.
89. Kubesov, A. and Rosenfeld, B. A., "On the Geometrical Treatise of Al-Fārābī", *Arch. Intl. Hist. Sci.* **22** (1969), 50.
90. Kutta, W. M., "Geschichte der Geometrie mit constanter Zirkelöffnung", *Deutsche Akademie der Naturforscher* **71** (1897), 69–101.
91. Lay-Yong, L., "The Chinese Connection Between the Pascal Triangle and the Solution of Numerical Equations of any Degree", *Historia Math.* **7** (1980), 407–24.
92. Levey, M., "Abū Kāmil", DSB I, 30–2.
93. Levey, M., "Some Notes on the Algebra of Abū Kāmil Shujā᾿: A Fusion of Babylonian and Greek Algebra", *L'Enseignement mathématique* **4** (1958), 77–92.
94. Levey, M., *The Algebra of Abū Kāmil*, University of Wisconsin Press, Madison, 1966.
95. Levey, M. and Petruck, M., *Kūshyar ibn Labbān: Principles of Hindu Reckoning*, University of Wisconsin Press, 1965.
96. Levi Della Vida, G., "Due nuove opere dei matematico al-Karağī (al-Karḫī)", *RSO* **14**, 249–64.
97. [= 6, 3].
98. Lindberg, D. C., ed., *Opticae Thesaurus Alhazeni Arabis...*, Sources of Science, no. 94. Johnson Reprint, N.Y., 1972. [Risner's edition]
99. Lohne, J. A., "Alhazens Spiegelproblem", *Nordisk Matematik Tidskrift* **18** (1970), 5–35.
100. London, F., "Die geometrischen Constructionen dritten und vierten Grades, ausgefürht mittelst der geraden Linie und einer festen Curve dritter Ordnung", *Zeitschrift für Mathematik und Physik* **41** (1896), 129–52.
101. London, F., "Uber cubische Constructionen", *Jahresbericht der deutschen Mathematiker-Vereinigung* **4** (1894/95), 163–5.
102. Luckey, P., "Der Lehrbrief über den Kreisumfang", *Abhandlungen der deutschen Akademie der Wissenschaften zu Berlin. Klasse für mathematik und allgemeine Naturwissenschaften.* 1950 [1953], no. 6.
103. [= 6, 11].
104. [= 6, 12].
105. Luckey, P., "Ṯābit b. Qurra über den geometrischen Richtigkeitsnachweis der Auflösung der quadratischen Gleichungen", *Bericht über verh. d. K. Sächs. Gesellschaft, Math-Phys. Kl. Leipzig* **93** (1941), 93–114.
106. Luckey, P., "Zur islamischen Rechenkunst und Algebra", *Forschungen und Fortschritte* **24** (1948), 199–204.

107. MacDuffee, C. C., *Theory of Equations.* John Wiley & Sons, N.Y., 1959. [1954]
108. [=5, 69].
109. Mahoney, M. S., "Another Look at Greek Geometrical Analysis", *Archive for History of the Exact Sciences* **5** (1968), 318–48.
110. Medovoy, M. I., "Negative Numbers in the Work of Abū'l Wafā'", (Russian), *Istorikomat. issled.* **11** (1958), 593–8.
111. Medovoy, M. I., "On the Arithmetic Treatise of Abū'l-Wafā'", (Russian), *Istorikomat. issled.* **13** (1960), 253–324.
112. Meyerhof, M., "Die Optik der Araber", *Zeitschirft für opthalmol. Optik mit Einschluss der Instrumentenkunde* **8** (1928), 16–29; 42–54; 86–90.
113. [=3, 27].
114. [=4, 46].
115. Naẓīf, M., *Al-Ḥasan ibn al-Haitham : His optical studies and discoveries,* (Arabic), 2 parts, Nūrī Press, Cairo, 1942–43.
116. [=1, 102].
117. [=2, 26].
118. [=2, 27].
119. [=2, 28].
120. Petersen, J., *Methods and Theories for the Solution of Problems of Geometrical Construction,* in *String Figures and Other Monographs,* Chelsea Publishing Co., N.Y., 1969.
121. Quraishi, M. F., "How to Determine the Side of a Regular Heptagon Inscribed in a Circle", *Ibn al-Haitham : Proc. of the Celebration of the* 1000*th Anniversary,* Hamdard National Foundation, H. M. Said, ed., Karachi, 1969, pp. 135–45.
122. Rashed, R., "Al-Karajī", DSB VII, 240–6.
123. Rashed, R., "La construction de l'heptagone régulier par Ibn al-Haytham", *J. History of Arabic Science* **3** (1979), 309–87.
124. [=6, 23].
125. Rashed, R., "Optique Géometrique et Doctrine Optique chez Ibn Al-Haytham", *Arch. Hist. Exact Sciences* **6** (1969), 271–98.
126. [=6, 24].
127. Rodet, L., "L'Algèbre d'Al-Khârizmi et les Méthodes Indienne et Grecque", *Journal Asiatique* (*ser.* 7) **11** (1878), 5–98.
128. Rosen, F., ed. and tr., *The Algebra of Mohammed ben Musa,* Oriental Translation Fund, London, 1831.
129. Rosenfeld, B. A., "The Algebraic Treatise of As-Samaw ʿal", (Russian), *Istorikomat. issled.* **20** (1975), pp. 125–49, 379.
130. [=3, 36].
131. Rosenfeld, B. A. and Youschkevich, A. P., *Omar Khayyam,* (Russian), Nauka, Moscow, 1965.
132. Rosenfeld, B. A. and Youschkevitch, A. P., eds., *The Key to Arithmetic* (*Russian*)*: al-Kāshī's Miftāḥ al-ḥisāb,* Moscow, 1956.
133. [=4, 63].
134. Ruska, J., "Zur ältesten arabischen Algebra und Rechenkunst", *Sitzungsber. Heidelb. Akad. Wiss. Phil.-Hist. Kl.,* 1917. 125pp.
135. [=4, 67].
136. Sacerdote, G., "Il trattato del pentagono e del decagono di Abu Kámil Shogiá ben Aslam ben Muhammed", pp. 169–94 in *Festschrift zum achtzigsten Geburtstage Moritz Steinschneider's,* O. Harrassowitz, Leipzig, 1896. [Reprint, G. Olms, Hildesheim, 1975]
137. [=6, 31].
138. [=5, 92].
139. Samplonius, Y., "Die Konstruktion des regelmässigen Siebenecks nach Abu Sahl al-Qûhî Waiğan ibn Rustam", *Janus* **50** (1963), 227–249. [Dold-Samplonius]

140. Sánchez Pérez, J. A., *Compendio de Álgebra de ABENBEDER*, Madrid, 1916.
141. [= 1, 123].
142. Sayili, A., *Logical Necessities in Mixed Equations by ʿAbd al Ḥamîd ibn Turk and the Algebra of his Time*, Turk Tarih Kurumu Basimevi, Ankara, 1962. [Rurk Tarih Kurumu Yayinlarindan VII, Seri, No. 41]
143. Sayili, A., "The Trisection of the Angle by Abû Sahl Wayjan ibn Rustam al Kûhî (Fl. 970–988)", in H. Guerlac, ed., *Proc. Intl Congress History of Science X Ithaca*, 1962, Hermann, Paris, 1964, pp. 545–6. [See also the Turkish version in *Belleten* **26** (1962), 693–700.]
144. Schoy, C., "Behandlung einiger geometrischen Fragepunkte durch muslimische Mathematiker", *Isis* **8** (1962), 254–63.
145. [= 2, 29].
146. Schoy, C., *Die Trigonometrischen Lehren des persischen Astronomen Abuʾl-Raihân Muḥ. ibn Aḥmad Al-Bîrûnî*, eds., J. Ruska and H. Wieleitner, Orient-Buchhandlung Heinz Lafaire K. G., Hannover, 1927.
147. Schoy, C., "Drei planimetrische Aufgaben des arabischen Mathemakiters Abûʾl-Jûd Muḥammad ibn al-Lîth", *Isis* **7** (1925), 5–8.
148. Schoy, C., "Graeco-Arabischen Studien", *Isis* **8** (1926), 21–40.
149. Schramm, M., *Ibn Al-Haythams Weg zur Physik*, Franz Steiner Verlag, Wiesbaden, 1963.
150. Schramm, M., "Steps Toward the Idea of Function", *History of Science* **4** (1965), 70–103.
151. Sesiano, J., "Diophantus of Alexandria", DSB XV, 118–22.
152. Sesiano, J., "The Arabic Text of Books IV to VII of Diophantus' *Αριθμητικά* in the Translation of Qusṭā ibn Lūqā", *Ph.D. Thesis*, Brown University, 1975.
153. Sesiano, J., "Un mémoire d'Ibn al-Haytham sur un problème arithmétique solide", *Centaurus* **20** (1976), 189–95.
154. [= 1, 128].
155. Sirazhdinov, S. Kh. and Akhmedov, A., "Some mathematical and astronomical problems in Canon al-Masudi of al-Bīrūnī", (Russian), *Obshchestv. Nauki v Uzbekistane* 1973, 50–64.
156. Smith, H. J. S., *The Collected Mathematical Papers*, ed. by J. W. L. Glaisher, 2 volumes, Chelsea Publishing Co., N.Y., 1965. [1894]
157. Stamatis, E. S., ed. (following J. L. Heiberg), *Euclidis Elementa*, 5 volumes in 6, B. G. Teubner, Leipzig, 1969.
158. [= 1, 139].
159. [= 1, 140].
160. Steinschneider, M., "Zur Geschichte der Uebersetzungen aus dem Indischen in's Arabische und ihres Einflusses auf die arabische Literatur", *Zeitschrift der deutschen morgenland. Gesellschaft* **24** (1870), 325–392.
161. Suter, H., "Das Rechenbuch des Abû Zakarîjâ al-Ḥaṣṣâr", *Bibliotheca mathematica* (*ser.* 3) **2** (1901), 12–40.
162. Suter, H., "Die Abhandlung des Abu Kamil Shuja b. Aslam über Funfeck und Zehneck", *Bibliotheca math.* (*ser.* 3), **10** (1909/10), 15–42.
163. [= 2, 35].
164. Szabó, Á., "Zum Problem der sog. geometrischen Algebra in Euklids Elementen", *Prismata* (Festschrift für Willy Hartner), ed. Y. Maeyama and W. G. Salzer, Steiner Verlag, Wiesbaden, 1977, pp. 373–93.
165. Tannery, P., ed., *Diophantus Alexandrinus: Opera Omnia*, 2 volumes, Teubner, Stuttgart, 1974. [1893]
166. Tekeli, S., "Muḥyi ʾl-Dīn al-Maghribī", DSB IX, 555–7.
167. Terzioglu, N., "Das achte Buch zu den ʿConicaʾ des Apollonius von Perge. Rekonstruiert von Ibn al-Haysam", *Publications of the Mathematical Research Institute, Istanbul. no.* 5 (1974). (*)

168. [= 4, 89].
169. [= 2, 37].
170. Tllasev, Kh., "On some mathematical and astronomical manuscripts of Naṣir al-Dīn al-Ṭūsī...", (Russian), Izv. *Akad. Nauk. UZSSR* ser. *fiz.-mat.* **4** (1972), 63–5.
171. [= 6, 39].
712. Toomer, G. J., "Apollonius of Perga", DSB I, 179–93.
173. Toomer, G. J., "Diocles", DSB XV, 115–18.
174. Toomer, G. J., ed., *Diocles on Burning Mirrors*, Springer, N.Y., 1976.
175. Tropfke, J., "Zur Geschichte der quadratischen Gleichungen über dreieinhalb Jahrtausend", *Jahresbericht d. deutschen Mathem.-Vereinigung* **43** (1933), 98–107; **44** (1934), 26–47, 95–119.
176. Uspensky, J. V., *Theory of Equations*, McGraw-Hill, N.Y., 1948.
177. Vernet, J., "Al-Karadjī", *The Encyclopaedia of Islam*, vol. IV, p. 600.
178. Viète, F., *Opera Omnia*, Introduction by J. E. Hofmann, Georg Olms, Hildesheim, 1970. [Leiden, 1646].
179. Vogel, K., Review of U. Cassina's "Sulle equazioni cubiche di Al-Biruni", *Mitt. zur Geschichte der Medizin, Naturwiss. und der Technik* **39** (1940), 318.
180. [= 3, 48].
181. van der Waerden, B. L., "On Pre-Babylonian Mathematics I–II", *Arch. Hist. Exact Sciences* **23** (1980), 1–25; 27–46.
182. [= 3, 49, vol. I].
183. Weinberg, J., "Die Algebra des Abū Soǧāʾ ben Aslam", *Dissertation*, Munich, 1935.
184. Wiedemann, E., "Ueber geometrische Instrumente bei den muslimischen Völkern", *Zeitschrift für Vermessungswesen* **29** (1910), 585–92, 617–25.
185. Wieleitner, H., "Die Erbteilungsaufgabe bei M. b. Musa Alchwarsmi", *Zeitschrift für math. naturwissen. Unterricht* **53** (1922), 57–67.
186. Winter, H. J. J. and ʿArafat, W., "The Algebra of ʿUmar Khayyām", *J. Royal Asiatic Society of Bengal*, **16** (1950), 27–78.
187. Woepcke, F., "Analyse et extraits d'un recueil de constructions géométriques par Aboûl Wafâ", *Journal Asiatique* (*ser.* 5) **5** (1855), 218–56, 309–59.
188. Woepcke, F., "Discussion de deux méthodes arabes pour déterminer une valuer approchée de sin 1°", *J. de Math. Pures et Appliquées* **19** (1854), 153–76; 301–3.
189. Woepcke, F., *L'Algèbre dʾOmar Alkhayyâmî*, Benjamin Duprat, Paris, 1851.
190. Woepcke, F., "Note sur des notations algebriques employees par les Arabes", *Comptes Rendus Acad. Sci.* (*Paris*) **39** (1854), 162–5.
191. Woepcke, F., "Notice sur des traductions arabes de deux ouvrages perdus dʾEuclide", *Journal asiatique* **18** (1854), 217–47.
192. Woepcke, F., "Notice sur différents morceaux tirés de manuscrits arabes et relatifs à lʾhistoire des mathématiques", *Comptes Rendus Acad. Sci.* (*Paris*) **31** (1850), 715–17.
193. Woepcke, F., "Notice sur un manuscrit Arabe dʾun traité dʾalgèbre par Aboul Fath Omar Ben Ibrahim Alkhayâmî, contenant la construction géométrique des équations cubiques", *J. für die reine und angew. Math.* **40** (1850), 160–72.
194. Woepcke, F., "Sur la construction des équations du quartrième degré par les géomètres Arabes", *J. Mathematiques Pures et Appliquées* (*ser.* 2), **8** (1863), 57–70.
195. Woepcke, F., "Sur un essai de déterminer la nature de la racine dʾune équation du troisième degré, contenu dans un ouvrage de Léonard de Pise découvert par M. le Prince Balthasar Boncompagni", *J. Mathematiques Pures et Appliquées* (*ser.* 1), **19** (1854), 401–6.
196. Woepcke, F., "Trois Traités Arabes Sur Le Compas Parfait", *Notices et Extraits des Mauscrits de la Bibliothèque Impériale* **22** (1874), 1–115.
197. Yagegari, M., "The Binomial Theorem: A Widespread Concept in Medieval Islamic Mathematics", *Historia Math.* **7** (1980), 401–6.
198. [= 2, 38].

199. Youschkevitch, A. P., "Omar Khayyam and his algebra", (Russian), *Trudy inst. istorii estestvozaniya. Akad. Nauk SSSR* **2** (1948), 499–534.
200. [= 6, 30].
201. [= 4, 61].
202. Zeuthen, H. G., *Die Lehre von den Kegelschnitten im Altertum*, with foreword by J. E. Hofmann, G. Olms, Hildesheim, 1966. [Reprint of 1886 edn.]
203. Zeuthen, H. G., *Geschichte der Mathematik im* 16 *und* 17. *Jahrhundert*, Johnson Reprint, N.Y., 1966. [Teubner, 1903]
204. Clagett, M., *Archimedes in the Middle Ages*, volume 4, pt I and pt II, The American Philosophical Society, Philadelphia, 1980.
205. [= 3, 16].
206. Halley, E., ed. and tr., *Apollonii Pergaei: De sectione rationis libri duo ex arabico ms...* E Theatro Scheldoniano, Oxonii, 1706.
207. Pelzer, A., *Abréviations Latines Médiévales*, 2nd edn., Publications Universitaires, Louvain, 1966.
208. Archibald, R. C., ed. and tr., *Euclid's Book on Divisions of Figures* (περι διαιρέσεων *βιβλιον) with a restoration based on Woepcke's text and on the Practica Geometriae of Leonardo Pisano*, Cambridge University Press, Cambridge, 1915.

UNIVERSITY LIBRARY
DURHAM

3 0104 00494294 6

ACCESSION No.

This book should be returned on or before the last date shown below, unless it is previously recalled by the Library when it must be returned immediately.

Fines are incurred on overdue books (see Library Regulations).

13

25